ANG CHEW FU, or KANTON
from
Nieuhof

BARBARIANS AND MANDARINS

BARBARIANS AND MANDARINS

NIGEL CAMERON

Thirteen Centuries of Western Travellers in China

HONG KONG
OXFORD UNIVERSITY PRESS
OXFORD NEW YORK
1989

Oxford University Press

Oxford New York Toronto
Petaling Jaya Singapore Hong Kong Tokyo
Delhi Bombay Calcutta Madras Karachi
Nairobi Dar es Salaam Cape Town
Melbourne Auckland

and associated companies in
Berlin Ibadan

First published by John Weatherhill, Inc. 1970

First issued, with permission,
by Oxford University Press 1989
Published in the United States by
Oxford University Press, Inc., New York

Library of Congress Cataloging-in-Publication Data

Cameron, Nigel.
Barbarians and mandarins: thirteen centuries of western travellers in China/Nigel Cameron.
p. cm.
Includes bibliographical references.
ISBN 0-19-585005-X (U.S.): $40.00
1. China—Description and travel—To 1900.
2. China—Relations—Foreign countries. I. Title.
DS707.C36 1989
303,48'251—dc20 89-25534
CIP

British Library Cataloguing in Publication Data

Cameron, Nigel
Barbarians and mandarins: thirteen centuries of Western travellers in China.
1. Western World. Relations with China, history
2. China. Relations with the Western World
I. Title
303.4'821821'051
ISBN 0-19-585005-X

Printed in Hong Kong by Skiva Printing & Binding Co., Ltd.
Published by Oxford University Press, Warwick House, Hong Kong

For Freda Wadsworth
and Chinese friends

CONTENTS

List of Principal Illustrations *9*

List of Tables *10*

Preface *11*

PART ONE. PRELUDE
A Nestorian and a Franciscan

1 Alopen and the Illustrious Religion *17*

2 The Priest Who Was a Heavy Man *28*

PART TWO. CHINA OBSERVED
A Merchant, a Saint, and a Tourist

3 Marco Polo *63*

4 John of Montecorvino *90*

5 Odoric of Pordenone *107*

PART THREE. VISITORS TO THE GREAT MING
Pirates, Casuists, Scientists, Diplomats

6 The Way from Portugal *123*

7 Poor Pirès *131*

8 Matteo Ricci and the Reluctant Dragon *149*

9 The European Rivals *195*

PART FOUR. SUITORS OF THE MANCHU

Ambassadors, Intriguers, Jesuits,
a Scotsman, and an American

10 A German, a Russian, and Two Dutch Merchants 221
11 The Grand Alliance: Ferdinand Verbiest and the Emperor K'ang-hsi 237
12 The Depths of Diplomacy: Spathary, Verbiest, and K'ang-hsi 250
13 Ripa the Neapolitan and Honest John Bell 263
14 Foreign Lords and Foreign Mud 288
15 Dr. Peter Parker and the American Effort 329

PART FIVE. CHINA IN CHAINS

Vandals, Reformers, Heroes, and a Poet

16 Lord Elgin and Wang's Mother 345
17 Mrs. Archibald Little: Unbind Thy Feet! 361
18 Defenders of the Faith, 1900 371
19 The Writers from the West 400
20 The Past, the Present, and the Future 417

Bibliography 421
Acknowledgments 429
Index 431

PRINCIPAL ILLUSTRATIONS

PORTFOLIOS

Sketches by William Alexander *305*

Photographs by Henri Cartier-Bresson *403*

MAPS AND PLANS

A Chinese Plan of Karakorum *54*

The Mongol Empire in 1290 *69*

Plan of Shangtu *71*

A Chinese Map of Zayton *103*

Sketch of the River at Canton *135*

A Chinese Map of China *160*

Plan of Chao-ch'ing *165*

Ricci's Route to Peking *168*

The Approaches to Canton *214*

Plan of Canton, 1857 *319*

Sketch Map of the Summer Palace Area *354*

Skeleton Map of Peking, 1900 *380*

Plan of Peking Cathedral *385*

REFERENCES TO OTHER ILLUSTRATIONS ARE INCLUDED IN THE INDEX

TABLES

Itinerary of Friar William of Rubruck *32*
A Comparison of the Jesuit and Chinese Curriculums *164*
Some Events in Europe and China, 1610–44 *198*
Principal Western Embassies to Peking *236*

PREFACE

FIRST AND FOREMOST, this is a book about people, about travelers and their motives and their adventures—and about their reluctant hosts at journey's end. Diplomatic and political histories, filled as they are with broad sweeps and heavy laden with grand significances, often seem to lose sight of the fact that history is made by people and, when written, must be read by people, that but for the farmer in his field there would be no governments and no prime ministers to govern them, that but for early travelers making their ways into unknown lands there would be no ambassadors extraordinary and plenipotentiary—for there would be no contacts between country and country, people and people, for them to maintain. History then, if we look at it in this way, is, in the words of Carlyle, "the essence of innumerable biographies." Here in this book we are dealing with the prime stuff of history—the people in whose actions its essence resides—in the belief that along the way we shall follow for so many centuries from Europe and America to the Middle Kingdom there is a significance that leaps the gulf of time and bears on the present day.

There is adventure enough in these travelers' tales. In most of them I have allowed the travelers to speak for themselves so far as possible, drawing upon many disparate and obscure and often out-of-print sources in the attempt to reveal a meaningful pattern of adventures—one that at the same time would be clear enough to follow and would also be representative. Here in this pattern we can enjoy the vantage point of time and look over the strange man-made edifices of the centuries with a perspective that was denied the travelers themselves, caught up as they were in an immediate and often hazardous present.

Writing in 1922, Bertrand Russell, one of the latter-day travelers, said: "A European lately arrived in China, if he is of a receptive and reflective disposition, finds himself confronted with a number of very puzzling questions. . . . Chinese problems, even if they affected no one outside China, would be of vast importance, since the Chinese are estimated to constitute about a quarter of the human race. In fact, however, all the world will be vitally affected by the development of Chinese affairs, which may well prove a decisive factor, for good or evil, during the next two centuries. This makes it important, to Europe and America almost as much as to Asia, that there should be an intelligent understanding of the questions raised by China, even if, as yet, definite answers are difficult to give. . . ."

Prophetic words indeed. The puzzling questions of China are even more acute today than they were in 1922—and still more difficult to answer. And, as events have turned out, their decisive nature seems to be much more imminent than those two centuries Lord Russell imagined. In this book I have attempted what is perhaps a new approach to the questions and their answers, an attempt to fill what writers of prefaces are wont to call a long-felt need.

In plain words, perhaps the reader "of a receptive and reflective disposition" may here be able to trace—without excessive expenditure of time and without the bafflement that would be occasioned by reading in full the often contradictory literature on the subject—the course of the Western wooing and assault (for it was both in varying degrees) of China. The East-West relationship is still far from workable. How to make it work?—this is a question to which possibly there may be answers as the picture unrolls down the years in the manner of a Chinese scroll—but an untypical scroll in that conflict rather than serenity, "barbarians" as well as Chinese, fill the scene.

I have purposely said almost nothing about the personalities involved in Sino-Western relations under the Republic, nor after the rise of the Communists. Important as these later developments are, they do not really form part of our story. The fundamental pattern of relations between the Western powers and China had clearly emerged in all its essentials by the time of the Boxer Rebellion in 1900. Events since that time have largely tended to be variations on the already enunciated theme. (We may note, as a simple example, the close likeness between Chinese Communist attitudes to the West, now inclusive of Russia, and the attitudes of the early Ming dynasty so long ago.) Tentatively it might be said that the attitudes displayed in the encounters between West and East which we shall examine have turned out to be essentially similar to those producing the same or similar problems that face us—and face the Chinese too—today.

This book is not intended to be a history of China, so the reader will find in it no more than the minimum of formal history necessary for the comprehension of the events with which we are dealing.[1] But perhaps the bare outline of the succession of dynasties, the character of a handful of representative high-ranking Chinese, and the press of Chinese event that is given in relation to our Western travelers will suffice for background. The reader

1. The better histories of China are included in the Bibliography. And the story of Peking—where much of the East-West dialogue took place—has already been treated at some length in my and Brian Brake's *Peking: A Tale of Three Cities.*

may take comfort in the fact that most of our Western travelers arrived in China knowing considerably less about Chinese history and customs than he himself will know when he has done with this book. For the most part it was on these men, in all their variety and in all their variegated states of ignorance and knowledge, that the story of East-West relations hung, at least until something like a hundred years ago. After that time governments tended to take the center of the stage, changing the pattern of relations. This change is reflected in the closing chapters.

By any standard, the story of the East and West face to face is a fascinating one. There is wit enough in it, and humor, suffering and endurance, courage, tenacity, and also cowardice—on both sides. There are great men, ideas and ideals great and less great, old and new. There are—a point worth noting—two of the major civilizations of the world for protagonists, the Chinese and the Western. But it is also a tragic tale of confrontation and clash arising from what is perhaps one of the fundamental misapprehensions of history—the idea that one race or people is innately superior to another. Very early in the story, when Western traders realized that their guns were all they needed in order to enforce their will on China, the West began to assume its own superiority toward the Chinese, and to Oriental peoples in general. Once set on this course, their feelings of superiority expanded to the dimensions of a cult and were accepted as axiomatic. Right up to the time when the present Chinese regime came to power, there were lamentably few Westerners who in their hearts thought of the Chinese as much more than an ancient nation sunk irretrievably in an outworn culture and its dark abuses.

The Chinese, for their part, had always—quietly, even smugly—felt superior to the rest of the world for reasons we shall see. And when the Western part of the world (as distinguished from China's less culturally developed immediate neighbors) came to her shores in numbers, the activities of its representatives tended to confirm the age-old Chinese opinion that China was indeed the seat and source of the sole valid culture in the world. From times far earlier than any described in this book—times long before the first recorded European set foot in the Middle Kingdom—the character *yi*, "barbarian," was the normal Chinese word applied to all non-Chinese peoples. When the first Europeans at last reached China, the Chinese saw little reason to make fish of one foreigner and fowl of the other. From the apex of that extraordinary pyramid of power—the emperor and mandarins in the capital and the august governors of the provinces—right down to the base, to the ordinary people with their feet in the immemorial soil, the Chinese continued to call *all* foreigners barbarians.

Westerners went on calling the Chinese heathens and pagans, and considering them very much a people to be pitied. (And to pity is, often enough, to despise.) Heathens and barbarians—West and East: the epithets carried almost the same derogatory connotation. Who, then, were the barbarians? And who the mandarins? What happened in the long confrontation? These are questions I have hoped to suggest in the title of this book, to trace and disentangle in the pages that follow.

Many quotations are used from the writings of the travelers themselves as well as from other sources. Every effort has been made to identify the precise source of each substantial quotation, without burdening the book with excessive paraphernalia. When all quotations from a particular author come from a single volume, the author's name alone is sufficient for finding the source in the Bibliography, where full bibliographical details are given. Two or more works by a single author are distinguished by abbreviated titles as well as the author's name.

PART ONE

PRELUDE

A Nestorian and a Franciscan

CHAPTER ONE

ALOPEN AND THE ILLUSTRIOUS RELIGION

In China, in the year 635 of the Christian era, a brilliant civilization flourished. Under the ruling T'ang dynasty at its capital, Ch'ang-an, near the mid-reaches of the Yellow River, a city already ancient and steeped in the Chinese past, a remarkable new phase of Chinese life had recently developed. One of the greatest of all the Chinese emperors sat on the Dragon Throne and guided the destinies of the country which Chinese, then and ever after, chose to call the Middle Kingdom, the hub of the world. It was already over eight hundred years since China had emerged for the first time as a coherent unit from the everlasting battles of petty warring states. It was over two thousand years since a recognizably Chinese civilization had emerged from unlettered darkness in that same region surrounding the middle Yellow River. While Europe, at the far western extremity of the Eurasian land mass, in that same seventh century, was still fumbling for a way out of its Dark Ages, China seemed to have already discovered her way. Indeed, that special Chinese way—the dynastic system of rule based on the moral code of Confucius—was to endure after the T'ang through an immense span of history, right up to the twentieth century.

In China, in Ch'ang-an, in the great houses of nobles as well as in royal palaces and pleasure gardens, it was an age of intricate refinement. It was an age, too, of lustrous poetry, of painting, of subtle philosophers, of pampered and painted women gliding from pavilion to lacquered pavilion beneath the moon. It was an age of discovery, a time when, suddenly, the arts, the humanities, took off into the skies and exploded like Chinese fireworks. And at that time, with the possible exception of India, there was nothing in the whole world, East or West, remotely comparable. China was undoubtedly the most advanced civilization in the world of that day.

Into this life at Ch'ang-an, through the crowded streets, past the palanquins of great men, under the shadow of intricate, painted eaves of palaces

THE NESTORIAN STONE as it is today in Sian. Over nine feet high and three feet wide, it is still in perfect condition. Note the elaborately carved dragon at the top, surrounding the stone's title. (Photo by Marc Riboud)

and great buildings, came a man from the West. His name was Alopen and he came from Syria.[1] Of Alopen himself, of what kind of man he was, why he traveled across Asia and precisely what happened to him in China, we unfortunately know absolutely nothing. But, almost by accident, we do know the outlines of what he accomplished in China.

The accident was the discovery of a stone monument, and since we know so little of the man himself, we shall have to depend upon this stone for what little of narrative this chapter can contain. But Alopen is important to our story, setting the scene as it were. So for a few pages let us forego swift events and consider Chinese attitudes toward this first known messenger from the West.

In 1625, just a thousand years after Alopen arrived in China, some workmen were preparing the foundations for a new building near Ch'ang-an when they uncovered a large stone tablet engraved with Chinese characters and also with another script that was not immediately decipherable. The city governor of Ch'ang-an had the bulky stone taken to a temple, where, but for another happy chance, it might have remained gathering dust in

1. Though in Asia Minor, Syria was then more allied with the West, with which it shared both Christianity and the Mediterranean.

obscurity, its little mine of information unsuspected by the world. Fortunately, a recent Chinese convert to Christianity happened to see the stone in the temple and took a rubbing of its inscription, which he sent to a friend in Hangchow. This friend, also a Christian convert, was a scholar called Li Chih-tsao. Having read the characters, Li made the following comment: "I was living in retirement in the country . . . when my friend . . . had the kindness to send me a copy of the Tablet of the T'ang, saying to me: 'It has the title, Praise of the Monument Recalling the Propagation in the Middle Kingdom of the Illustrious Religion.' We had not heard of this religion before. Could it be the holy religion that Ricci had come to preach from the farthest West?"[2]

Li was mistaken. The great Jesuit father Ricci (of whom we shall hear much more later) was a Roman Catholic, and the stone described the arrival and progress in China of an entirely different brand of Christianity called Nestorianism.[3]

Presently, the Chinese inscription from the stone reached the Jesuits in Peking, and one of them, Father Semedo (who also figures in our story later) made the first translation of it into a European language. The name Alopen occurred in the inscription along with the Chinese name Ta-ch'in, which designated roughly the area we now call Syria.

Alopen, then, is the first Western visitor to China whose name is known.[4] The stone credits him with founding a Nestorian church in China. Most historians have inferred that he was a monk, although there is only one contemporary statement to this effect. However that may be, the story of the Nestorians in China in the T'ang dynasty is a fascinating one, partly because it contains some of the earliest news we have of Westerners in the Middle Kingdom, and partly because in it we find even at this very distant date the elements of later events which overtook other Westerners when they came to China.

While it is quite possible that Alopen was indeed a Nestorian priest, we can just as credibly visualize him as a pious Nestorian merchant. At the time when he must have set off from Syria, the Arab armies, crusading out of their own land after the death of Mohammed, had not yet reached Syria. Persecution, then, could not have been his reason for leaving his own country, although it was probably the reason why later Nestorians set out for the East.

2. Quoted by Cary-Elwes.

3. The Nestorian brand of Christianity was a much more subtle and complicated matter than can be briefly told here. Baldly stated, however, it originated in the heresies of Archbishop Nestorius (died about A.D. 451), who taught that two distinct entities, the divine and the human, were coexistent in Christ.

4. There were doubtless many Western travelers to China before Alopen, but their names are unknown today. For instance, a still earlier Chinese record speaks of an ambassador from the Roman emperor Marcus Aurelius who arrived in the Chinese capital in A.D. 166.

It was an age of caravans. On horseback and on camels, thousands upon thousands of Western merchants had for centuries before Alopen been struggling their way across the vastness of the Asian continent along the silk routes. From the Mediterranean, with its nucleus of Western culture, those clusters of hardy and courageous men had been taking the long, long road toward China—ever since the times when wealthy Roman women discovered the alluring effects of Chinese silk and bought so much of it that the Roman senate became alarmed at the drain of gold from the coffers of the empire.

Alopen may have been a devout Nestorian layman, perhaps the spokesman of a party of merchants. In that capacity he would have had access to high officials of the T'ang court and administration, whose outlook—as we know from Chinese writing of the time—was unusually tolerant of foreigners. Ch'ang-an itself undoubtedly contained a whole community of Westerners. In many excavated tombs of Chinese nobles and rich families there have been found large numbers of figurines, placed there to attend the dead in the afterworld, whose faces and costumes are certainly Middle Eastern, Armenian, Greek, and Central Asian. The upper classes of T'ang

T'ANG POTTERY FIGURE OF AN ARMENIAN holding a wine skin. The caricaturing intention of the potter is obvious. Height, 36 cm. (Royal Ontario Museum, Toronto)

China seem to have preferred to choose their servants from the non-Chinese who came to Ch'ang-an with the caravans and elected to stay and enjoy the benefits of the civilization they found there. There are grooms, maids, dancers, barbers, cooks, messengers, musicians, and others among the figurines. So Westerners—often in their own, to the Chinese, delightfully foreign styles of dress—must have been a common enough sight in the streets and houses of the capital when Alopen arrived there. He most certainly did not find himself an isolated representative of a foreign land, as did many later Westerners in China, amid entirely alien people.

Not fancy but real probability allows us to imagine him passing time in conversation and discussion with cultured Chinese, men who for the first time in many a century were inclined to turn outward from the strict confines of their own traditional Chinese way of thought, to seek and enjoy the achievements of peoples from distant lands. And it is highly likely (to go along with this speculation for the moment) that the interest in Alopen's talk of his Nestorian faith was so vivid in Ch'ang-an that in the course of time messages sent by Alopen, perhaps backed by official inquiry, brought a group of Nestorian priests to the capital. Such a process was to occur many times in the future in China. In this way we may explain with a certain historical naturalness the possible manner of the foundation and early growth of the Nestorian church in T'ang China.

It was not long before the new religion was well enough established to find favor with the emperor—the illustrious T'ai Tsung. The inscription on the Nestorian Stone puts the case this way: "The emperor . . . caused this doctrine to be translated in his palace, and seeing the law to be true, he powerfully commanded that it should be divulged throughout the kingdom, and presently after, he sent forth an edict which contained the following: 'The true law hath no determinate name. The ministers therefore go about in every part to teach it unto the world. . . . In the kingdom of Ta-ch'in this Alopen, being a man of great virtue, hath brought from so remote a country doctrine and images, and is come to place them in our kingdom. Having well examined what he proposeth, we find it excellent and without any outward noise [perhaps meaning that the Nestorian faith had an innate humility and seemliness] and that it hath its principal foundation from the creation of the world . . . wherefore I have thought it convenient that it should be published throughout our empire.' "

Such is the translation made by Father Semedo in 1655 and published in his history of China. But the actual text of T'ai Tsung's edict, which is luckily preserved in the T'ang official records, is a much more cautious document. The measure of state and imperial support afforded to the Nestorians was in fact counterbalanced by an equal, and judiciously diplomatic, support given to the Buddhists.

T'ai Tsung had come to the throne in the autumn of A.D. 626 after an

EMPEROR T'AI TSUNG of the T'ang dynasty, after an engraved stele formerly at Chao Lin, Shensi province. (From Favier)

unusually thorough palace orgy of butchered relatives. But, from such an unpromising beginning, his reign went on to become one of the great peaks of Chinese culture. The circumstances were ripe for such greatness. For the first time in five hundred years, under T'ai Tsung's leadership China subdued its most violent neighbors, exacting both obedience and tribute from half of Central Asia. Direct rule from Ch'ang-an extended as far as the Pamirs, and Tibet was neutralized by the convenient marriage of an imperial female relative to the chieftain of the Lhasa region—a fact which went into Chinese folklore and has remained there ever since as a lament. The emperor's biographer quite justifiably puts into T'ai Tsung's mouth some brave words on the subject: "By taking my three-foot sword in my hand, I have subjugated the two hundred kingdoms and made quiet all within the four seas, while the far-off barbarians have come one after another to make their submission!"[5] Apart from anything else, this is a statement of what was to be for many a century China's basic foreign policy.

Among the results of T'ai Tsung's success were broad affluence in Ch'ang-an, an apparently solid and stable internal situation throughout China, and an intellectual and artistic ferment reminiscent in many ways of the cultural climate of the Italian Renaissance at a later date. It was a luminous age in Ch'ang-an. The turmoil of ideas occasioned by the arrival in China of philosophies, religions, techniques, and knowledge of social and sexual mores from the four corners of subjugated Asia and beyond was unique and disturbing and exciting. Certainly there had been nothing like it in China since the Han dynasty, when Buddhism arrived from India. In life and art, under T'ai Tsung there were flavors of Persia, of India, of Central Asia. And now, with the coming of the Nestorians, there was added a whiff of the Mediterranean.

For Alopen and the Nestorians the contrast must have been an acute one. Just as that Chinese age was opening its flower, they arrived from a Europe whose legacy of civilization from Greece and Rome was in near-total

5. Quoted by Grousset.

eclipse. Syria, perhaps, did not present quite so somber a picture as did Europe proper, but Alopen must nevertheless have found his eyes widening in surprise as he surveyed the Chinese scene. For at that precise time the culture of the world flowed to China. And in the Middle Kingdom itself it was the "Golden Age" of T'ang poetry and T'ang pottery, of painting and the arts of music and dance, an age of high sophistication, of economic expansion, and of wit and intelligence unrivaled in China before that time. This revitalized cradle of Chinese civilization boasted a refinement such as Europe had not known since Rome fell, an ambiance long-since forgotten in the West. There, in Ch'ang-an, stood temples and palaces such as had not been built in the West since the Parthenon and the country villas of great Romans rose into the Mediterranean air. And there were gardens, acres of pleasances subtly contrived so that nature obeyed the curious canons of Chinese art. There also, if we may turn momentarily to more practical things, were evidences of the first crops of Chinese sugar cane, the earliest wine from Chinese-cultivated grapes, the spread of tea-drinking as a national rather than an aristocratic addiction. And even in these facts there was an element of sophistication which was, at that time in the seventh century, peculiar to China.

The Nestorians prospered and their church had numerous adherents—probably for the most part drawn from the Western expatriates in the capital, but doubtless also boasting a few inquiring minds from the ranks of the Chinese upper classes. But that happy situation was not to last for long. Auspicious as a beginning, it proved the start of a checkered career. T'ai Tsung was succeeded by his son Kao Tsung, who, according to the Nestorian Stone, "spread the religion over the ten provinces. . . . Monasteries occupied every big city and families enjoyed brilliant happiness!"[6] The description, with its implication of instant joy at the touch of the Nestorian philosopher's stone, sounds too good to be true, and in fact it can be countered by information engraved on a Buddhist stone only slightly later in date (A.D. 824): "Among the different foreigners who have come to China," this lapidary document states coldly, "there are the Manichaeans, the Nestorians, and the Zoroastrians. All the monasteries of these sorts of foreigners in the empire together are not enough to equal the number of Buddhist monasteries in one small city."

The population of China at that time was probably around fifty-four million, so we may guess that remarkably few Chinese had turned into Nestorians by virtue of that instant joy. China was a hard nut to crack. Very soon it was to turn into a nutcracker and pinch the Nestorians painfully. The emperor Kao Tsung was a lesser man than his father. His character was, besides, eroded by a grand passion for his late father's favorite con-

6. Translation from Moule's *Christians in China.*

cubine. He retrieved her from the Buddhist nunnery to which all the harem of T'ai Tsung had been consigned at his death, and set her up again in the palace. According to the contemporary poet Lo Ping-wang, who hated her, this "vixen—her eyebrows arched like the antennae of a butterfly" tried every stratagem to gain imperial power for herself.[7] By a grisly series of murders she eventually accomplished her aim. As the notorious empress Wu, she turned her elegant but treacherous smile on the Buddhists. Thus encouraged, and "taking advantage of these circumstances," as the Nestorian Stone delicately puts the facts, "the Buddhists raised their voices against the Luminous Religion."

After the death of Empress Wu it was the turn of the Taoists—"inferior scholars" is the stone's euphemism for them—to attack. All in all, to be a Nestorian must then have been a hard path to tread. Alopen, fortunately for him, could not have lived long enough to suffer this reverse. First the Nestorians had suffered from Kao Tsung's wayward sexual fancies, and then from the dislike of a homicidal strumpet on the throne. Now other sects were hacking at their ankles.

Their eclipse was temporary. With Hsuan Tsung, who reigned from 712 to 756, the sun shone again on the Nestorians. "The consecrated rafters which had been bent for a time were once more straightened," as the Nestorian Stone, with consummate and very Chinese tact, puts the matter. Something of the quality of the T'ang is typified by this emperor. Hsuan Tsung was a patron of the theater, and of literature. He it was who established an actors' school in the pear garden of his palace—and the Chinese, with their innate traditionalism, have ever since referred to stage people as "sons of the pear garden." And toward the end of his reign there was set up the Han-lin-yuan (literally, Forest of Pencils Academy), a very early equivalent of the Académie Française, which turned out to be one of the most enduring of all Chinese institutions. Both the arts and religion flourished in the long reign of the mild Hsuan Tsung, but toward its end China found herself confronted with growing problems on the far frontiers as a long struggle developed for domination of Central Asia.

The fortunes of the Nestorians were intimately linked with this struggle, for, with Islam already overflowing from the Middle East, Chinese power was being challenged. The Nestorian return to favor in Ch'ang-an was partially motivated by Chinese desire that they exert their influence on sister Nestorian churches in the Middle East and elsewhere to join the Chinese side of the struggle.

The Nestorian Stone in its oblique way records something of all this, and even names a priest who came just then from Syria. The priest was ordered to celebrate Christian service in the T'ang princes' palaces. Once

7. Quoted by Grousset.

more, relations between Chinese court and Christian church were cordial. But not for long. Chinese armies suffered defeat after defeat on the frontiers. After the disaster at the Talas River in 751, when they were routed by a Moslem army, the T'ang administration saw that the Nestorians could not help them. And, once the emperor had been deposed, the Nestorians were deprived of their shelter. Their plight, like that of the T'ang dynasty itself in those days, was a sorry one. For the T'ang was entering its terminal phase, a time of atrophy which had afflicted many a dynasty before and was to be paralleled by similar witherings in later dynasties in the future.

The process was accompanied by unusually virulent antiforeign sentiments, and by anticlerical ones too—for priests and religions were mostly of non-Chinese origin. The great T'ang poet Po Chu-i sums up the situation in a few lines. Lamenting the decay of T'ang refinement, the absence in the capital of the dancers and musicians of old, he says:

"No more do I hear the sound of music-making, but only bells and chimes:
On the gates of temples in golden words the Imperial Patent shines.
For nunneries and wide Buddhist courts plenty of room is found;
Here moss under the brilliant moon, acres of vacant ground;
But in the crowded homes of ordinary folk there is hardly space to live.
It makes me think of how at P'ing-yang a great house went up,
Seizing and eating I don't know how many people's homes.
And when the two princesses died, even their palace turned into a temple.
So now I begin to fear every home will soon turn into a Buddhist church."[8]

The end of the T'ang and the eclipse of the Westerners and their religion were both very near. The edict of A.D. 845 extinguished the brief Nestorian candle of Christianity in China. Gone were the days when the emperor smiled. Wu Tsung, in his edict, underlined the need for action, the futility of contemplation and decorative spirituality. The dynasty was in desperate straits.

"Under our three famous dynasties," he said, "no one ever heard of Fo [Buddha]. It is since the Han and Wei [from about the time of Christ onward] that the sect began to spread in China. Since then the foreign elements have established themselves without our people being sufficiently on their guard . . . and the State suffers in consequence. . . . In all the cities, in the mountains, there are nothing but priests of both sexes. The number . . . of the monasteries grows daily. Large numbers of workmen are employed making statues . . . in all kinds of materials. A great deal of gold is wasted in embellishing them. People forget their traditional rulers in order to serve under a master priest. . . . Could anything more pernicious be imagined? . . . Today an infinite number of priests . . . live upon . . .

8. Translation by the author and Pih Tze-wing.

NESTORIAN-CROSS RELICS. *Left,* on monuments found in various parts of China in the present century. *Right,* on bronze seals found about 1929 in the region of Tenduc in the great loop of the Yellow River. (From *Nine Dragon Screen*)

the sweat of others, expend the time of an infinite number of workmen in building . . . and furnishing at great expense their magnificent edifices. Need we seek further for the cause of exhaustion of the empire?. . . "[9]

Tendentious though the argument was as a cause of his dynasty's failure, Wu Tsung had some reason for his edict. He goes on to castigate his forebears and to give some astonishing statistics. "As for our dynasty . . . its founders . . . concerned themselves with ruling . . . by wise laws. . . ." But T'ai Tsung was too soft toward the Buddhists. Wu Tsung has determined to "dry up this vile source of errors which flood the empire." Then comes the command for dissolution: The more than 4,600 monasteries shall be destroyed, and the 260,500 men and women who inhabit them shall return to the world and pay their proper taxes. Their lands are to be confiscated, and the 150,000 slaves of the monasteries are to be returned to the bosom of the Chinese population. As for the Nestorians, together with another sect, there are only about three thousand of them. They too are commanded to return to the ordinary world, "so that in the customs of our empire there shall be no mixture."

No mixture. The idea was to be repeated again and again in the future as the Chinese strove to protect what they visualized as their own purity, their Chinese-ness, of which they were so proud.

The great days of T'ang power and T'ang culture were over. The tremulous flame of Nestorian Christianity, of Western ideals, was snuffed out. Along with all other religions, Nestorianism was made the scapegoat. But not all the sacrifice of its institutions (pernicious though they may have

9. The edict is translated by Henri Havret in the series *Variétés Sinologues.*

been in a financial sense), not all the suppression of the refinements of life lamented by the poet, could save the T'ang. One of the great ages of Chinese life had reached its natural end.

What happened to the Nestorians is not known. They vanish with all their effects from the Chinese scene for several centuries, although they were disseminated over Central Asia. In China, even today, a Nestorian cross comes to light now and then, often in an unexpected place. Father Ricci, seven or eight hundred years after the T'ang, was puzzled from time to time when he discovered people in China who made the sign of the cross but did not know what it meant. The completeness of the dissolution of the Nestorian monasteries and churches, which Alopen and his followers had set up in Ch'ang-an, is well shown by the tale of an Arab traveler to China in the ninth century. "In the year 377 (A.D. 987) behind the church in the Christian quarter of Baghdad," says the writer who reports the matter, "I fell in with a certain monk of Najran, who seven years before had been sent to China by the Catholics with five other ecclesiastics, to bring the affairs of Christianity in that country into order. He was a man still young, and of a pleasant countenance, but of few words, opening his mouth only to answer questions. I asked him about his travels and he told me that Christianity had become quite extinct in China. The Christians had perished in various ways; their church had been destroyed; and but one Christian remained in the land. The monk, finding nobody whom he could aid with his ministry, came back faster than he went."[10]

Not even the most ardent partisan can trace any residual influence of the two Nestorian centuries on the life and thought of China. Unlike its bigger rival, Buddhism, Nestorian Christianity did not survive except as the vaguest of memories here and there. China, Chinese thought, Chinese institutions, other sects, other rulers came and flourished and passed away in the great mill of China. As the T'ang poet Li Po says: "The waves follow one after the other in an eternal pursuit." The infinitesimal, transient Nestorian wave that lapped the shore of the great T'ang in China simply broke on that shore with its small splash, and disappeared.

WINGED BABY OF PAINTED COTTON. Until a few decades ago these dolls were on sale in rural China to customers predominantly non-Christian. They appear to be derived from Western angels or cherubs. (From Ecke and Demiéville)

10. Quoted by Roberts.

CHAPTER TWO

THE PRIEST WHO WAS A HEAVY MAN

IT WOULD be foolish to approach the journey of Friar William of Rubruck with anything but humility. For it was with this fundamental attitude that he approached his own life, the hazards of his trip, and the pitiless Mongols, whom he so ardently wished to bring to his God. Although he never reached China proper, never crossed the Great Wall, he did have close contact with the rulers of China, and for this reason, as well as for the magnitude of his travels along roads to be followed by later visitors to China, his story has a very real bearing on our own.

Unfortunately, the only description we have of Friar William is his own remark that at the time of his travels he was "a very heavy man."[1] No one else thought to record a detail of what he looked like. As to his birth and death, we are ignorant of the dates, but it seems he came from a village in French Flanders. He emerges among the crossed swords of popes and emperors that characterized his times, when the Saracens were occupying the Holy Land and the West was crusading to rescue it from them, and when the Mongols had already overrun most of eastern Europe, decimating its population and striking terror into the hearts of popes and kings alike. He emerges, solid as that rock on which his faith was founded, and never wavers from the goals he set for himself.

We know little of his appearance, but of his character we have an excellent picture in his own words—in those dog-Latin sentences in which he set down the report of his journey for the foolish saint, Louis IX of France, who had sent him across the wastes of Asia. Unwittingly, he gives a remarkably full account of himself, one that leaps clearly out of the astounding tale he has to tell of that pagan wilderness of barbarity and ice through which he went toward China.

The story of Friar William properly begins in Paris in the mid-thirteenth

1. Throughout this chapter the quotations from Friar William are taken from Rockhill's *Journey of William of Rubruck.*

century, at the court of Louis IX, about four years before the friar set out across Asia. Here it seems possible that he met two travelers, Friar John of Pian de Carpini and his companion Benedict the Pole, who had recently returned from a papal mission into Mongol territory. Perhaps it was at this time that Friar William's Franciscan heart first filled with the urge to go into the wilderness and attempt the prodigious task of converting the heathen Mongols. John of Pian de Carpini, after all, seems to have known St. Francis of Assisi himself, and what more profound incentive could a Franciscan have to strike out into the pagan unknown than from the fountainhead of the saint's recent presence?

But the time had not yet come for Friar William: the good Louis of France, blazing with his own brand of impetuous Christianity, was set on a crusade, and with him in 1249 went his queen Margaret and the friar. The expedition to Egypt was hardly a success, and after defeat at the battle of Mansurah, in the Nile Delta, and captivity, the royal court turns up in the summer of the following year with the raggle-taggle of an army on the Palestinian coast. It was here that detailed news reached Louis of Christians among the Mongols, and of a Christian Mongol chief called Sartach, son of the ferocious general Batu, whose armies had ravaged eastern Europe and Russia. Such was the unreliability of the information, and the delight of the saintly king at hearing it, that he and Friar William seem to have taken it for granted that Christian missionaries would now be favorably received in Central Asia by the Mongol khans. How far from the truth this was, Friar William was soon to discover. But he was not yet ready to start out.

With a thoroughness quite unlike the religiosity of his master the king, Friar William first set about learning all he could of Mongols and of Asia. Opportunities for expanding his knowledge were not lacking, and, more than any traveler to Asia before him, and more than most who followed in later times, he took full advantage of them. Probably the foundation of his knowledge had been laid in Paris by John of Pian de Carpini and Benedict. And he had carefully read and memorized the geographical writings, dating from the third and seventh centuries, of Solinus and Isidorus of Seville, from both of whom he quotes. His is not, however, the slavish quoting that blindly takes for granted the accuracy of the source: almost always as he later stumbled on peoples and places and customs about which he had read, he carefully checked his reading with his own observations.

While waiting to set out, Friar William undoubtedly met returning embassies—that of Friar Ascelin, sent by the pope and accorded only humiliation by the Mongols they encountered, and that which King Louis himself had dispatched from Cyprus some years before, equally a failure. This latter mission returned with a letter from the Mongol female regent

Ogul-Gaimish that must rank as one of the rudest missives ever written by one ruler to another. "If you do not send me gold and silver each year by your envoys," she replied, in effect, to Louis, "I shall simply destroy you and them." But it must be added that she also sent a cloth made of asbestos, which astonished Louis and his court in that it resisted burning.

Undeterred by the threatening letter, Louis pursued his schemes to convert the Mongols and range them on his side in the reconquest of the holy places of Christendom. With the returning travelers there also came much hyperbolic and apocryphal information, which undoubtedly also reached Friar William's ears. Mixed in with facts, such as that the Mongols came from the steppes to the east of impassable mountains, were half-truths of a windy immensity matched only by the extent of that vast territory, then still vaguely termed Seres, stretching away across Asia. He heard tell of the people of Gog and Magog, previously thought to be located in the Caucasus, who were now alleged to live even farther east. He heard of Genghis Khan, the great chief of the Mongols, as having been converted to Christianity after a vision in which the Lord promised him power over the fabled Prester John. He heard too of hundreds of Christian chapels mounted on carts, and of European prisoners toiling as slaves for the Mongol hordes. And he heard those grisly and all-too-true stories of the mounds of bleaching human bones, of the devastated cities of southern Russia, which the Golden Horde of Batu's Mongol army had left in its wake.

But which of these tales were fact and which fable? Friar William had no means of telling. How to sort out fact from fable in the thirteenth century (and Friar William was very capable of skepticism, as we will see) must have been a puzzle even for an astute man. For King Louis and for the friar the salient fact was the presence of Sartach, the Christian Mongol chief, somewhere in the Asian wastes. And on this deceptive pinpoint of light in the Mongol dark both king and friar seem to have fixed their gaze.

Apart from this mixed bag of unreliable information, Friar William had at his command some knowledge of Arabic, probably picked up in the Middle East. On his Central Asian journey he even managed to acquire a smattering of Mongol, and both tongues proved of use to him.

To understand the reason for the friar's journey, we have to recall the dire fear in which the Europe of those days existed—fear of extinction in a very literal sense at the hands of Mongol armies. Already European Russia, Poland, Hungary, and most of eastern Europe had been ravaged, their cities razed, their peoples decimated, their fields burned, and their farms destroyed by the fury of an enemy entirely unknown to the West before its sudden irruption. The process of expansion by the sword, begun far away to the east in the Mongol heartland by Genghis Khan, had licked out and, like flame on flesh, had seared the body of Europe wherever it touched.

In Western Europe, one of the effects of Mongol devastations was, at first, to open even wider the rift then existing between the pope and Europe's temporal rulers. The death struggle in which the pope was engaged with the emperor Frederick II was in no way healed by their common danger. At first neither recognized the extent of the danger; and when they did, neither could bring himself to sink his differences with the other for the common good. Had the Mongols—the Golden Horde under the ultimate generalship of the brilliant Batu—pushed on, there is no doubt that all Europe, together with its ancient way of life stemming from Greece and Rome and Christendom, would have been overwhelmed as easily as a sand castle by the incoming tide.

Why did the Mongols' European adventure come to an abrupt end just when all Europe lay at its feet? The ostensible reason was that the great khan Ogotai (third son of Genghis Khan and ruler of all the Mongol clans) died in 1241 and that in order to elect a successor it was customary for all the chiefs, descendants of Genghis, high-ranking officers of state, and great generals to repair to Karakorum (in what is now the Mongolian Peoples' Republic) to form a sort of parliament for this purpose. With what seems a nomad lack of perseverance in what they had willy-nilly begun —the conquest of the world—they dropped everything and made their way to Karakorum, to the power struggles that took place in those round felt tents outside the capital itself.

Thus, although Ogotai died in 1241, the same year as did Pope Gregory IX, Europe did not know of his death until some time later, did not know that the terrible scourge had finally been lifted. Both the new pope and his predecessor's archenemy, the Holy Roman emperor Frederick II, still trembled for their lives and domains. In 1248, Innocent IV was still convinced that Europe was in danger from renewed Mongol incursions and, with more reason, that Frederick too might overwhelm him. So he sent John of Pian de Carpini and Benedict the Pole to Paris in an effort to deter King Louis from setting out on his projected crusade to Egypt and to enlist his help against Frederick if and when the need arose.

It was at this time that Friar William probably met the two Asian travelers, who had actually seen the election of Ogotai's successor, Kuyuk, in Asia in August, 1246. Kuyuk Khan was to reign just under two years, dying in April, 1248; and by the time Friar William reached Karakorum a new khan, Mangu, was on the throne. The haphazard conquest of Europe had long been given up by the Mongols, whose energies now turned in an eastern direction, on China. But Europe, deeply divided against itself and obtuse in regard to the character, institutions, and possible intentions of the Mongols, still shivered in its shoes at the thought of further waves of cavalry from the East.

Friar William was evidently a much cleverer and in many ways a more realistic man than his master Louis. If Louis imagined he could enlist the

"Christian" Mongols' aid in wresting the holy places from the grasp of the Saracens, Friar William appears to have had fewer illusions on the subject. It would seem, indeed, that while faithfully carrying out the commands of Louis in the political field, he himself had a much more purely evangelistic intention.

Sailing out from Constantinople toward the Black Sea on a May day in 1253, Friar William had with him a pathetic little party of four. His companion was a friar as frail as William was stout of body and heart, Bartholomew of Cremona. There was also Gosset, "the bearer of the presents"—those entirely inadequate "fruits, muscadel wine, and dainty biscuits" which Friar William in his simplicity felt were the most a monk under vows of poverty could take to offer to chieftains gorged with the treasures of half of Europe and most of Asia. The remaining two of the party were Homo Dei (the "man of God," perhaps a Latin version of the Arabic

ITINERARY OF FRIAR WILLIAM OF RUBRUCK

		1253
May	7	Left Constantinople
	21	Landed in Crimean Peninsula
June	3	Met the Mongols for the first time
	5	Reached Scatay's camp
	8	Left Scatay's camp
July	20	Reached the Don
	31	Reached Sartach's camp
August	3	Left Sartach's camp
	5	Reached the Volga and Batu's camp
September	16	Left Batu's camp
	27	Reached the Ural River
December	3	Reached Ala Kul
	27	Reached Mangu's camp
		1254
January	4	Audience with Mangu
March	29	Left for Karakorum
April	5	Arrived at Karakorum
May	24	Audience with Mangu
August	18	Left Karakorum
		1255
August	15	In Tripoli (Italy)

See map, page 69, for route

CASTLE OF SOLDAIA, or Sudak, on the Crimean Peninsula, where Friar William began his long trek into Mongol lands. (From Yule's *Cathay*)

"Abdulla," which means the same), who acted as dragoman; and "the boy Nicholas whom I had bought at Constantinople by means of your [King Louis'] charity." The boy was probably a Coman, a native of southern Russia, through which they were to pass. Besides these, Friar William had his vestments, chalice, holy oil, censers, a Bible and missals, a cross, and the psalter "which my lady the Queen had presented me with. . . ." And of course he bore letters from the emperor of Constantinople (Baldwin III, a cousin of King Louis), and from Louis himself, who, having been insulted by a Mongol queen, insisted that Friar William pretend to be merely a messenger and not in any way an envoy to the presumed Christian chieftain Sartach and others.

Reaching the north shore of the Black Sea at Kersona (Sevastopol on the Crimean Peninsula) and disembarking a little to the east, at Soldaia, where later one of Marco Polo's uncles owned a house, Friar William and his party started out on their journey on horseback. In three days they encountered the Mongols who had driven the inhabitants from the peninsula thirty years before.

"And when I found myself among them it seemed to me of a truth that I had been transported into another century," the friar says with surprise. And then he breaks off his narrative to give King Louis a summary of the strange customs of the Mongols. His reasons for surprise, however, appear as soon as he has made the king conversant with the elements of Mongol

A MONGOL PRINCESS AND HER SUITE crossing a river in a camel cart. Northern Sung painting, Ta Kuan period (early twelfth century). (Museum of Far Eastern Antiquities, Stockholm)

character and usage. "When, therefore, we found ourselves among these barbarians, it seemed to me, as I said before, that I had been transported into another world. They surrounded us on their horses. . . . The first question was whether we had ever been among them before. Having answered that we had not, they began to beg most impudently for some of our provisions. We gave them some of the biscuit and wine that we had brought with us from the city [Constantinople], and when they had drunk one flagon they asked for another, saying that a man enters not a house with one foot only. . . . Then they asked whence we came and where we wanted to go. I told them that we had heard that Sartach was a Christian, and that I wanted to go to him, for I had your letters to deliver. . . . They made most diligent inquiry whether I was going of my own free will or whether I was sent. I answered that no one forced me to go, nor would I go if I did not want to, so I was going of my own free will, and also the will of my superior. I was most careful never to say that I was your ambassador." Friar William was at a disadvantage in explaining his position, and the Mongols called him an impostor. It was not a good beginning.

"It is true that they took nothing by force; but they beg in the most importunate and impudent way for whatever they see, and if a person gives it to them it is so much lost, for they are ungrateful. They consider themselves masters of the world, and it seems to them that there is nothing that anyone has the right to refuse them." But the Mongols agreed to take him to a chieftain named Scatay, for whom he had letters from the emperor of Constantinople. "In the morning we came across the carts of Scatay carrying the dwellings, and it seemed to me that a city was coming toward me."

Friar William's surprise was like that of Macbeth when Birnam Wood advanced upon him. It was one thing to hear of a city on wheels, quite another to experience it. But his description of Mongol yurts as of many other sights, is vivid and direct. "Nowhere have they fixed dwelling-places, nor do they know where the next will be. They have divided among themselves . . . the land which extends from the Danube to the rising sun; and every captain . . . knows the limits of his pasture lands, and where to graze winter and summer, spring and autumn. . . . They set up the dwelling in which they sleep on a circular frame of interlaced sticks converging into a little round hoop on the top, from which projects above a collar as a chimney. And this framework they cover with white . . . or black felt. The felt around the collar they decorate with various pretty designs. Before the entry they suspend felt . . . embroidered with designs in color . . . making vines, trees, birds, and beasts. . . . And they make those houses so large that they are sometimes thirty feet in width. I myself once measured the width between the wheel-tracks of a cart, twenty feet; and when the house was on the cart it projected beyond the wheels on either side five feet at least.

MONGOL YURTS. This nineteenth-century drawing faithfully follows Rubruck's description. The tent on wheels drawn by many oxen is complete even to the embroidery. A camel-cart follows it, while in the background yurts are seen in various stages of construction. (From Yule's *Travels of Marco Polo*)

I have myself counted to one cart twenty-two oxen drawing one house. . . . The axle of the cart was as large as the mast of a ship, and one man stood at the entry of the house on the cart, driving the oxen."

Then, meticulously, he goes on to describe how the women make "turtle-backs of twigs" covered with felt and tallow against rain, embroidered and charming—but utterly strange—in which to store their bedding. And these are "tied tightly on high carts drawn by camels, so that they can cross rivers without getting wet."

In those first few days with the Mongols, Friar William is both shocked and fascinated. His is the surprise of a European, a man at home at the court of France, in the intellectual society of Paris. But in the thirteenth century a wonder was a genuine wonder, and even in Paris mental palates were not glutted with the world's strange things. There was nothing remotely resembling those dwellings in Europe. Moreover, as those of us who have had the good fortune to see the Mongols at home will know, Friar William's description is often as accurate today as it was then. It is one of his merits that, faced with the shock of the unknown and the knowledge that his statements were not likely to be checked, he never exaggerates on matters of observation, and also that he seldom misinterprets them.

Much stranger experiences lay ahead of him, and much more terrible. From the moment he met the Mongols, there was never an instant, until he left them almost two years later, when he was not totally in their power. From many an expression he uses one feels his taut sensations of claustrophobia, of being shuttled here and there, ever deeper into the cold heart of Mongol country, at the behest of men to whom he was merely another European among millions whom their armies had destroyed in lands overrun. The value of human life at that time was not high in Europe itself; with the Mongols, as Friar William quickly sensed, it was naught. Yet during his whole journey the friar must have kept almost daily notes, or he could never have recalled the events of exact days, as he does, when his account was later put to paper in Italy. In the circumstances this was something of a feat in itself; but he feels the result is inadequate. Of Mongol encampments he says aoplogetically to King Louis: "I would not know how to describe them to you unless by a drawing, and I would depict them all to you if I knew how to paint." And then he goes on to set down one of the best descriptions of Mongol life, beliefs, food, customs, laws, that has ever been written—one which in many ways was as valid of Mongols until recently as it must then have been.

"When they set down their dwelling houses they always turn the door to the south, and after that they place the carts with coffers [containing their possessions] on either side near the house . . . so that it stands between two rows of carts as between two walls. . . . A single rich Mongol has quite a hundred or two hundred carts with coffers. Batu [whom he was later to

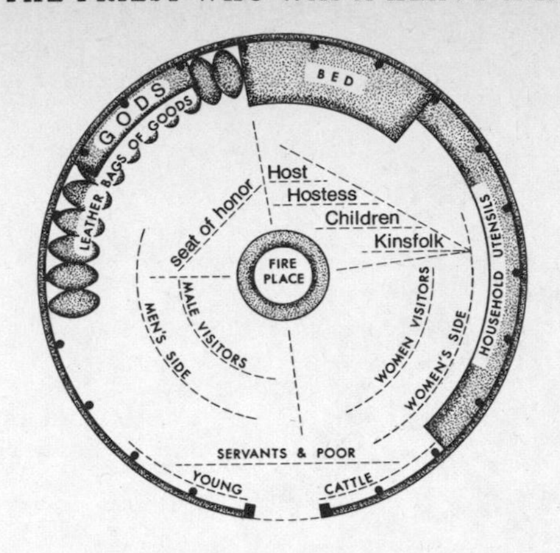

PLAN OF A MONGOL KHAN'S YURT. (Redrawn from Rockhill)

meet] has twenty-six wives, each of whom has a large dwelling, exclusive of the other little ones which they set up after the big one, and which are like closets, in which the sewing girls live; and to each of those large houses are attached quite two hundred carts. . . . The *ordu* [encampment] of a rich Mongol seems like a large town, though there will be very few men in it. One girl will lead twenty or thirty carts, for the country is flat . . . and should it happen that they come to some bad piece of road, they untie them, and take them across one by one. So they go along slowly as a sheep or an ox might walk."

When the yurt is erected, with its door facing south, "they set up the couch of the master on the north side. The side of the women is always the east side, that is to say on the left of the master, he sitting with his face turned toward the south." The men are placed on the west side; "men coming into the house would never hang up their bows on the side of the women." To see how exact and precise this and following statements of Friar William are, we can compare them with the sketch made by the traveler Rockhill in the 1890's. In over six centuries little has altered, and Friar William's words bring the scene to life. Over the chief's head, suspended on the tent wall, "is always a felt image like a doll, which they call the brother of the master; another similar one is above the head of the mistress . . . and higher up between them is a little lank one who is . . . the guardian of the whole dwelling. . . . Beside the entry on the women's side is yet another image with a cow's tit for the women, who milk the cows. . . . On the men's side there is a mare's tit for the men, who milk the mares."

The chieftain, reclining on his couch, has with him several wives, "she with whom he has slept that night sits beside him in the day, and it becomes

COSMAS CHURN AND MONGOL IDOLS. (From Yule's *Cathay*)

all the others to come to her dwelling that day to drink, and the court is held there that day, and the gifts which are brought that day are placed in the treasury of that lady." Strange as it is to Friar William, he nonetheless appreciates the workability of this system of maintaining both psychological and material comfort among the ladies. There is nothing at all narrow about him.

But let us follow him on his travels. "For two months from the time we left Soldaia to when we came to Sartach, we never slept in a house or a tent, but always in the open air or under our carts, and we never saw a city, but only the Comans' tombs in very great numbers." (Comans were South Russians, decimated by the Mongols—the people from whom came his "boy" Nicholas.) The truth of that part of what the friar had heard in Palestine was beginning to be apparent. Departing from this camp, "it seemed to me that we had escaped from the midst of devils. On the next day we came to their captain."

Scatay, to whom the Mongols now led him, "was seated on his couch with a little guitar in his hand, and his wife was beside him. It seemed to me that her whole nose had been cut off, for she was so snub-nosed that she seemed to have no nose at all; and she had greased that part of her face with some black unguent, and also her eyebrows, so that she appeared most hideous to us." He presented his gifts—a basket of those "dainty biscuits" and a flagon of wine, "excusing myself, being a monk and not allowed by my order to own gold or silver or costly robes." Scatay was unable to read the letter from the emperor of Constantinople (which was written in Greek) and sent it back to the Crimea for translation. "He asked us if we would drink *cosmas,*

A MONGOL WOMAN SADDLING A YAK. Note the blunted horns and tethered legs of the animal. The scene, photographed in 1957 in northwest Kansu province, is one which would be familiar to Rubruck, whose description of yaks is accurate. (Photo by Brian Brake)

or mare's milk"—the Nestorian Christians and other European elements enslaved by the Mongols and carried with them refused to drink it, considering themselves no longer Christians if they did. But Friar William was not taken aback. He had already tried the stuff the day before, when "at the taste of it I broke out in horror and surprise. . . . It seemed to me, however, very palatable, as it really is." Often enough in the future he was to be glad of this drink, when it stood between him and starvation. *Cosmas* was made in this way: "They stretch a long rope on the ground, fixed to two stakes" and to it "tie the colts of the mares they want to milk. Then the mothers stand near their foals and allow themselves to be quietly milked." When a mare objects, her foal is brought to her, allowed to suckle a while, and then taken away.[2] The milking continues. The milk, "which is as sweet as cow's as long as it is fresh, they pour into a big skin or bottle, and they set to churning it with a stick . . . as big as a man's head at its lower extremity and hollowed out." The milk "begins to boil up like new wine and to sour and ferment, and they continue" until the butter is extracted. "Then they taste it, and when it is mildly pungent, they drink it. It is pungent on the tongue like *rapé* wine . . . and when a man has finished drinking, it leaves a taste of milk of almonds on the tongue, and it makes the inner man most joyful, and also intoxicates weak heads, and greatly provokes urine." Among more weighty matters this is an example of the incidental delights of Friar William's story.

Later on in his book the friar, in another of his vivid little verbal snapshots, describes the animal we call a yak, which was much in use then among the Tangut people in what is now the province of Kansu in western China. "These people have very strong cattle with very hairy tails like horses, and with bellies and backs covered with hair. They are lower on their legs than other oxen, but much stronger . . . have slender, long curved horns, so sharp that it is always necessary to cut off their points." The cows, he continues with a touch of dry humor, "will not let themselves be milked unless sung to."

From Scatay's *ordu,* Friar William and his party were sent on to Sartach in charge of a guide. By oxcart the journey took almost two months. And it was during this time that the friar began to experience the inhumanity of the Mongols. Because he had brought nothing of value to give them, for the first leg of the trip (to a camp five days distant) he was provided with only one goat, several skins of cow's milk, and a little *cosmas*. This was supposed to feed the party. "And so we set forth due north [Friar William's sense of direction is not his strong point—it was northwest that they must have gone] and it seemed to me that we had passed through one of the gates of hell.

2. For a description of this same technique of milking mares in present-day Mongolia, see Lattimore's *Nomads*.

The men who conducted us began robbing us in the most audacious manner, for they saw that we took but little care. Finally . . . vexation made us wise." It was the penalty they paid for having brought nothing but those dainty biscuits to offer the Mongols. They were despised.

But on they went, "with the sea [of Azov] to the south and a vast wilderness to the north . . . over thirty days in breadth; and in it [South Russia] is neither forest nor hill nor stone, but only the finest pasturage. Here the Comans used to pasture their flocks." And he quotes Isidorus on the extent of the country. "It was all ravaged by the Mongols," who, when they can extract no more gold from the inhabitants, "drive them off into the wilds, them and their little ones, like flocks of sheep. . . ."

Thus "we travelled eastward, seeing nothing but the sky and the earth . . . and the tombs of the Comans. . . . As long as we were in the desert it fared well with us, but such misery as I had to suffer when we came to inhabited places, words fail me to express." For they could find nothing in the way of food to be bought with whatever little money they had, and at every place their guide, to ingratiate himself, wanted them to give presents to the headman. And as they had nothing to give, the headman did not trouble himself to provide even the essentials for their further progress. "To add to this, when we were seated in the shade under our carts, for the [midsummer] heat was intense, they pushed in most importunately among us, to the point of crushing us, in their eagerness to see all our things. If they were seized with a desire to void their stomachs [he means, presumably, bowels], they did not go farther away than one can throw a bean; they did their filthiness right beside us while we were talking together. . . . Above all this, however, I was distressed because I could do no preaching to them; the interpreter would say to me: 'You cannot make me preach, I do not know the proper words.'. . ." After a while, when Friar William had learned "something of the language, I saw that when I said one thing he said a totally different one, according to what came uppermost in his mind. So . . . I made up my mind to keep silence." But it went much against the grain to do so.

They crossed the Don (described with nostalgia—"as broad as the Seine at Paris") with the help of the local ferrymen stationed there by the Mongols. With each of its two wheels in a separate boat their cart was paddled across the water. And at last, with much hardship they came to the camp of the putative Christian, Sartach, whom William so ardently desired to meet. But, irrepressible observer that he was, William has put down on the way his comments on the geography of the route and its surroundings; the pasture habits of the Mongols; the beauty of the wooded country; the plight of its inhabitants, whom the Mongols had pushed in front of them to be slaughtered at the head of their armies in the invasion of Poland and Germany.

They were now more than four hundred miles from the Crimea, where they had disembarked, "three days from the Etilia"—the Volga. It was the last day of July, 1253, and the heat lay like a blanket on the land. Friar William little knew it then, as he reached this place due north of what is now Stalingrad, but he was to journey farther, more than two thousand miles into the Mongol unknown, before reaching a destination for which he had not thought of aiming.

Sartach's camp was much bigger than Scatay's, as befitted a more important officer. On the day following his arrival there, Friar William was instructed to put on his priestly vestments for the presentation to Sartach. He did so, and "took the bible you [King Louis] had given me, and the beautiful psalter which my lady the Queen had presented me with, and in which were right beautiful pictures. My companion [Friar Bartholomew] took the missal and the cross, while the clerk [Gosset] put on a surplice and took the censer. And so we came to Sartach's dwelling and they raised the felt which hung before the entry so that he could see us." What he felt at this moment, Friar William does not tell us, but it is apparent from the way he proceeds that he was conscious that this was the great moment to which his long and arduous journey had led him.

The attendants cautioned the party not on any account to touch the threshold of the great tent, this being a strict taboo (a point worth noting, as we shall see later). And, as the Mongols and all Sartach's wives pressed round them with intense curiosity, they went into the chief's presence chanting "Salve, Regina!" Coiac, a Nestorian Christian and an important man in Sartach's camp, took the censer and other holy objects Friar William had brought and handed them to Sartach, who "took a good look" at the illuminated psalter "as did his wife who was seated beside him." Sartach wanted to know if it was Christ's image on the cross. "I replied that it was. Those Nestorians . . . never make the image of Christ on their crosses. . . . Then he caused the bystanders to withdraw so that he could better see our ornaments." It is easy to imagine Sartach's wives and retainers falling over each other in their rude curiosity, and the gesture of irritation with which Sartach waved them back. Then the friar handed over the letter of King Louis, which was immediately translated into Mongol from the Arabic and Syriac of the original.

Friar William must have waited for the answer with hopes reasonably high. The following day it came. "The Lord King hath written good words to Sartach," Coiac the Nestorian reported to him, "but they contain certain difficulties concerning which he would not venture to do anything without the advice of his father; so you must go to his father." And Coiac added: "You must not say our lord is a Christian. He is not a Christian but a Mongol." And Friar William says in some disgust: "They have risen so much in their pride that they may believe somewhat in Christ, yet they will not be called Christians, wishing to exalt their own name of Mongol above all others."

He was sad. It was obvious that those stories of a Christian Sartach were false, and that there was little hope of spreading the faith in Sartach's *ordu*. The great moment had proved empty. Wretched, but with infinite perseverance, he set out again, this time to Batu, Sartach's father, somewhere in the distances east. In the four days with Sartach no food had been offered to him, and his hosts forced him to leave all his vestments and holy objects behind, saying: " 'What! You have brought these to Sartach, and now you want to take them to Batu!' I had one comfort; as soon as I discerned their greed I abstracted the bible. . . . I did not dare take out the psalter of my lady the Queen, for it had been too much noticed on account of the gilded pictures. . . . Of Sartach I know not whether he believes in Christ or not . . . and it even seemed to me he mocked the Christians." But, observes Friar William shrewdly, "he is on the road of the Christians"—of all those who come from the conquered lands and have to pass through Sartach's *ordu* on the way to the court of Batu. "Those who bring him the best presents are sent along most expeditiously."

And so, crossing the Volga and heading south for a hundred miles, they came to the *ordu* of the great general Batu, the scourge of a whole continent, whose face was covered with red spots. With the loftiest condescension, Batu received Friar William and made him kneel in a sort of kowtow. The friar,

RUBRUCK KOWTOWING BEFORE BATU. An all but completely fanciful engraving of the early eighteenth century. Only the circular form of the tents is correct. Palm trees certainly never grew in Central Asia. (From Vander)

finding himself in this posture, temporarily forgot himself in his very natural excitement and, as he says, for a few moments imagined he was in the presence of God himself, praying on his knees!

It was in sentences couched as for God that he began his address to Batu. But Batu allowed his entourage to make fun of the travelers and then dismissed them from his presence. It was five weeks before he made up his mind what to do with these strangers—or perhaps he just forgot about them for that space of time. And then he commanded they go on to Mangu, the khan of all the Mongols, leaving Gosset behind with him. King Louis had made a great mistake in asking by letter that Friar William be allowed to stay and preach to the Mongols. It would have gone much better for Friar William had he merely asked this himself when his hosts had had a chance to sum him up for themselves. But with such an introduction they suspected he was some sort of spy, potentially dangerous. And as such they tended to treat him.

During the five long weeks that it took for Batu to make up his mind, the little party traveled with the *ordu,* but, unlike the Mongols, they went on foot. "Sometimes my companion would say to me almost with tears in his eyes: 'It seems to me that I shall never get anything to eat.' The market always follows the *ordu* of Batu, but it was so far away from us that we could not get there"—which tells us something of the size of the moving horde that swarmed over the country in the wake of the general. It was only the friendship of some of Batu's Hungarian slaves, who gave Friar William and his group food, that kept them from starvation.

"At last, about the feast of the Elevation of the Holy Cross (September 14) there came a rich Mongol to us, whose father was chief of a thousand. 'I am to take you to Mangu Khan,' he said. 'The journey is a four months' one and it is so cold that stones and trees are split by the cold. Think over whether you can bear it.' " Doubtless the Mongol little relished a winter journey that far and hoped somehow to dissuade the friar. "I answered him," William said, " 'I trust by the grace of God we shall be able to bear what other men can bear.' "And the Mongol, after inspecting their clothing, brought sheepskin gowns, breeches, felt stockings, boots, and hoods.

From the Volga, in mid-September, they went eastward, crossing the southern tip of the Ural Mountains and the Ural River not far from where it flows into the head of the Caspian Sea. This, as Friar William reminds his king, is the country from which came in former times that other scourge from Asia—the Huns. The cold was coming on and there was seldom anything like enough to eat; but all the time, even in the dire straits in which he found himself, he is commenting on the languages of each region, the customs, and what he could learn of the history or geography. Each day "we went, as well as I could estimate, about the distance from Paris to Orleans"—a distance of some sixty miles, which hardly seems likely when traveling by packhorse.

But seeing the conditions the friar had to suffer, one can forgive such small exaggerations. It probably seemed at least that long to them by the end of each day. "They always gave me a strong horse" at the posting stations the Mongols had established over Asia, "on account of my great weight, but I dared not enquire whether he rode easily or not. . . . Consequently we endured great hardships; often the horses were tired out before we reached the stage, and we had to beat and whip them up. . . . Times out of number we were hungered and athirst, cold and wearied. They only gave us food in the evening [following normal Mongol eating habits]. In the morning we had something to drink or millet gruel, while in the evening they gave us meat . . . and some pot liquor. When we had our fill of such meat broth, we felt greatly invigorated; it seemed to me a most delicious drink and most nourishing." On Fridays, even if he had food, the good father fasted until evening when, "though it distressed me sorely," he had to eat some meat or not have strength to go on the next day. "Sometimes we had to eat . . . nearly raw meat, having no fuel to cook it." Reaching camp after dark, they could find no cattle dung or briars to make a fire. It is not hard to re-create his feelings, the cheerless thought of camping in those wilds, cold and hungry, without a fire, and with only raw meat of very doubtful freshness to fill a yawning stomach. To Mongols this sort of life might be normal among the young, but even the older ones traveled when they could in their carts; and to a European the conditions must have been all but insupportable.

"At first our guide showed profound contempt for us . . . disgusted to have to guide such poor folk." But after a little the haughty Mongol softened and began to use them to show off, taking them to other Mongols' tents on the way, where "we had to pray for them." And Friar William comments ruefully again that if only he had had a proper interpreter, he might have done some good. Every one of those Mongols was highly surprised at the temerity of the little party in venturing among them without gold or other tribute, but they were curious about the parts of the West which they had not conquered, wanting to find out if the pope were really five hundred years old as they had heard. Among the Mongols, stories of western Europe seem to have been quite as implausible as were those concerning eastern Asia current in the West.

The four-month journey in the stone-splitting cold occupies a considerable section of the friar's book. In its way, with its simplicity and straightforward sentences, above all by its revelation of tolerance toward all men and toward the outlandish and often reprehensible attitudes he found in many of them, this passage must be one of the most remarkable in the literature of travel. It is the writing of a man of profound humility and great moral and physical stamina. Friar William stands very far above the rest of the Europeans who journeyed across Central Asia in the dark times of Mongol ascendancy.

His sentences, in that dog Latin of no pretensions, lead simply and plainly, one into the next, with the naturalness of a stream flowing through fields. "In that desert I saw many wild asses called *culam* and they greatly resemble mules: our guide and his companion chased them a great deal, but without getting one, on account of their great fleetness. The seventh day we began to see to the south some very high mountains, and we entered a plain irrigated like a garden, and here we found cultivated land. On the Octave of All Saints (November 8) we entered a town of Saracens . . . and its captain came out of the town to meet our guide, bearing mead and cups. For it is their custom that in all towns subject to them [the Mongols], they come out to meet the messengers of Batu and Mangu with food and drink. At that season of the year there was ice on the roads in those parts, and even earlier we had frost in the desert. . . . And there came a big river down from the mountains, which irrigated the whole country wherever they wanted to lead the water, and it flowed not into any sea, but was absorbed in the ground forming many marshes. There I saw vines, and twice did I drink wine."

His *culam* is the Mongol *kulan* or *hulan;* and Rockhill, whose translation we are following, here remarks: "I have often chased them on horseback, but even when wounded they could get away from the best pony I have ever seen." The river, named Talas by the friar, is absorbed in the sands of the Myun Kum (west of Lake Issik-kul and 150 miles northeast of Tashkent) in what is now South Kazakhstan in the USSR. High though the region is—up to three thousand feet—grapes have been grown and wine made there for two thousand years.

Here, Friar William—a mere Western speck in the Mongol universe—was about halfway on the road to Karakorum. That he felt his isolation is often evident, but his energy and attention never flag. Now and then he gives a sample of Mongol history, an incident which agrees well with the account in the Mongol *Secret History,* of which he could not have known. On Mongol customs he is always just and objective. He realizes the Mongols are not in all ways a barbarian people. The careful documentation he accords their customs and mores is tacit acknowledgment of this. He records the close-knit fabric of their society, the fact that nomad life depends as much on a high degree of social organization as on an understanding of pasturage and stock management. He appreciates those skills in the Mongols. And one feels that, had he read it, he would equally have appreciated the thirteenth-century *Secret History,* whose strange poetry, with its alliteration at the beginning of lines, tell the stories of past race heroes. Friar William never makes that mistake which the Chinese have made throughout their history of regarding everyone other than themselves as barbarian.

Soon Friar William, Bartholomew, and the others entered the domains of Mangu Khan—whose subjects everywhere "clapped their hands before our guide because he was the envoy of Batu. . . . They show each other this honor. . . . The subjects of Batu, however," he observes astutely, "are the stronger [presumably at this geographical point on the fringe of Mangu's territory] so they do not observe the custom so carefully." Crossing the great river now called Ili, somewhere south of its delta where it flows north into Lake Balkhash, he describes the wide plains of the area in phrases like those of a Russian writer describing the steppes. The towns had been destroyed by the Mongols so that they might graze their herds, but he stayed perforce in one surviving town, waiting the pleasure of his guide to continue. In twelve days of rest there—and welcome rest it must have been—he was not inactive, and later he marshaled the information he amassed there in digest form for King Louis. Little though that monarch may have appreciated the details, the section contains most comprehensive documentation on matters previously only fragmentarily known or not known at all.

From mid-September, 1253, for three months (not four as the guide had predicted), they traveled on through the shriveling cold of the Asian winter toward Karakorum and the unguessable fate that awaited them at the hands of Mangu Khan. Half-starved, led as though prisoner by unfeeling guides who despised him, it was a searching endurance test, and equally an example of that moral courage—more impressive than its physical counterpart—which lends to human beings *in extremis* their peculiar greatness. For here was a man who did not set out into the cold hell of Asia for profit, nor was he going specifically at the behest of his superiors. He was not a Marco Polo on a trading venture, nor an ambassador in the ordinary meaning of the word. In reading Friar William's story one cannot escape the fact that really, fundamentally, and with all his heart, he went across Asia for an ideal. His sublime assurance that he was doing right, the only possible right he could do, is implicit in every page of his book.

A great man, his tolerance is wide, his forgiveness free. The sole passage of sustained and deep condemnation occurs in a description of the habits of Nestorian priests in Asia. In it we can see into what degradation the Illustrious Religion of the T'ang days in Ch'ang-an had fallen as it spread over Asia. "Living among [the peoples of Asia] though of alien race are Nestorians and Saracens all the way to Cathay. In fifteen cities of Cathay there are Nestorians, and they have an episcopal see in a city called Segin [probably Ch'ang-an]. . . . The priests of the idols [Buddhists] all wear wide saffron-colored cowls. There are among them, as I gathered, some hermits who live in forests and mountains, and who are wonderful by their lives and austerity. The Nestorians know nothing. They say their offices and have sacred books in Syriac, but they do not know the language so they chant like those monks among us who do not know grammar, and they are absolutely depraved. In the first place they are usurers and drunkards; some . . . even have several wives. When they enter church they wash their lower parts like Saracens [some Moslems still do]. The bishop rarely visits these places, hardly once in fifty years. When he does they have all the male children, even those in the cradle, ordained priests, so nearly all the males among them are priests." (It is interesting to compare the fact that, until the last decade or so, more than one-third of the male Mongols in the Mongolian Peoples' Republic were in monasteries.) "Then they marry, which is clearly against the statutes of their fathers, and they are bigamists, for when the first wife dies they take another. They are all simoniacs, for they administer no sacrament gratis. They are solicitous for their wives and children and are consequently more intent on the increase of their wealth than of their faith. And so those . . . who educate . . . the sons of rich Mongols, though they teach them the Gospel . . . through their own evil lives and their cupidity estrange them from the Christian faith. For the lives that the Mongols themselves lead are more innocent than theirs." This is much more in anger than in sorrow—the reverse of his usual critical approach.

It would be easy to make a list of facts about Asia which Friar William is the first to mention in Western literature, among them the identification of the ancients' Seres with Cathay, the two having for long been thought to be separate places. But it is rather the accuracy of his intelligence, gathered as it was in circumstances daunting in the extreme, which fascinates. Apart from sorting out very acutely the ethnography of the regions he traversed, he tells much of the Buddhists with their shaved faces and heads, their characteristic robes, and puts down recognizably the incantatory phrase *Om Mani Padme Hum,* still familiar in Tibetan worship. He describes the rosaries of Buddhists with their "one or two hundred beads"—still in use in the lamaseries of Mongolia and Tibet. He shows how the Mongol script was taken over from the Uighur, describes Tibetans, and speaks of the Tanguts of the Hsi Hsia kingdom, which was conquered by the Mongols on their way to the conquest of China proper.

Meanwhile the friar is disputing with the idol worshipers in words ringing with sincerity. "They asked me, as if in derision: 'Where is God?' To which I said: 'Where is your soul?' 'In our body,' they said. I replied: 'Is it not everywhere in your body, and does it not direct the whole of it, and nevertheless, is invisible? So God is everywhere, and governs all things, though invisible, for He is intelligence and wisdom.'" Then, just as I wanted to continue reasoning with them, my interpreter got tired and would no longer express my words, so he made me stop talking."

Then he expresses his joy at finding "a village entirely of Nestorians. We entered their church singing joyfully and at the tops of our voices 'Salve, Regina!' For it was a long time since we saw a church." No more is needed to paint the picture of their feeling of abandonment in the wilderness. The land was vast and fearsome as they came to the lake of Ala Kul. "There opened a valley which came out of high mountains and there amid the mountains was visible another big sea . . . and there blows nearly continuously such a wind through that valley that persons cross it with great danger, lest the wind should carry them into the sea. . . . It was extremely cold so we turned our sheepskins with the wool outside."

The last few days of the journey to Mangu's *ordu* were blighted by the information that Batu's letter to Mangu said King Louis's letters contained requests for Mongol troops to join the crusade against the Saracens. The friar feared to contradict Batu's reported words, reasoning in his own mind that it must have been the Hermenians ("great haters of the Saracens") who translated the letters and inserted the request. He said nothing and waited his time.

At last, two days after Christmas, 1253, they came to Mangu's camp, a few days' journey from Karakorum. Summoned for examination into the presence of Bulgai, a Nestorian and a great power in the Mongol government, he went barefoot as was the custom of his order, to the amazement of the bystanders, who asked if he had no use for his feet since he would inevitably lose them. After close questioning he was returning to his lodging when he noticed a tent with a cross over it. Inside was an altar, embroidered pictures of Christ, the Virgin and John the Baptist, and a Hermenian monk "swarthy and lank . . . dressed in a tunic of roughest haircloth . . . and over it a stole of black silk lined with fur." The monk had been there three months: God, he said, had enjoined him repeatedly to convert Mangu to Christianity. Friar William's hope of finding a kindred spirit was dashed by the chatter of this self-seeking man. The following morning "the tips of my toes were frozen so that I could not thereafter go barefoot."

A week later Mangu commanded their presence. But before they went to him they were questioned on the sort of obeisance they would make. "We are priests in the service of God," Friar William replied. "Noblemen in our country do not . . . allow priests to bend the knee before them. Nevertheless we want to humble ourselves to every man for the love of God. . . . We will do as it shall please your lord, this only excepted, that nothing be required of us contrary to the worship and glory of God." This was apparently satisfactory—a fact worth noting when the vexed question of the kowtow rears its disputatious head later between rulers of China and Europeans at audience with them.

Now it seemed once more, as it had at Sartach's camp, that the great moment of the journey had arrived for Friar William. He and Bartholomew went to Mangu's huge tent. The flap of the doorway was held up; there they stood, just after Christmas, singing a Nativity hymn. After being searched for knives, they went in, leaving the interpreter by a bench with the *cosmas* on it. Under the tent walls of gold cloth, surrounded by his ladies and his courtiers, with a fire of dung blazing before him, "Mangu was seated on a couch, dressed in a skin spotted and glossy like seal. He is a little man, of medium height, aged about forty-five, and a young wife sat beside him;

and a very ugly full-grown girl called Cirina." In this scene of wild splendor they finished singing of the birth of Christ, and Mangu offered them a choice of drinks—wine, rice wine, mead, clarified mare's milk. "My lord," said the friar, "we are not men who seek to satisfy our fancies about drinks: whatever pleases you will suit us." Alas, the interpreter was not of like mind and, placed as he was close to the butlers and the *cosmas,* he was soon drunk.

Then Mangu appeared to forget about the visitors—a favorite trick of later emperors in Peking when they wished to impress. He sent for his falcons, "which he took on his hand and looked at. And after a long time he bade us speak. Then we had to bend our knees." Which they did, without feeling they had lost irretrievably their European dignity as did later visitors to Peking at the mere mention of the kowtow.

Mangu had his interpreter and the friar his, "such as he was and already drunk." Friar William first praised God, who had brought them from so far to Mangu, to whom God had given such power on earth. He prayed for Mangu's long life. "Then I told him: 'My lord, we had heard of Sartach that he was a Christian, and . . . rejoiced greatly, principally my lord the king of the French. So we came to him, and my lord the king sent him letters by us in which were words of peace. . . . And he begged Sartach to allow us to remain in his country, for it is our office to teach men to live according to the law of God. Sartach sent us, however, to his father Batu, and Batu sent us to you. We pray . . . permission to remain in your dominion. . . . We have neither gold nor silver nor precious stones to present to you, but only ourselves to offer you to serve God. . . . At all events give us leave to remain here till this cold has passed away, for my companion is so feeble that he cannot with safety to his life stand any more of the fatigue of traveling on horseback.' "

Mangu replied oracularly: "As the sun sends its rays everywhere, likewise my sway and that of Batu reach everywhere, so we do not want your gold or silver." And that, Friar William says, was the first and only sentence of Mangu's that the drunken interpreter gave him whole and intelligible. " 'Twas by this I found out he was drunk, and Mangu himself appeared to me tipsy."

It is hard to suppress a feeling of anger as the good friar recounts this shabby treatment. He had come too far and with too great pains to be thus treated. But his tolerance was, as ever, profound. Mangu was displeased they had gone to Sartach first, and not come directly to him—something obviously impossible. Friar William protested that the mention of gold and silver did not imply he thought Mangu lacked these things but that he would have been glad to honor the khan with "things temporal as well as spiritual."

And soon the audience ended—as pointless a passage of words as can be imagined; but Mangu relented a little later (perhaps when he was sober), and his interpreter came to tell Friar William that they would be allowed a two-month stay till the thaw and that meanwhile they might go to Karakorum if they wished. But the friar declined the invitation, preferring to stay with Mangu.

One of Friar William's virtues is that he is never dull. Just after this depressing audience, he begins: "A certain woman from Metz in Lorraine, Paquette by name, who had been made a prisoner in Hungary, found us out, and she gave us the best food she could." Like many thousands of Europeans, Paquette had been carried away by the Mongol armies on one of their incursions into eastern Europe and was a slave in the *ordu* of the ugly Cirina's mother. "And she told us of the unheard-of misery she had endured before coming to the *ordu*. But now she was fairly well off. She had a young Ruthenian husband, of whom she had three right fine-looking boys. . . . Furthermore she told us that there was in Karakorum a certain master goldsmith, William by name, a native of Paris."[3]

Friar William's spirits must have risen at this mention of a namesake from his own city—a man, moreover, who was a friend of the great Bulgai, the Grand Secretary. And doubtless his spirits rose even further when he heard that William had adopted a son who was a good interpreter. Paquette confirmed what he must have sensed himself, that the opinion in Mangu's camp was that the friar and his party were acceptable but that the drunken interpreter had made fools of them. So the friar wrote at once to William, but was answered that the son could not come for another month as Mangu had entrusted him and his father the goldsmith with work which was not quite finished. What this work was, we shall soon discover.

Once more, in February, 1254, Friar William was summoned to Mangu's presence. Going into the big tent, he was at first puzzled at the sight of a servant removing charred shoulder blades of a sheep from a fire. In a remarkable passage, he explains correctly this divination process, practiced long since by the Chinese, in which the directions of the cracks that appear in the charred bones are used to determine whether or not a given action should be taken. The Mongol method seems to have been a primitive version of the Chinese oracle-bone technique in which the bones were inscribed with questions before being put in the fire and, after they had been divined, often with the answers as well, by way of record.[4]

Once more they were cautioned not to touch the threshold of the tent, but after Mangu had examined their Bibles and dismissed them, "my com-

3. William Boucher (sometimes Bouchier) served Kuyuk and Mangu at Karakorum between 1246 and 1259, during which time his brother still lived on the Grand Pont in Paris. Before it collapsed in 1296, the Grand Pont was flanked by shops, as is the Ponte Vecchio in Florence to this day, most of them belonging to money-changers and goldsmiths. There is an interesting book on William Boucher by Olschki.

4. Scapulimancy was also known in Europe from early times. In England it was called reading the "speal bone." Until recently it was thought that the Chinese characters on oracle bones were the earliest form of calligraphy, but earlier forms are now known.

panion who had turned his face toward the Khan bowing to him . . . hit the threshold" with his foot. Immediately poor Bartholomew was seized and taken away to Bulgai. "When I looked back and did not see him coming, I thought they had detained him to give him lighter clothing, for he was feeble and so loaded down with furs that he could scarcely walk." If there is just a hint of dry humor in this sentence, it quickly passes. Returning Bartholomew to the party, Bulgai asked whether they knew not to touch Mangu's threshold. "My lord," said Friar William, "we had no interpreter with us; how could we have understood?" Bulgai pardoned Bartholomew, "but never thereafter was he allowed to enter any dwelling of the Khan." Poor Bartholomew! He was a good man, but frail, tremulous, and inclined to complain. Friar William's loyalty to him is constant, though now and then tinged with a trace of irritation.

The two friars remained with Mangu's *ordu* for three months, attempting by every means to see how they could spread their faith among the Mongols. One stumbling block in the way of such progress was the cringing Nestorian monk with whom they shared accommodations, whose character must have been apparent to the Mongols and which doubtless rubbed off on Friar William, since to them he was merely another Christian priest. It is for this cloth weaver (as Friar William discovered the monk was when he passed through Hermenia on his return to Europe) that he reserves his invective. This craven soul was always up to some trick or other. At one point he swore he was about to baptize Mangu himself. "And I begged that . . . I might be an eye-witness of it. And this he promised me." Summoned to court on the day appointed, the friar's hopes rose, only to be dashed when he found no such ceremony was to take place. "So we went back by way of the monk's, who was ashamed of the lie he had told us. We came back to our cold and empty dwelling" ill-supplied with food. "It would have been quite sufficient . . . but there were so many suffering from want of food who . . . had to be given to eat. Then I experienced what martyrdom it is to give in charity when in poverty."

Another time the Nestorian monk rashly swore he would cure the illness of Mangu's wife or else cut off his head. And when the woman got no better, Friar William came to his aid with holy water to add to the powdered rhubarb the Hermenian was administering. Again, when the monk dressed up in clothes like a bishop, the friar seemed to comply and treated him as such in order to demonstrate solidarity, though loathing every moment of the masquerade. When Friar William and Bartholomew were again near starvation and had to beg for food from Mangu, William forgave the monk for keeping in the chapel "under the altar, a box of almonds and raisins and prunes, and many other fruits, which he ate all through the day whenever he was alone. . . . Nevertheless we kept to his company for the honor of the cross."

If he made no progress in evangelizing the Mongols during his stay with Mangu's camp, Friar William amassed much important information. For the first time in Western literature we hear of the kingdom of Korea and the empire of the Southern Sung Chinese, still unsubdued by the Mongols—facts he picked up from talks with a Tibetan priest returning home from Cathay.[5] He notes that all the rivers run in the general direction east, a fact of great importance in Chinese history; he checks the stories of Solinus and Isidorus (stories of dog-headed people in India, and others who lived on the odor of wild apples) and finds no one ever heard of such monstrosities; he tells of herds of monkeys being trapped by putting out wine on which they get drunk and are thus easily caught, a story that can be identified in a Chinese written source. "They also told me as a fact (which I do not believe) that there is a province of Cathay, and at whatever age a man enters it, that age he keeps." This, too, is Chinese in origin—a variant of fables centering on the Kun-lun mountains where grows the Peach of Immortality.

Then, in a most important passage no more than a few lines long, Friar William describes Chinese paper money (of which we will hear Marco Polo speak later), and also the nature of Chinese writing (which Marco Polo never mentions). "The Cathayans write with a brush such as painters paint with, and make in one figure the several letters containing a whole word." This is the first statement in Western literature on the subject, and although he was wrong about the "several letters" (Chinese being analphabetical), he had understood the nature of the ideograph as essentially different from other forms of writing.

Early in April, accompanied by the adopted son of William Boucher, who had finished his work in Karakorum and come back to report to Mangu, Friar William arrived in Karakorum itself. He entered the town "with raised cross and banner . . . and the Nestorians came to meet us in a procession." The same day he at last met William Boucher, the Parisian goldsmith, who "took us with great rejoicing to his house to dine with him" and his French wife. Another guest was an Englishman called Basil who, like Paquette, had been captured in Hungary. It is easy to conjure up the happiness and camaraderie of that evening on the steppes of Asia when the little group of Europeans, speaking French and doubtless exchanging their stories of Europe and homes now lost forever, passed a few hours in entirely un-Mongol conviviality.

On the following day Friar William went to the palace, for Mangu Khan had arrived. Bartholomew, because he had trodden on the threshold, had to stay behind. This is undoubtedly the climax of the Franciscan's marvelous

5. The word Cathay came into English in this way: The Khitan, a pre-Mongol people, established a state in northern China at the expense of the Sung dynasty and ruled later as the Liao dynasty, 937–1125. The Chinese word is Ch'i-tan, the Mongol Kitat, the Russian Ḳitai, and the Arabic Hitai. Hence the English Cathay.

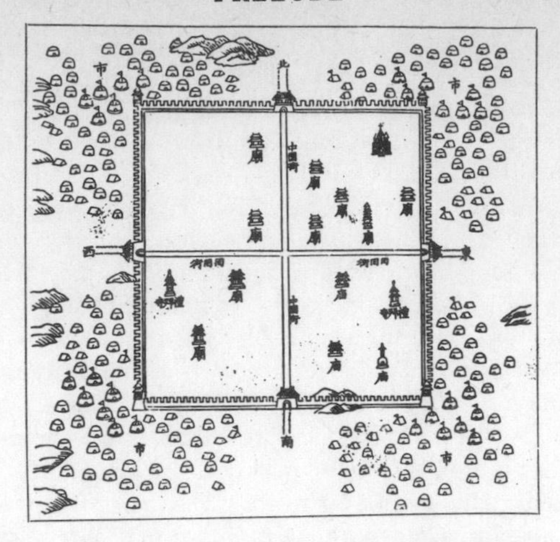

A Chinese Plan of Karakorum. (From Favier)

tale—the first description of the Mongol capital of Karakorum and of its palaces, of the style and state of the great Mangu Khan in the heart of his immense realms. Here Friar William saw, in the fullness of his power, the man who was ruler of the largest empire ever to have come under one sovereign, an empire, moreover, which was to be even further expanded under Kublai Khan a little later.

Friar William describes very well what he saw, but for him these were temporal things. It was not at all what he came to do or to see, but to the last he remained hopeful, and pressed his Christian point of view. His description of Karakorum[6] is intensely interesting, but we can fill it out a little with facts which the friar probably did not know.

Under Genghis Khan, Mangu's grandfather, the town had been made into a capital on the advice of Yeh-lu Ch'u-ts'ai, a scholarly descendant of the former ruling Liao (Khitan) dynasty in China, a man who was later to be made "auxiliary secretary of the Boards of Right and Left" and was the virtual governor of Peking. Yeh-lu Ch'u-ts'ai's advice to the Mongols had been that since they were now a great power, they required a fixed place (as opposed to a moving *ordu*) which tribes would know as the administrative center, and to which they could bring tribute. Genghis, though personally an almost compulsive nomad, consented to this. He had no idea of settling there and is said to have remarked that perhaps his children would live in stone houses, "but not I." How accurate this prediction was, he could hardly have dreamed. To call Karakorum a city is, however, hardly correct. In the

6. The name is derived from Mongol *kara* (black) and *kuren* (a camp or its surrounding fence).

midst of the rolling steppe it had no excuse to turn into an urban center in the strict sense. It was, rather, a focus, never very big—as we hear from Friar William himself—and its usefulness diminished as soon as the capital moved to Peking.

"Of the city of Karakorum," he says, "you must know that exclusive of the palace of the Khan it is not as big as the village of St. Denis [now part of Paris], and the monastery of St. Denis is ten times larger than the palace." The town was divided into two sections (a type of layout we will meet later throughout the history of Peking), one for the Saracens and Mongols and the other for Chinese, all of whom were artisans. "Besides these quarters there are great palaces which are for the secretaries of the court"—and for the officials of the swollen Mongol administration. "There are twelve idol temples of different nations" for ambassadors and numerous others at Mangu's court. There were also two mosques "in which is cried the law of Mahomet," doubtless by the muezzin, as it is today, and one Christian church to serve the many waifs and strays from eastern Europe and elsewhere who had been swept along by the Mongol armed tide. "The city is surrounded by a mud wall [as was Peking in Kublai's time] and has four gates. At the eastern is sold millet . . . and at the western one, sheep and goats . . . at the southern, oxen and carts . . . and at the northern, horses." So, later in Peking, were the markets mostly outside the square, walled town.

Within the walls lay Mangu's palace, "enclosed within a high wall like those which enclose monks' priories among us. Here . . . he has his drinkings twice a year: once about Easter . . . and once in summer when he goes westward. The latter is the greater feast, for then come to court all the nobles, even though distant two months' journey: and then he makes them largesse of robes and presents, and shows his great glory. There are many buildings long as barns in which are stored his provisions and treasures. . . . The palace is like a church, with a middle nave and two sides beyond two rows of pillars, and with three doors to the south." Mangu was seated high up at the northern end on a platform approached by two stairways between which stood his cupbearer. To his right (the west) were the men, and to his left the women. And the whole palace was aligned north-south. Mangu, says Friar William, "sits up there like a divinity," and we may well imagine the awe with which he was regarded by everyone— a feeling enhanced by the arrangement of the audience hall which the friar describes.

But there is a further device to impress the audience, the product of that work which had kept Boucher's adopted son from coming to see the friar sooner than he did:

"In the entry to the great palace [presumably the courtyard before its three southern doors], it being unseemly to bring in there skins of milk and other drinks, master William the Parisian had made for Mangu a great silver tree. At its roots are four lions of silver, each with a conduit through it,

THE MIRACULOUS FOUNTAIN OF MANGU KHAN made by William Boucher the Parisian and set in the palace at Karakorum. (From Vander)

all belching forth white milk of mares [in fact the dregs from the making of the clarified mare's-milk drink, and fit, according to Mongol custom, only for slaves]. And four more conduits are led inside the tree to its tops, which are bent downward, and on each of these is a gilded serpent [dragon, in all likelihood] whose tail twines around the tree. From one of these pipes flows wine, from another clarified mare's milk, from another *bal*, a drink made with honey, and from another rice mead which is called *terracina*. And for each liquid there is a special silver bowl at the foot of the tree to receive it.

"Between these four conduits at the top, master William made an angel holding a trumpet, and underneath the tree he made a vault in which a man can be hid. And pipes go up through the heart of the tree to the angel. In the first place he made bellows but they did not give enough wind.

"Outside the palace is a cellar in which liquors are stored, and there are servants all ready to pour them out when they hear the angel trumpeting. . . . When drink is wanted, the head butler cries to the angel to blow his trumpet. Then he who is concealed in the vault . . . blows with all his might in the pipe . . . and the angel places the trumpet to his mouth, and blows it right loudly. The servants who are in the cellar, hearing this, pour the different liquids into the proper conduits which lead them down into the bowls . . . and then the butlers draw from the bowls and carry the drink into the palace."

The magical effect of this fountain can readily be appreciated. The bystanders not in the know merely heard the butler call, the trumpet blow, and then saw the wine flow from the top of the tree. "It is in this way," as Olschki remarks in his book on Boucher, "that this forgotten prototype of modern prosaic drug-store fountains turns out to be one of the most attractive and significant monuments of thirteenth century Asia." At first sight, he goes on, the contrivance may seem just an eccentric result of a baroque imagination. But its constituent parts reveal a complex apparatus of symbolism. The inhabitants of Mangu's realm included people from most of the European nations and those of Asia too, most of the religions of those territories and its superstitions. To Buddhists the tree must inevitably have recalled the sacred *bo* tree under which Buddha attained his enlightenment in India; to Christians it must have brought to mind the Tree of Paradise, at whose roots originated the four rivers; and Moslems must have been reminded of that tree which was supposed to have grown from a staff of the Prophet.

But it is doubtful if Mangu meant to make a symbol of religious freedom. It may be that he had in mind the tales of early Mongol times as told in the *Secret History*. There is at least one passage in which, celebrating the election of a new ruler, the Mongols "feasted under the Great Tree. They were all feeling very happy and danced round the tree until their feet wore a deep trough in the ground." The angel on top seems to have been a purely Christian symbol, doubtless put there by William Boucher himself in a sort of calculated defiance of his position as a slave in the wilderness of pagan and idolatrous peoples. If so, he got away with his gesture.

In Karakorum, Friar William assumed the spiritual lead among Europeans there. Boucher had made him an "iron to make wafers" and vestments to replace those taken from him, just as he had made an image of the Virgin and put it in a Christian chapel (on a cart in conformity with Mongol custom), which was a place of secure worship for those same Europeans. The goldsmith almost died of some illness while Friar William was in Karakorum. Too late, the latter discovered that the patient had been doctored by the meddling Nestorian monk with a potion of his emetic rhubarb. Going to the monk in high indignation, he said: "Either go as an apostle doing real miracles by the grace of the Word. . . or do as a physician in accordance with medical art."

It is with just this sort of common sense that Friar William picked his way through controversies and jealousies at the Mongol court. By the month of May he was wondering to himself what to do about the future. The two months Mangu had granted him to stay had long elapsed. He decided to ask Mangu if after all he and Bartholomew could stay on in Karakorum to preach the faith.

Reminded of the whole affair, Mangu at once sent officials, and another round of questioning began, the Mongols being still unsatisfied that William was not in fact an ambassador of Louis and some sort of spy. Eventually Mangu himself sent for the friar. In a confrontation reported with a simple mastery of narrative, Mangu first stated his religious opinions. "We Mongols believe that there is only one God, by whom we live and by whom we die, and for whom we have an upright heart." Then, evincing a good knowledge of Christianity, he went on to question Friar William: "God gives you the Scriptures and you Christians keep them not." He lighted on the fact that Christians fought and wrangled with each other, that they would do almost anything for money. He made it clear that he did not refer to Friar William but to Westerners in general. Then he said: "God gave us diviners and we do what they tell us, and we live in peace." The argument, as Friar William well knew, was fallacious; but he could reply nothing since it was forbidden to answer Mangu without permission.

Mangu then decreed that the friar should return to Europe, asked if he would take letters to Louis, and in kindly enough fashion, gave William something to drink. Then he dismissed him. "And then I went out from before him, and after that I went not back again. If I had the power to work by signs and wonders like Moses, perhaps he would have humbled himself."

There is an infinite regret in these passages. For William knew that there was little he could do to convert the Mongols. He obtained permission for Bartholomew to remain at Karakorum—since the latter could not face the long journey home—and with many heart-searchings, prepared himself to go back across the continent. Master William the goldsmith, who had helped him greatly, sent a girdle ornamented with a precious stone to his former king, Louis. "So we separated with tears, Bartholomew remaining with master William, and I alone with my interpreter and my guide and servant going back. . . ."

A sad day. Against his own judgment he was leaving Bartholomew. He was leaving that little circle of friends, the goldsmith, his wife and son, and others, to their fate in the capital of the Mongol wilderness. And perhaps too, Friar William was leaving a place in some ways become dear to him, for he gives hints of his sensitivity to the beauties of place and landscape. He may have felt, as have others, the marvelous beauty of Asian mornings when the sun comes up over that boundless earth and the scattered, pathetic little groups of its people begin the boisterous day. For a while he had been, in his way, part of that odd polyglot community inlaid among the rough pebbles of the Mongols. For a time he had had a chance—as he saw it—of making a Christian mark on them. But now he knew for certain that he had failed in his primal aim; had failed in the wide steppes, and in the tents, and in the palace of Mangu Khan. It must have been, though he does not stress the fact, with a downcast heart that he left Karakorum.

One thing that certainly did not cross his mind then or afterward was how great in itself his journey had been, nor how much that lucid chronicle he wrote in Tripoli, after his return to Europe, could have contributed to the sum of Western knowledge of Asia. *Could have* contributed: for it is strange that, apart from the use of it made by Roger Bacon in his *Opus Majus,* the narrative was afforded hardly a mention from the time of Friar William until the mid-nineteenth century. Perhaps because he was sent by a king and not a pope, historians of the Franciscan order, such as the great Wadding, do not even mention it, while devoting much space to the less important writings of Carpini, who preceded Friar William into Asia. Only now when we have at hand all the known writing of travelers to Asia and to China in those early times, can we attempt to evaluate William of Rubruck's journey as he tells it.

It is a great work, the story of an unassuming, humble, but great human being. Though he had failed in his real task (because it was impossible to succeed), Friar William nonetheless gave the world a treasury of entirely new and important information. None of it appears to have been acted on, either by King Louis or anyone else. But as we shall see in tracing the stories of later travelers to China, this is merely an early example of a process of disregarding that was to be the norm rather than the exception. It is sad but true, and often ludicrous, that until little more than a century ago most of the information brought back from China was effectively lost—in the sense that it was hardly correlated at all and future travelers to China by sea or by land either had no knowledge of the existence of such information or made no intelligent use of it. Thus the story of Europeans in China is, for the most part, a tale of the curiously ignorant in confrontation with the utterly surprising—the facts of Chinese life they could have read about but never did. Christian priests at the court of Kublai Khan in Peking (shortly after Friar William's time in Karakorum) appear to have had no knowledge at all of his story; just as—even later—the Jesuits at the courts of Ming emperors obviously had not the faintest inkling of any of their Christian predecessors, either Nestorian in the T'ang or Franciscan in the Yuan dynasty.

Here, before we run too far ahead, we must temporarily leave this strange phenomenon. And here, too, we must leave William of Rubruck as he makes his hazardous way back though the camp of Batu and down the Volga; down the western shore of the Caspian, which he is the first to understand as a lake and not a gulf; and so to lower Turkey, which he calls Hermenia; and to the Mediterranean. Eventually he came to Italy in August, 1255—a year after leaving Karakorum—and even later reached his beloved Paris.

There, after a meeting with Roger Bacon, Friar William's bulky figure and humble greatness drop finally out of history. We know nothing more of him. Perhaps his most fitting epitaph may be the words of the great Sir Henry Yule in his *Cathay and the Way Thither:* "The generation immediately preceding his [Marco Polo's] own has bequeathed to us in the Report of the Franciscan Friar William of Rubruck . . . the narrative of one great journey, which in its rich detail, its vivid pictures, its acuteness of observation and strong good sense, seems to me to form a Book of Travels of much higher claim than any one series of Polo's chapters; a book, indeed, which has never had justice done to it, for it has few superiors in the whole Library of Travel."

This seems a not unreasonable evaluation.

A MONGOL ARCHER drawn by Pisanello. (Louvre Museum, Paris)

PART TWO

CHINA OBSERVED

A Merchant, a Saint, and a Tourist

CHAPTER THREE

MARCO POLO

"I HAVE NOT told the half of what I saw," Marco Polo is said to have declared shortly before his death. The statement was made in answer to friends who wanted him to retract or at least tone down what seemed to them, and to the public at large in the Italy of that time, the gross exaggerations in the narrative of his long sojourn in China. Marco Polo apparently refused the request. However wild many of his opinions and pronouncements about the cities and landscapes and manners of that far country—and from our twentieth-century perspective we must certainly convict him of some inflation of fact—he was being perfectly truthful in saying he had not told half of what he saw. Obviously the experiences of seventeen years of life and travel in China could never have been compressed into one volume of moderate size.

But from another point of view Marco Polo's statement is really an unintentional comment on his work. For the book he wrote is in fact no more than a half-truth about the China of his day. It is only one side of a piece of sculpture in the round.[1] Important and fascinating though his description of the Middle Kingdom under the Mongol ruler Kublai Khan will always remain, Marco Polo's view of China, and of the great figures of China whom he knew, is wholly frontal. Hardly had he arrived in Peking than the young Marco was elevated to a position of privilege from whose height he could indeed survey a very wide field, but which at the same time largely prevented him from understanding what ordinary Chinese life was like. Below the social and official level on which he found himself, and of which he gained a fairly shrewd understanding, his picture of China is a mass of generalizations and easy assumptions. His assertion made in later life has, therefore, an ironic truth which Marco Polo probably did not realize.

Given the man he was, given the times in which and for which he was writing, we should perhaps not expect his tale to be other than it is. Books were still written largely by, for, and about the great ones of the earth, of

1. Most quotations from Marco Polo in this chapter follow the Everyman edition, but in a few instances Yule's translation has been preferred for its greater felicity.

MARCO POLO. No authentic portrait of the man is known to exist. This is an Italian engraving of the seventeenth century. (From Yule's *Cathay*)

whom, in China, Marco Polo became a minor example. The mass of people in that country, tilling its fields, contributing their anonymous lives of work and striving to the national existence, were seen merely as the large, multifariously moving bulk of a mechanism geared to the high purposes of exalted figures of state for whose motions and intentions they supplied the wherewithal. It is in this context, first of all, that we must look at Marco Polo and at what he wrote.

That "half" which Marco Polo told about China in his narrative has made it one of the world's most popular books.[2] The fame and the romance that have clustered round the author's name have obscured the man, just as they have inflated parts of his achievement and ignored others. But this is the way of legends. Behind this, as behind other legends, reside facts. To take them out and look at them may absolve us from the seeming impiety of tampering with what modern Chinese would undoubtedly call a "culture hero."

It is hard to imagine any more exciting adventure that could befall a young man of seventeen than to start out by ship from Venice on a journey to Peking. Even today this would be thrilling, and when the young Marco set sail with his father and uncle, he must surely have been the envy of his friends. Unfortunately no one has left any word of description of Marco, either of the youth or of the man, and the traveler himself is almost equally

2. Everywhere, that is, except in China, where heroes are all of indigenous growth.

reticent on this subject. But there is no reason to suppose he was not an average young man, son of a prosperous merchant family, accustomed to reasonable wealth, to the life of a unique maritime trading center with its come-and-go of galleons laden with the exotica of those days—the spices and scented Oriental woods from India and Ceylon, the red eyes of rubies from Burma, the peacocks' feathers, the furs from Asia, the luxurious silks from Cathay, whose dimly understood identity he was soon to know in all its strange and fabulous detail.

Years before, in 1260, when Marco was about six, his father and uncle had left Venice on what turned out to be a journey with an unpremeditated destination. On their trading venture they had been immobilized in Bokhara for three years by local wars, and eventually they accepted an invitation to accompany a party returning to Peking, since this seemed the safest way out of their predicament. In the newly established capital they were well received by Kublai Khan and eventually sent back on an embassy to the pope from the Mongol ruler of China. By the time they returned, Marco had grown into a young man of fifteen. Marco and his father must have been almost strangers to one another, for nine years had elapsed since Nicolo left, and it was now the year 1269. The pope to whom Kublai had sent the elder Polos was dead when they reached Europe, and his successor not yet elected. Two years elapsed before they made up their minds what to do—and in those two years the young Marco must have learned much from his father which no other youngster of his times had had the opportunity of learning. Soon he was to find a use for this knowledge.

But before we follow the three Polos as they leave Venice on their second journey in 1271, it will be a good idea to examine briefly the European background against which their journey must be seen, and in which they (particularly Marco, who was to write the story of that journey) had grown up.

One's first thought on reading how the elder Polos dared once again to cross the Asian wastes to Kublai is that they were both foolhardy and courageous to tempt the fates in the shape of unpredictable Asian tribal rivalries and wars between minor khans, of which they already had some experience. But in fact, they were assured of Kublai's protection from the moment they entered his domains.

Venice of the 1270's was only one, though for the moment the most powerful, of a number of armed mercantile principalities in Europe among which there was a complete absence of unity save for their recognition of papal supremacy in matters of religion. Even this last factor was a comparatively new thing, since from the long feud between popes and sovereigns of the Holy Roman Empire, the papacy had only recently emerged triumphant. Meanwhile, in France, King Louis IX, whom we have already met on his ill-organized crusades, had given up the goal of taking back Syria from the Saracens and retired to more sober government of his own country. Here he had come into his own, and by the time of his death in 1270, a year before the Venetians set sail from Venice, France was the sole stable kingdom in an unruly Europe. In sharp contrast to France were divided Germany, and England under the devout but extravagant Henry III, who, before he relinquished his French possessions, had been forced to renounce most of his royal prerogatives and powers at home (the Provisions of Oxford, 1258).

Far to the south, Portugal under Alfonso III had achieved its definitive shape as a nation but was showing little sign of future maritime excursions. In Spain power was also much divided. Further east, Hungary, Poland, and European Russia were still almost literally licking the wounds inflicted by Mongol armies, and not for a long time would they emerge from the catastrophe of those incursions.

The age, indeed, was dark enough. Even in Italy the Polos left behind them an inflammable situation of commercial rivalry between the several strong mercantile cities of the country. Of these cities Venice, their home, was for the time being in complete control of the Aegean, lording it over Constantinople, and assured of supremacy in the Black Sea, which she had made almost as safe for her ships and traders as were the lagoons of her own little archipelago. The most serious rival to Venice was Genoa, which later assumed the former Venetian control of the eastern Mediterranean and was to venture in the person of Columbus (who was inspired by what the young Marco was to write of China) into Atlantic waters in search of the sea route to India. Disunited Italy was itself a dangerous place to live in if one wished to travel from city to city, but in Venice at least, and to some extent elsewhere, there was a measure of prosperity from the profits of markets opened up by the Mongol conquests as if in compensation for the trade recession experienced toward the end of the Crusades.

It was in this setting—European and Venetian—that Marco Polo had grown up, a background that was to serve him well in some respects in China, while in others—when he came to give an account of that country—it proved less than adequate and had most curious effects. After the return of his father and uncle, Marco undoubtedly learned much of their business in distant Asia, and it was possibly then also that he got a grounding in the Mongol language and perhaps also in Persian. The former, he tells us, was known to his father and uncle. But, just as important, the older men imbued him with their knowledge of trading, for even on his outward journey and before he reached China, Marco tells us much of the materials of trade en route.

The two years' delay in Venice before the party set out for China again was due to the Polos' desire to fulfill the ambassadorial duties with which Kublai had charged them. His letters to the pope which they carried asked for no fewer than one hundred men "of the Christian faith . . . learned in the seven arts [Rhetoric, Logic, Grammar, Music, Geometry, Arithmetic,

Scenes from the Polos' Travels. This painting from an illuminated manuscript of Marco Polo's book dated about 1400 depicts the Polos' embarkation at Venice, with a view of the city itself in the background and, in the foreground, scenes from the subsequent voyage. The ships are the type of trading vessel common at the time as carriers of Venetian goods in the Mediterranean. (Bodleian Library, Oxford)

and Astronomy] and qualified to demonstrate that idols were of the devil and the law of Christ better than the religion the Mongols knew." Kublai had also asked for some holy oil from the sepulcher of Christ at Jerusalem. But the pope was dead when Nicolo and Maffeo returned from Peking to Acre; so they gave Kublai's letters to Tedaldo Visconti of Piacenza, papal representative in the Levant, and, as things would shortly turn out, the next pope. His advice was that they await the election of the pope; but after two years had passed, they decided that to tarry at home longer would at least be undiplomatic. So Marco came to Acre on his first trip out of Venice and

The Polo Caravan to China. An illustration in a corner of the Catalan Map of 1374 prepared for Charles V of Spain. (Bibliothèque Nationale, Paris)

the three Polos again conferred with Tedaldo. This time he gave them letters to Kublai and sent them on their way, without those hundred men he had requested. But the party had not gone far before it was recalled, by Tedaldo, who had now become Pope Gregory X. The new pope now bestowed on them Letters Apostolic and two Dominican friars, in cautious response to the request for one hundred. The Polos had promoted the idea that the Mongols might prove susceptible to Christianity, but the opinion of the shrewd Gregory was evidently much less sanguine. The two unwilling Dominican friars reached Lajazzo with the Polos, invented all manner of excuses for not going further, and turned back. This was no great loss.

The Polos must have known how irrevocably the Mongols of East Asia were committed to Buddhism, even if they did not know that those of West Asia were already committed to Islam. And they must certainly have known that Kublai really wanted those hundred Christians for secular and not religious reasons. Only recently had he managed to conquer a section of China, so recently that it was too soon for him to trust Chinese with senior positions in his service. Most of his administration was in fact run by non-Chinese, and this was the function he intended for the hundred intelligent Christians.

The three Polos left Lajazzo toward the close of 1271, going in a southerly direction to the port of Ormuz on the Persian Gulf, apparently intending to take ship there for China. But just conceivably there was a better reason. Marco takes the trouble (as usual in his book) while at Ormuz, to list the "spices, drugs, precious stones, pearls, gold tissues, elephants' teeth, and various other articles of merchandise" which arrived in that flourishing port from all parts of India. These, he says, "are disposed of to a different set of traders by whom they are dispersed throughout the world." It seems possible that the reason the merchant Polos went to Ormuz was to purchase goods of a suitable exoticism to take with them.

In fact, from Ormuz they turned inland across country to Balkh (in

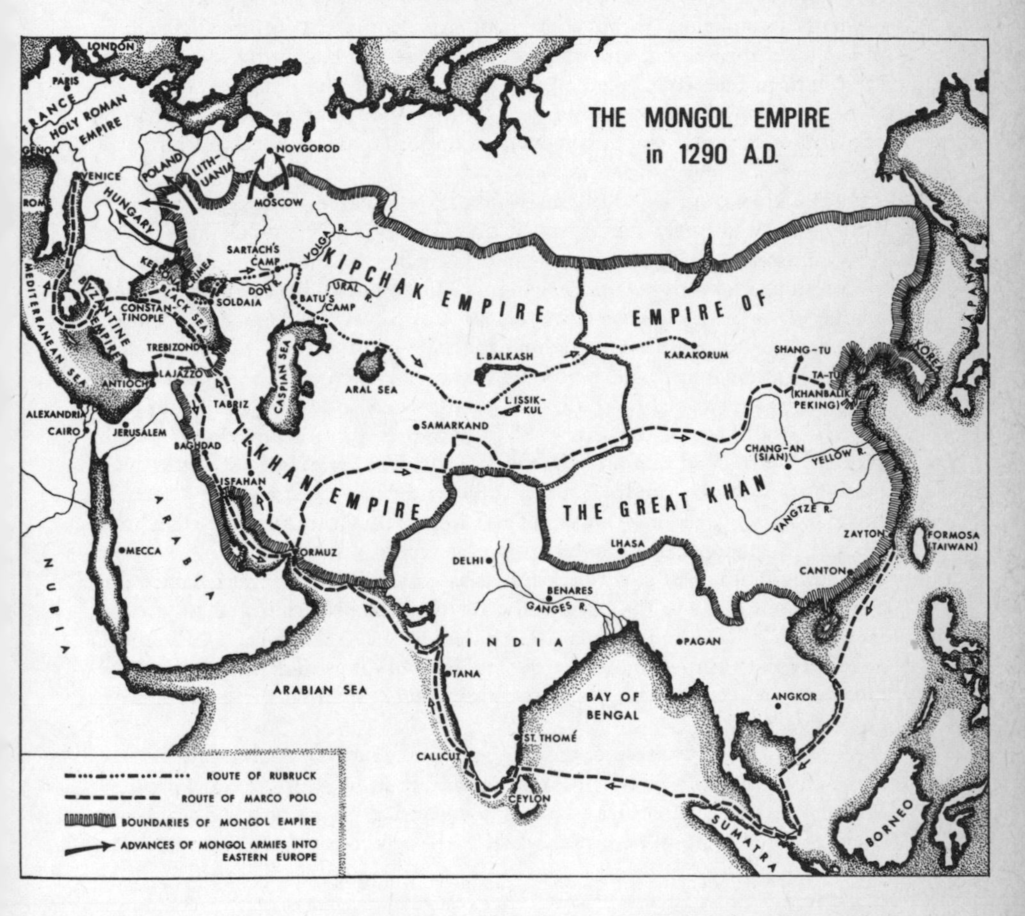

The Mongol Empire in 1290. Map showing the routes of Rubruck and Marco Polo.

modern Afghanistan) and thence over the Pamir Mountains to Kashgar on the edge of the Takla Makan Desert; and onward by a well-worn trade route to the southern edge of the desert—to Khotan; and from there, still following the southern trans-Asian route, to the lake of Lop Nor. Here, close to the wonderful painted and sculptured Buddhist caves at Tun-huang (though they did not know of it, and would have made little of the place), their route deviated from the usual one. Instead of bearing southwest to Lanchow on the upper Yellow River, they seem to have followed the northern side of the Great Wall of China and, meeting the big loop of the Yellow River, to have followed it into the country of the descendants of Prester John. Marco Polo never mentions the Great Wall—one of the many incomprehensible omissions in his book—but he says quite a lot about Tenduc, as he called this region, and about the Christian (Nestorian) King George, "the sixth in line from Prester John." Now this is a very interesting statement. As we shall see in the story of John of Montecorvino, King George, or perhaps his successor, played an important part in the Christian endeavors of this later traveler to Peking.

As the Polos approached Kublai Khan, the great ruler sent riders forty-days' journey to meet them; and it must have been in some state, in the early summer of 1275, that they neared his palace at Shangtu—the Xanadu of Coleridge's famous poem. Hereabouts Marco noted five species of cranes, among other birds conserved for the sport of the great khan. And—one of those slender bridges that sometimes span the centuries—a traveler writing in 1925 comments: "The writer will always remember the surprise and delight he experienced en route to Shangtu when, a day's march from that city, he saw three of the five sorts of cranes so graphically described by Marco, and realised that after all the scene had changed but very little since that spring day six hundred and fifty years ago, when the great Venetian traveler first sighted the towers of the imperial summer city."[3] Reaching the site of Shangtu, the modern traveler continues: "The city . . . itself [now only mounds of earthwork traceable with difficulty] must have been a marvellous sight to the eyes of the few Europeans privileged to visit it. Its position in the central plain encircled by beacon-crowned hills must have served to throw the numerous palace buildings and its reported one hundred and eight temples into high relief against the surrounding greenery."

Marco Polo was impressed, though he is somewhat vague about the layout of the city. Probably he was more impressed than he seems in a description written years later, after he had seen Peking and also his favorite Hangchow. In the city of Shangtu, he says, Kublai had "caused a palace to be erected,

3. L. Impey, "Shangtu, the Summer Capital of Kublai Khan," in the *Geographical Review*, 1925.

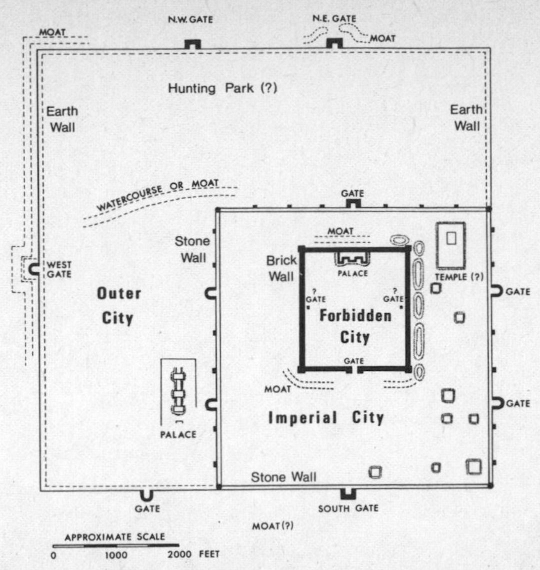

PLAN OF SHANGTU. Here Marco Polo first met Kublai Khan. Comparison between this plan and that of Chinese cities reveals how closely the Mongols' ideas on the layout of an imperial city followed those of their great neighbor China. (Redrawn from Impey in the *Geographical Review,* October, 1925)

of marble and other handsome stones, admirable as well for the elegance of its design as for the skill displayed in its execution. The halls and chambers are all gilt . . . and it presents one front toward the interior of the city, and the other toward the wall." This, doubtless, is Coleridge's "stately pleasure-dome." From the plan which the modern traveler Impey drew after his exploration of the old site, it appears that Marco is talking of the Imperial City, rather than of the palace itself.

The walls surrounding the whole, Marco goes on, were sixteen miles in circuit with but a single entry, and the enclosed area was only accessible through the palace. Here he was wrong, since Impey found four gates in the outer walls. But Marco was a privileged person and doubtless never saw them, having come into the city through the khan's palace. It is this kind of approach to China which makes him fascinating because he had inside information, and inaccurate because the level from which he observed was too exalted. Friar William would have known the facts, because no one would have sent an escort forty days to fetch him in state. He would have walked up to, and probably around, the walls of Shangtu, seeking entry in his humble way. As to Marco's estimate of sixteen miles for the circuit of

Kublai Khan Hunting, by Liu Kuan-tao, a court painter of the Yuan dynasty. The emperor is dressed in a white ermine cloak and rides a dark horse. It seems likely that this painting was based on a sketch actually made during the hunt. (National Palace Museum, Taipei)

those walls, six is more like the true figure. The reputation for exaggeration which in later years is said to have earned him the title "Il Milione" in his native Italy was not without foundation.

With a certain verve Marco turns to a long description of the hunting preserve, where the "grand Khan" repaired once a week. "Frequently when he rides about there, he has one or more small leopards carried on horseback behind their keepers; and when he pleases to give direction for their being slipped, they instantly seize a stag, or goat, or fallow deer, which he gives to his hawks, and in this manner he amuses himself." The leopards were doubtless cheetahs, similarly used in India to this day. "In the center of these grounds, where there is a beautiful grove of trees, Kublai has built a royal pavilion, supported on a colonnade of handsome pillars, gilt and varnished. Round each pillar a dragon, likewise gilt, entwines its tail, whilst its head sustains the projection of the roof. . . . The roof is of bamboo, likewise gilt, and so well varnished that no wet can harm it. The bamboos . . . are split into two equal parts [lengthwise] so as to form gutters, and with these the pavilion is covered. . . . The building is supported on every side [like a tent] by more than two hundred very strong silken cords." From those sturdy yurts described by Friar William the Mongol rulers had now indeed moved far into a realm of Chinese refinement.

Marco is happy at the chase. It arouses his innate sporting instincts in a way that produces more lively description than his usual. He tells us a lot about the ten thousand mares of Kublai—probably not an exaggeration, since the Mongols always possessed huge numbers of horses—about Kublai's sacrifice of mares' milk, about the astrologers who miraculously prevented the tempests from affecting the palace while winds raged unabated all round it. As the story continues we find that Kublai, like all his ancestors and all his successors, shows a preference for being surrounded by every sort of religious and mystical service, as if to avail himself of each to stop the gaps of fate. Once more echoes of Friar William obtrude in Marco Polo's writings—though Marco, of course, had never heard of the Franciscan.

In Shangtu, Kublai received the Polos. Marco refers to the meeting in the first chapter of his book and the narrative is in the third person, a further divider between the writer and the reader. "Upon their arrival, they were honorably and graciously received by the grand khan, in a full assembly of his principal officers. When they drew nigh to his person they paid their respects by prostrating themselves on the floor." William of Rubruck's approach to the question of the kowtow was one of simple logic: now we have Marco's—that of a merchant only too willing to please. Perhaps the historian Wieger is too harsh when he says that Marco's career in China was not quite an honorable one; but one can see from a sentence here and there that he was at least a very accommodating young man where Kublai was concerned.

The great Kublai at once commanded the Polos to rise "and to relate to him the circumstances of their travels with all that had taken place in negotiation with his holiness the Pope. To their narrative, which they gave in the regular order of events [an attitude to which the lost reader of Marco's book could wish Marco had adhered when he dictated what is one of the more disordered of the world's classics], and in perspicuous language, Kublai listened with attentive silence." Doubtless he was interested to hear how his request for those hundred Christians had fared. In that hope he was disappointed, but he was given the presents the Polos had brought from the pope, the holy oil they had brought from Jerusalem, and the Letters Apostolic, which were probably no great comfort to a king who needed trained men more than resounding phrases. But at least the two elder Polos had returned, bringing one new recruit, on whom Kublai at once remarked.

"Nicolo made answer, 'This is your servant and my son.' To which Kublai replied: 'He is welcome, and it pleases me much.' " Marco was at once enrolled among the attendants of honor, and Kublai ordered a great feast to celebrate the Polos' return. "Marco was held in high esteem . . . by all belonging to the court. He learned in a short time and adopted the manners of the Tartars [Mongols] and acquired a proficiency in four different scripts." What these scripts may have been is an interesting point—probably Mongol, Uighur, Persian Arabic, and Tangutan. It is most unlikely he ever learned more than a smattering of spoken Chinese or more than a few written characters, if we may judge by his poor efforts to reproduce Chinese names and his preference for their Mongol equivalents. But ignorance of the Chinese language is typical, rather than the reverse, of Westerners who have stayed long in China. Owen Lattimore's sage comment, "No foreigner ever learned to read and write Chinese without boasting about it,"[4] hits the nail squarely on its head. Marco did not learn it. He had little need to, since the official language was Mongol and he became a member of the Mongol administration.

Nothing in Marco's book tells us of his arrival in Peking. We plunge abruptly into the midst of the new palace Kublai had recently built in his new city there, a city named Ta-tu, on the site of modern Peking, which was completed about 1290. But his description of the city is excellent.

First, he introduces and identifies Kublai, his hero. "In this book it is our design to treat of all the great and admirable achievements of the grand khan now reigning, who is styled Kublai Khan; the latter word implying in our own language lord of lords, and with much propriety added to his name; for in respect to number of subjects, extent of territory, and amount of revenue, he surpasses every sovereign that has heretofore been or that now is in the world; nor has any other been served with such implicit obedi-

4. Lattimore, *Inner Asian Frontiers.*

ence by those whom he governs. . . ." Perhaps significantly, he further introduces Kublai at the moment when the emperor returns to the capital and entertains the Christians in his entourage. "Being aware that this [Easter] was one of our principal solemnities, he commanded all the Christians to attend [in the palace] and to bring with them their Book. . . . After causing it to be repeatedly perfumed with incense in a ceremonious manner, he devoutly kissed it."[5]

But, adds Marco, Kublai observed in the same way the festivals of the Jews, Moslems, and Buddhists. He tells how his own father, Nicolo, at the audience when the older man and Maffeo were sent by Kublai to the pope, had "ventured to address a few words to Kublai on the subject of Christianity." To which Kublai had rejoined: "Why should I become a Christian?" and added that Christians were unable to do anything miraculous, whereas the "idolators" (Buddhists, shamans, and others) had the power to make cups in the middle of the palace hall "come to me filled with wine . . . spontaneously, and without being touched by human hand." (And yet later, in Europe, the elder Polos apparently did nothing to refute the idea that Kublai was anxious to be converted to Christianity: so much for their intellectual honesty.)

Having made his point, Kublai then swore he would change to Christianity when those hundred men who were to come from Europe proved themselves more versed in the miraculous arts than idolators, or could prove that such things were inventions of the devil. One can hardly credit that Marco Polo believes his own statement: "From this . . . it must be evident that if the Pope had sent out persons duly qualified . . . the grand khan would have embraced Christianity."

Kublai was a man "of middle stature . . . his limbs well formed . . . in his whole figure a just proportion. His complexion is fair, and occasionally suffused with red which adds much grace to his countenance. His eyes are black and handsome, his nose is well shaped and prominent." And then Marco, perhaps recalling incidents and court gossip from those days twenty years before the time he was dictating his story, lingers more than a little over a description of how concubines were selected for Kublai. If it is possible to reconstruct any entirely personal trait in Marco's character or life in China, it must be from those long moments that occur in his book where he hovers, as it were, over the remembered beauty of women. At these moments he becomes almost a human being instead of a disembodied observer.

The choosing of Kublai's harem is, however, interesting in itself. Every

5. Kublai's mother, a Kerait princess from the region of Lake Baikal, came of a people who had long been Nestorian Christians. The official history of the period tells us that she was buried at what may be translated as the Temple of the Cross.

year or so officers were sent into the Uighur country (where Marco Polo has already noticed the attractiveness of the girls as he passed that way) to collect four or five hundred of the most beautiful young women "according to the estimation of beauty . . . in their instructions." The hair, the face, the eyebrows, mouth, lips, and the symmetry of features are examined, and the girls' worth estimated in carats (probably of gold). The khan wants only those of at least twenty carats' worth, and these are brought to Peking, where they undergo another examination, after which thirty or forty of the girls are selected. Then they are sent to the wives of certain nobles who "observe them attentively during the course of the night, to ascertain that they have not any concealed imperfections, that they sleep tranquilly, do not snore, have sweet breath, and are free from unpleasant scent in any part of the body." Those who survive this intimate scrutiny are allotted in parties of five, "one of which attends during three days and nights in his majesty's interior apartment, where they perform every service that is required of them, and he does with them what he likes." The parties take it in rotation to serve in this way, while yet another group, perhaps less luscious, is stationed in the outer apartments ready to bring refreshments should the great khan need them.

The chapters in which Marco Polo describes the palace at Peking, the condition, life, and attainments of Kublai, are perhaps only rivaled for interest by that section dealing with the city of Hangchow. The Peking chapters lay an impressive foundation for all that was to happen in that city during the remainder of the Mongol dynasty when Montecorvino and Odoric and others were there.

Here, in Peking, Marco begins, "on the southern side of the new city, is the site of Kublai's vast palace. . . . In the first place is a square enclosed with a wall and a deep ditch, each side of the square being eight miles in length [more accurately, only about six miles], and having [centrally] an entrance gate for the concourse of people coming thither from all quarters." Within this outer perimeter is a series of smaller square enclosures, each with its boundary wall. The outer one is filled with troops (Kublai was evidently taking no chances with the newly conquered Chinese), the next with their arsenal and equipment, which was disposed in four symmetrical buildings, one on each side; and lastly, in the center was the sanctum of the palace itself, within a wall of "great thickness, its height full twenty-five feet. The battlements . . . are all white."

Between these concentric squares of walls, the whole area is planted with trees, and on the meadows beneath their branches roam stags and fallow deer, and "the animals that yield musk. . . . Within these walls . . . stands the palace of the grand khan, the most extensive that has ever been known. . . . It has no upper floor, but the roof is very lofty. The paved foundation or platform on which it stands is raised ten spans above the level of the ground, and a wall of marble two paces high is built on all sides, to the level of this pavement, within the line of which the palace is erected; so that the wall, extending beyond the ground plan of the building and encompassing the whole, serves as a terrace, where those who walk on it are visible from without. Along the exterior edge of the wall is a handsome balustrade with pillars, which the people are allowed to approach. The sides of the great halls and the apartments are ornamented with dragons in carved work and gilt, figures of warriors, of birds, and of beasts, with representations of battles. . . . The grand hall is extremely long and wide. . . . The exterior of the roof [the eaves] is adorned with a variety of colors, red, green, azure, and violet, and the sort of covering is so strong as to last for many years. The glazing of the windows is so well wrought and so delicate as to have the transparency of crystal."

This picture is probably an accurate one, presenting as it does a palace whose main outlines agree very well with the palace built in the succeeding Ming dynasty, when the Mongols had been chased back to their steppes. The only puzzles are the roof, which was certainly of tiles (why did Marco not say so?), and the glazing of the windows, which has been explained by some authorities as of transparent paper. But in fact there is nothing to stop us visualizing glass, made in China for over a thousand years before Marco Polo's time.

"In the rear of the body of the palace there are large buildings containing several apartments where is deposited the private property of the monarch, his treasure in gold and silver bullion, and also his vessels of silver and gold plate. Here likewise are the apartments of his wives and concubines; and in this retired situation he despatches business with convenience, being free from every kind of interruption."

What Marco omits is almost as interesting as what he tells. For he must surely have heard how Kublai had employed a Chinese-Arab architect and his son to plan and erect this grandiose and typically Chinese complex of buildings. The two men, Yeh-hei-tieh-er and Ma-ho-ma-sha respectively, were in all probability still alive or only recently dead in Marco's time. But Mongol-centered, swallowing whole whatever the Mongols told him, and unable to talk to Chinese, Marco fails to understand the significance of the utterly Chinese concept of Peking as a city. There was little in it that had anything to do with traditionally Mongol ideas (see plans of Peking, page 380, and Shangtu, page 71).

In the picture Marco Polo projects of Peking and its lord, the astonishing change which had come over the Mongols is implicit. No doubt his picture of Kublai does not apply to Mongols of the lower echelons, who were proving, even at this early date, totally incapable of the administrative greatness required to govern a country of high and subtle sophistication and culture. Nonetheless, the contrast between Karakorum and Peking, between Mangu and Kublai, is a vivid one. Here, in Peking, is no converted encampment of brash and showy riches new-found, but a city designed in conformity with an ancient Chinese model, fitted for the greatness of an emperor who now ruled the greatest empire the world has ever known. The record shows how conscious the Mongols were of their mantle of power, as of the necessity to make a setting in which to wear it.

"In the third year of Chih-yuan," says the official record of the times, "it was decided to locate the national capital at Yen [the present Peking]. In the eighth moon Kublai directed the technicians of all the bureaus concerned to assume the duties of constructing the palaces and issued the following command: 'As the Great Enterprise [of conquering China] has come to its conclusion, our national influence is in process of expansion. If the palaces and metropolitan adornments are not beautiful and imposing, they will not be able to command the respect of the empire.' Yeh-hei-tieh-er labored ceaselessly at the task until he had drawn up a grand scheme." The resulting buildings must have pleased Kublai by their gaudy magnificence. "All the imperial gates had cinnabar doors and vermilion pillars, and the red walls were decorated with a scroll design and green-glazed tiles. . . ." In one of the palace buildings, "the floral base was of blue stone pavement covered with a double fabric; red pillars decorated with gold, surrounded with carvings of dragons. The four walls were red and the windows in emerald green [perhaps the alleged glazing of Marco Polo's text]. Between the beams and ceiling it was painted in gold and studded with tiles. The staircase with its balustrade was painted red and covered with gilt-copper. . . ."[6]

The rude courts of Mangu Khan that Friar William described have indeed undergone a remarkable metamorphosis, and in this sumptuous confection sat Kublai Khan in the utmost perfection of state. He was almost a god. His bodyguard of twelve thousand horse, three thousand of which were always on duty in the palace yard, presented an extraordinary image of power; but they were not kept, Marco hastens to inform us, "for any apprehensions entertained by him, but as a matter of state." And he hurries on with his story, demonstrating more dramatically than anything else could how imperial dignity and Oriental near-divinity had focused on the grandson of the wild and woolly Genghis Khan. That fierce, shrewd, and brutish tribal chief had once said that perhaps his children would live in houses. And so, indeed, they did.

In his palace Kublai sat on his throne, elevated above everyone else, in the midst of that amalgam of Chinese and Mongol pomp, his face to the south (the direction required by both traditions). On his right and at a much lower level were his sons, and to his left his womenfolk and all the other women of the court—an arrangement which clearly reflects that of the Mon-

6. The first quotation comes from Ch'en Yuan's work, and the second from vol. 1 of Chuta Ito's *Architectural Decoration in China* (Tokyo, 1941–44).

gol tents of Central Asia (see plan of yurt, page 37). "The greater part of the officers and even the nobles . . . eat sitting on carpets in the hall; and on the outside stand a great multitude of persons who come from different countries and bring with them many rare and curious articles." Nothing delighted Kublai more, Marco Polo tells us, than to hear of far parts of his dominions and, of course, to receive their tribute in "curious articles." At the same time it is interesting to note that even in Mongol Peking the practice of receiving ambassadors from "abroad" (in the Chinese sense of the word, which always meant anywhere not in the central Chinese area) was well established, and that it mostly took place at the great feasts of the court. Had some of the ambassadors from the West in later days in Peking understood the meaning of these audiences, they might have saved themselves much heart-searching and no little dismay; or, alternatively, they might not have come at all until they and their countries were ready to accept the fact that, to the Chinese court, "ambassador" was a word meaning a tributary, a suppliant from a place outside the immediate sway of the emperor and hence of little importance. That suppliant nonetheless came to crave the imperial smile, the imperial favor and protection.

"In the middle of the hall," Marco Polo goes on, "there is a magnificent piece of furniture made in the form of a square coffer, each side of which is three paces in length, exquisitely carved in figures of animals, and gilt. It is hollow within" and contains a capacious jar made of precious metal calculated to hold a very large amount of wine. On each of the four sides of this object stands a smaller vessel, filled with one of four types of beverage—the same alcoholic drinks we have met at Karakorum. But the Mongol "fountain" here described is now a contrivance at once more simple and more Chinese. Boucher's fountain has given place in Peking to a much more simple object, for now there was much less need to impress the *hoi polloi*.

The Mongol capacity for alcohol is noticed by the Venetian. "At each door of the grand hall . . . stand two officers of gigantic figure, for the purpose of preventing persons from touching the threshold with their feet"; strangers are warned of the prohibition, and if anyone infringes it, "they take from him his garment, which he must redeem with money," or beat him. But at the end of a feast or court "as some of the company may be affected by the liquor, it is impossible to guard against the accident, and the order is not then strictly enforced." It is very evident that for all their newfound wealth and dignity, the Mongols still caroused in an entirely un-Chinese manner, just as Friar William had depicted them. Marco Polo, still very young when he first saw such scenes of inebriation, appears to have accepted them as normal to the great ones of the Chinese earth, although such drunkenness is not common among Chinese.

But these were mere flaws in the amber. The picture of Kublai in all his almost-Chinese majesty is an impressive one. Servants wear veils of embroidered silk over their mouths and noses "so that his victuals or his wine may not be affected by their breath." The page who presents the emperor's cup retires three paces and kneels, as do all those present, when Kublai raises the cup to drink; and simultaneously a "numerous band" begins to play and continues to do so until the emperor has finished drinking, "when all the company recover their posture." Tantalizingly, and unlike Friar William, the Venetian does not comment on the food at those feasts except to say its "abundance is excessive." We do not learn whether the Chinese cuisine had routed the inferior Mongol mutton and cereal. Unlike Marco, who shows little sign in his book of any interest in food (or wine for that matter—his main preoccupations being trade, somewhat naïve wonder at architectural and other "marvels," and, rather slyly, women), we would like to know what sort of fare was offered before the tables were cleared and jugglers, acrobats, comedians, and other performers appeared. This entertainment was Chinese, and similar shows at imperial banquets were later to produce the same sort of comments from Europeans more sophisticated than Marco, in courts of purely Chinese splendor at Peking in the eighteenth century.

The sole item of food mentioned by Marco Polo in this context is game, obtained by hunting parties organized at distances of not more than a forty-day journey from the capital and sent frozen (by the deep cold of the North Chinese winter) to Kublai's table during the months he stayed in Peking. Marco has already given a description of Kublai at the chase, and now gives another of the great man on a journey. Kublai is borne in a pavilion carried on the backs of four elephants, its structure carved, covered with leopard or tiger skins, and lined with cloth of gold. A rare note of confidential information finds its place here, for we discover that the real reason behind this clumsy if gorgeous mode of transport is that Kublai suffers from gout! The image is still further humanized by the information that Kublai generally has with him in this small pavilion twelve officers to amuse him and twelve falcons to let loose for hunting. If this is true, there must have been quite a crush beneath the leopard skins and cloth of gold.

But apart from this conceit, the progress of the emperor is made in Mongol fashion, and Marco's description of the camp is reminiscent of Friar William's sight of Mongol *ordu* that seemed to him like cities.

After the pomp of the court we come to the everyday life of the capital. Within the vast foursquare city laid out like a chessboard on the plain and surrounded by its fifty-foot-high walls, the regular grid pattern of its streets was teeming with people. Marco Polo had never seen such a city, and he is overcome by the huge number of people, both within and without the walls in the spreading suburbs, and also by the manner in which life was closely regulated by the strong police force of the army. The hours or "watches" were struck by an enormous bell in its tower (one of the few relics of Mongol Peking that exist to this day, the Bell Tower), and the populace obeyed the curfew of each night, only daring to leave their houses in the narrow residential roads, called *hutung,* to fetch a doctor for a sick person or a woman in labor, being required to carry a light to do so and to answer the questions of the guards at the wooden trellis-work screens that curtained off the ends of all these streets. The main streets were broad and ran straight as arrows across the city from east to west and from south to north, interrupted only by the forbidding enclave of the palace city standing like a strong room and a sacred shrine within the twenty-four-mile circuit of the outer walls.

Hardly surprising that Marco Polo was impressed; for at the time there was nothing in Europe, or anywhere else, remotely as grand, as carefully planned and regulated by a supreme authority. Those twelve suburbs corresponding to the twelve gates of the city contained a multitude of inhabitants "greater than the mind can comprehend. . . . And it is there that the merchants and others whose business leads them to the capital, and who on account of its being the residence of the court, resort there in great numbers, take up their abode. . . . In the suburbs there are . . . as handsome houses and as stately buildings as in the city, with the exception of the palace. . . ."

Here one feels the Venetian is at home, if a little overcome by Peking's superiority to Venice. Important functionary of the court,[7] favorite of the emperor, a merchant at heart, the lover of women we have some reason to suppose him to be, teller of tales—all these roles of Marco blend well in his description of the city and suburban scene. He delights in the prostitutes and estimates their number at twenty thousand, noting that they are not supposed to ply their trade within the walls of Peking ("unless it be secretly," he adds, obviously knowing very well what goes on); he delights quite as much in "the vast concourse of merchants and others," for whom the number of prostitutes is not "greater than is necessary," who bring from their own countries "everything that is most rare and valuable in all parts of the world . . . and more especially from India [a word which Marco uses to cover Southeast Asia as well as India proper], which furnishes precious stones, pearls, and various drugs and spices. . . . No fewer than a thousand packhorses and carriages, loaded with raw silk from the Chinese provinces, make their daily entry." And clustering round the colossal business center of Peking and its straggling suburbs had sprouted a rash of subsidiary towns "whose inhabitants live chiefly by the court, selling the articles they produce in the markets" of Peking.

But having said all this, our guide, writing after he had spent seventeen years in China and returned to Italy, gives us no sign that he comprehended what it all meant. He gazes at the spectacle of bustling trade and happy nights with a wide choice of delightful Chinese women, at the heraldic and

7. Here and elsewhere Marco Polo's words are being taken at their face value. Oddly enough, however, his statements about his high position are completely unsubstantiated by Chinese records.

immeasurably wealthy emperor and his court, at the corruption which he knew festered everywhere in the higher ranks (he even gives examples of it) —he gazes and gazes without being in any way aware of what the Mongol regime was plainly doing to China, to the ordinary Chinese person.

What was happening economically in China is important for us to know in bare outline, because later on, when the Europeans had established themselves in economic force round its coasts, something in many respects similar was to occur. In Mongol times there were at least two related aspects of the situation. First, under the Yuan, China became easier of access to the West than ever before, for the very simple reason that Mongol overlordship and control extended from the China Sea to Basra and almost to the Mediterranean, and from the Pamirs and Himalayas northward to Siberia. The great trans-Asian trade routes were wide open for the first time since the T'ang dynasty. Not only did foreign merchants come and trade in China in very large numbers, but many foreigners came to stay—temporarily or permanently, employed as was Marco Polo in quite high situations in the administration, with every opportunity to squeeze money out of China. Second, while this aspect was in some ways good, its net result in economic matters was a colossal drain of assets in silver and every imaginable precious substance, *from* China toward the rest of the world. The progressive opening up of China to foreigners resulted in nothing but the rapid impoverishment

MONGOL PAPER MONEY. A Mongol-dynasty woodblock for printing paper money preserved in the Bank of Japan, Tokyo, and a print made from the same block.

of the country and, as a side effect, the concentration of what wealth there was in China in the hands of very few people. The very things which assisted the T'ang dynasty to broaden its outlook—contacts with the rest of the world—resulted, under the alien Kublai, in Mongol aggrandizement and Chinese poverty.

As for paper money, although Marco Polo attributes its invention to Kublai, it had in fact been used by various dynasties following the T'ang and had always ended in extortion and inflation. It was by no means "the secret of the alchemists," the miracle of perpetual solvency, he imagined it to be. It was, in Mongol hands, little but an elaborate hoax, a facade of legality concealing the power to extort real goods from all and sundry and place them eventually in the imperial treasury. There are times when Marco's naïveté is better termed stupidity. Contrary to his image of a prosperous and wealthy China, Mongol times were those of steadily worsening financial problems, which were not eased by the colossal expenditure of the court and the equally colossal expense of road- and canal-building and military adventures that were undertaken. By 1290, when figures of taxpayers were compiled, the total number was only slightly over half of the number of taxpayers in the previous dynasty, the Sung. This in itself reveals the growing poverty of the people. Marco himself tells us graphically of the corrupt and ruthless Ahmed, a Moslem finance minister of Kublai, who was at last assassinated by a Chinese patriot in the palace. After the death of Kublai no fewer than eighteen thousand of his officials were degraded for peculation big and small. And with Kublai died whatever was vital in the Yuan dynasty. By then about ten percent of the total population of perhaps sixty-odd million were officially admitted to be starving.

Through the eyes of Marco Polo, however, we see a prosperous China, a vision through a distorting mirror—the mirror of a merchant, a man of small culture. His description of the efficient posting stations along the main roads all over China could hardly be bettered in its details. But he sees it all as a highly privileged man. Of the posthouses themselves he remarks: "They are large and handsome buildings, having several well-furnished apartments hung with silk and provided with everything suitable to persons of rank." Running through the way the system worked, the huge numbers of horses required for its upkeep, and the large body of persons employed, he mentions the prodigious expense of the system; and a question does just creep into the corner of his mind: How does the population supply sufficient numbers to run the posts and how are they fed? His answer is summary: "All idolators [Chinese] keep six, eight or ten women, according to their circumstances [this was not actually true of most], by whom they have a prodigious number of children . . . and with regard to food there is no deficiency of it, for these people . . . subsist for the most part on rice . . . and millet. . . ." He goes on to say how fertile Chinese soil is [a half-truth, dependent entirely on the area]. Lord Macartney, despite his nobility and position of privilege and his few months in China in the late eighteenth century, gives a picture which is altogether more in accordance with the facts of life in rural China.

But let us be fair. What he knew, what he understood, Marco Polo said well. A large part of his book is taken up with two long journeys he made at the command of Kublai through various parts of China and to the area now called Vietnam. The incredulity of his readers in Italy when he returned there at what he said of China is hardly surprising on many counts. Just his picture of the two great rivers of China—the Yellow and the Yangtze—must have seemed to them ludicrously exaggerated. No river of comparable size existed in the then-known world of greater Europe, and the volume of traffic he described on the Chinese rivers was inconceivable. Yet most of these passages in Marco Polo's book were reasonably accurate, as was his story of the black stones which the Chinese used for fuel—the first European description of coal, which had been in use in China from the fourth century if not earlier. Thus, also, his stories of the tribal areas or the western borders of China must have seemed just as fanciful to thirteenth-century Europe as did the tales of one-legged and one-eyed men to Friar William, who eventually made up his mind they did not exist. But doubtless many of the tribal customs described by the Venetian were actual—stories of virgins, for example, who were unwanted in marriage until sexually initiated—sound not at all strange to the ears of those who, even ten years after the Communist revolution, have traveled in the same areas. When one has seen a tribal priest in Yunnan communing with a herd of small black pigs under the conviction that they represent the incarnate souls of his flock of humans who have already died; when one has talked to a *sawbwa*, or hereditary chief, of another tribe in the same area whose right, regularly exercised, was to sleep with any girl on the first night of her marriage to one of his subjects; then Marco Polo's tales of similar regions and peoples fall into truer perspective. But one could not expect his compatriots in the thirteenth century to believe them.

What is missing in most of his stories is what he himself lacked—a background of average culture. For all his detail on the Yellow River he apparently did not know that it was in this area that Chinese civilization began and had its early flowering—something known to every Chinese of the times who had the rudiments of education. And when he comes to Hangchow, that lovely city of the Southern Sung emperors, made and remade under the surveillance of two of the most cultivated of them, a city in which Marco delighted, his tale is lively, detailed, enthusiastic; but in the end one is left wishing he had related the sophistication and atmosphere of the place to that tremendous culture of China of which the city was probably the most perfect urban manifestation.

With Hangchow he fell in love. He spent three years in a high position in Yangchow not far away and, as he says, frequently visited Hangchow, where he had made friends with a man who had been in the service of the last Sung emperor before that unfortunate ruler was deprived of his kingdom by the Mongols. Unquestionably Hangchow was the most beautiful city the Chinese ever constructed, far more beautiful than Peking, because Hangchow was built and rebuilt in the times of Chinese cultural apogee; whereas Peking, to be rebuilt in thoroughly Chinese taste by the Ming, did not become a capital until that peak of vitality and taste had suffered the concussive blows of Mongol conquest and rule, from which in many senses it was never completely to recover.

Hangchow, a city built like Venice, laced with canals, was bordered by the fabled beauty of the West Lake that had been sung in poetry for centuries before Marco. A city of huge public squares, of fine two-story houses, of an organized fire brigade, of extremely beautiful women, of public parks and pleasure boats, of sophisticated entertainments and finest silks, of public bathhouses and personal cleanliness, of excellent drainage, of palaces beyond compare—Hangchow must have deeply stirred Marco Polo with its likeness, and superiority, to his boyhood Venice. In one of the longest passages in his book devoted to a single subject, he describes it. Apart from the usual inflation of numbers (twelve thousand bridges seem excessive even in what was perhaps the world's largest city), most of his statements are borne out by Arab travelers writing just after his day. Ibn Battuta, a lovable man and perhaps the greatest traveler of all medieval times, writes of a canal in Hangchow where "there was assembled a large crowd in ships with brightly colored sails and silk awnings. . . . They began a mimic battle and bombarded each other with oranges and lemons." Marco tells of the pleasure boats, but if he saw such a mimic battle (or oranges and lemons either), he failed to mention them. It is not the kind of detail which appealed to him, and we have to wait until later European travelers of more refinement come to China to find a kindred spirit to the Arab visitor.

It was after seventeen years in China that the Polos decided the time had come to return home. Both father and uncle were now well advanced in years, and Kublai too was getting old. Marco feared that after the emperor's death he might not find favor in the eyes of his successor—a judicious observation of the way matters usually went in China. Although reluctant to lose the three men, who had proved so useful to him in many ways, Kublai at last consented to their departure, but not without assigning them one last task. They were appointed to escort the princess whom Kublai had chosen to send as consort to his great-nephew Arghun, the *ilkhan* or Mongol ruler of Persia, who in 1286 had lost his much-loved wife.

The party was sent off by sea. A fleet of thirteen four-masted ships, six hundred men (not including the sailors), the princess and her companions,

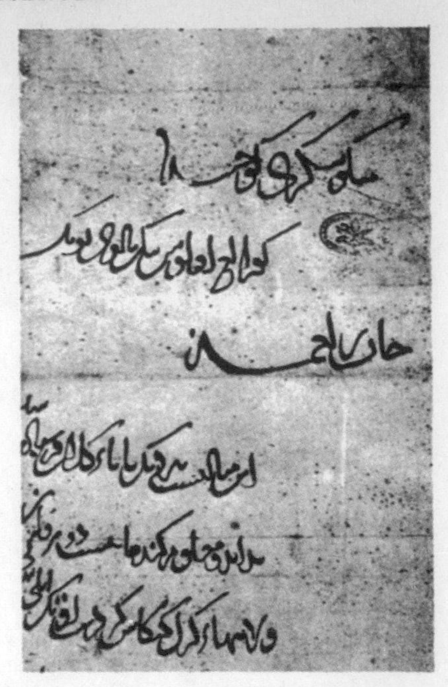

THE SIGNATURE OF ARGHUN KHAN, the Mongol ruler of Persia and the great-nephew of Kublai Khan. (Vatican Archives, Rome)

three Persian "barons," and the Polos made up the expedition sent by Kublai. They sailed early in 1291, and in a voyage lasting two years eventually reached Persia—only to find that Arghun was dead. The princess was given to his son, and the Venetians were received in great honor and forwarded on to Venice. On the last stage of the journey they heard that Kublai too had died. By 1295 they were back home in Venice. Marco was forty-two years old.

Here at least one might expect in his narrative a shred of reminiscence, a morsel of nostalgia for those years of fame and evident enjoyment that Marco spent in China, a word of regret for the death of Kublai. But there is nothing of the sort. We must conjecture what he felt, or read some of the probably apocryphal tales that have woven themselves round the Venetians' reception in Venice. Marco was soon commanding a galley in battle, in September, 1296, and was taken prisoner by the Genoese. It was during the three years of his imprisonment that he dictated his story to a fellow prisoner. On his return to Venice in 1299, or perhaps just before his imprisonment, he married. He made his will on January 9, 1324, and must have died shortly afterward, since in June of the following year his wife and daughters executed a document which implies that he was dead. He was buried outside San Lorenzo in Venice, but there is no trace of his grave.

There are, however, important traces of his book's influence. In spite of the perhaps exaggerated tales of his being discredited and nicknamed "Il Milione" by his contemporaries, it seems established that Marco Polo's work had an influence on Christopher Columbus, who came from Genoa, on the drawing of the famous Catalan Map of about 1375, and on Henry the Navigator (1394–1460). But popes and the missionaries they were sending out to China during the remainder of Yuan times after the Polos' return to Italy seem to have been ignorant of Marco's narrative. There is no sign that what it contained about Nestorian Christianity in China served any purpose at all in church circles. A vague impression of a wonderful land in the Far East, peopled by Christian subjects of the king called Prester John, seems to have had general currency, backed up by the lying Mandeville, who merely stole out-of-date information from others. Vasco da Gama left Portugal in 1497 with letters for Prester John; but this, though curious, was no more and no less natural than the fact that Marco Polo's tales of China tempted Columbus across the Atlantic to find it. Indeed, on reaching the West Indies, Columbus at first believed he had found China. It was not until the arrival of Matteo Ricci, the greatest of all the Jesuit travelers in China, that a man of intelligence and culture made use of what he read in Marco Polo's book. And by that time the Mongols had been hounded out of China, their very memory reviled, by the Ming Chinese. But, as we shall see, even in much later times, and even when Europe believed most of what was reported on China, the information was seldom used in any constructive way by travelers to that country, who still went out "blind" and returned with the usual garbled reports and miserable prejudices.

Marco Polo tells us that on his journeys about China he made notes "in order to gratify the curiosity of his master," Kublai Khan. When he was captured by the Genoese and imprisoned, he probably did not have those notes with him and therefore must have dictated his reminiscences from memory. But certainly he must have checked the written word later, since it is not likely that he could have recalled all that is in the book without the aid of notes. How, therefore, are we to explain the strange omissions, the discrepancies, the muddled chronology of the book? And to what is the entirely impersonal nature of his tale attributable?

The list of things never mentioned by Marco Polo, things with which he must have been familiar in China and must have missed in Italy on his return there, is a long and curious one that has occupied the attention of many commentators. Given a man of Marco's character, there is not much surprise in his failure to mention Chinese painting or Chinese literature. He never shows the slightest conception that a world of art—in China or elsewhere—exists, and no doubt what painting he saw and what poetry he heard of appeared to him merely as the incomprehensible work of Chinese, who were, after all, a subject people in his times. He admires Chinese architecture, enjoying its abundant decorative detail rather than appreciating its structural differences from European architecture. For all his stories of wonderful and gigantic things, there is not a single mention of the Great Wall, through one of whose gates he most certainly passed, and which runs within a morning's ride from Peking. Traveling in the south, he never mentions the tea which was grown there, and in fact never alludes to tea at all; yet he must have seen it drunk every day, and it was quite unknown in Europe. An enthusiast for the beauty of women in various provinces of China, there is not a word, among his many appreciations of woman, about the bound feet which were an essential feature in every attractive Chinese beauty. Talking much of silk, whose variety in China far exceeded that which filtered through to Italy in his day, he attempts no description of the ingenious looms on which it was woven. More serious is his omission of that Chinese instrument, the compass, with whose aid the ship in which he began his journey back to Europe was almost certainly steered. The Chinese were using the compass at least a century before the first knowledge of the instrument in the West, and as a mariner, it surely must have surprised and delighted him. Even in his favorite city of Hangchow, about which he repeats that he took detailed notes, he apparently never noticed that the populace were using sulphur matches to light their fires. The earliest match of any kind in Europe was invented in 1781.

Much more could be added to this brief list. The reasons for Marco Polo's omissions may have been multiple, rather than simple, and were perhaps also the reasons for his total suppression of anything like a personal statement in his book. First, as he said, he did not tell the half of what he knew (partly because there was too much to tell and partly from the fear of being disbelieved—a fate which overtook him anyway). Second, being at heart a merchant, and a Mongol-centered administrator and servant, his omission of many things peculiarly Chinese, or cultural, is understandable, more especially since he apparently knew little or no Chinese language. Third, the suppression of himself, of the details of his three years in high office in Yangchow, for example, perhaps has to do with the audience who would read his book in Italy. And his omission of personal details, such as any emotional or marital attachments he may have made, can be explained by a natural desire not to reveal too much in Italy, where he married soon after returning. Yet there are still a large number of peculiarly Chinese things unmentioned in the book. How is it that we are not told of printing (then unknown in Europe) and the three million copies of the Chinese calendar run off the press in Peking for the beginning of each new year and distributed all over the empire? For these and the hopeless muddle of Marco's chronology, there is no explanation at all.

The immense importance of Marco Polo's *Description of the World* lies in the richness of the detail with which he favors us, a richness which it must be said matches well enough that of his opportunities and that of the land in which he sojourned, whose history and culture he little understood. His temperament, his backgound, and the very position of affluence and influence that he attained so soon on his arrival in Peking are perhaps responsible for the most conspicuous lack in his writings. The absence of humanity, the missing qualities of compassion and humility, and the lack of empathy in regard to people in general make the book duller than it deserves to be.

For he is humorless as well as credulous, and his total picture of China hardly makes the civilization of that great land come alive. Had it been Friar William who journeyed in China, we might well have been given a less copious book, but surely a better one. For the friar, though not a poet by any means, yet had the vision of poetry, its capacity to tell something real and important in a few simple words. This, alas, no one can attribute to the Venetian merchant who grew up in the land of Kublai Khan.

CHAPTER FOUR

JOHN OF MONTECORVINO

Saints are hard to deal with. Their convictions burn them up. It is pointless to question the validity of their ideals. Each of their acts has the purity of something from the fire, and can hardly be put into words that measure up to the fact. There is an inevitability about them which, like that of a fire out of control, leaves little to be done but record it. John of Montecorvino was such a man, such a saint. The fire which was his person and his flaming soul lit smaller fires that burned on in Peking for a few decades after he himself burned out. And that is much more than can be said for most Europeans who went and lived there.

There are two villages called Montecorvino in the dry and needy south of Italy, and no one can now tell from which of them Friar John came. In one or the other, perhaps in that a few miles inland from Salerno, he was born, according to his own estimate, in 1247, several years before Marco Polo saw the light of Venice in the more prosperous north of Italy. Friar John appears to have come from the upper classes of society since he went to university (which one we do not know) and later entered the Franciscan order (at a date also unknown). A priest who followed him in Peking, John of Marignolli, gives us those facts, while another Franciscan, Friar Elimosina, describes the ordained John of Montecorvino as "a fervent devotee and follower of St. Francis, upright and austere."[1] Of the two schools of Franciscan thought and behavior current at the time, Friar John undoubtedly adhered to that which sought to imitate the actual life of St. Francis himself. And he seems to have felt, as St. Francis did, that he had received a divine command to imitate the life and evangelistic labors of Christ. Friar John's whole life points clearly to this.

Between the time of his ordination and his first mission abroad he apparently gained some renown as a preacher and a theologian. Opinions differ considerably on the date of his departure from Italy toward the Middle East, but it is certain that by 1289 he had been in the Levant for some

1. This and other quotations from the writings of Montecorvino's followers Elimosina and Marignolli are taken from Yule's *Cathay*.

years and that he was then legate of the Roman church at the court of the king of Armenia. He had already visited Persia, the country of Arghun Khan to which the Polos were shortly to import a bride for the ruler. He was, therefore, already well schooled in the obstacles and perhaps also the joys of missionary work.

There is not a scrap of evidence for it, but it would seem that, moving as Friar John did in very high circles in Persia, he must have met Mark and Sauma, two Uighur monks from Peking. Their curious story belongs properly to another volume. But, briefly, Mark and Sauma were two Nestorian Christians who left Peking and came to the Middle East overland, hoping to reach the holy places of Christ's life. The irony of their journey was that neither of them managed to see Jerusalem, but that Mark was elevated to the position of patriarch of the Nestorian church (as Mar Yahballaha), and Sauma to the office of his visitor-general. From them Friar John may have learned of conditions in China, and of Peking in particular. He must have heard of the Mongol religious eclecticism, the beliefs of superstitious Kublai Khan, of the prevalence of the Nestorian heresy and the absence of any Catholic endeavor in China. And he must also have met in Persia merchants returning from Central Asia and even from China itself. For the Mongol trade routes across the great continent were wide open, the flood of Westerners in China was at its peak and not to be paralleled in numerical strength until the nineteenth century, during another period of alien rule.

The evidence for all this is for the most part inferential. But what is certain is that when the pope, to whom Friar John delivered his account of the state of affairs in the Middle East in 1289, saw a man determined and strong in missionary zeal, with a heart set on the conquest of the souls of Asia, he took what was (for once in the sorry story of papal actions in an age of power politics) an enlightened step. He sent the right man to the right place. Equipped with a battery of letters to almost every known ruler through whose territory he might have to pass, John of Montecorvino was sent off toward China about the beginning of 1291.

His feelings can be reconstructed. No one has ever set out for China without a flutter of excitement—either then or now. And from this moment on, as if taking over when the mere prologue had been told, John of Montecorvino himself supplies the central portion of his life story. It is contained, concisely and almost incidentally, in the three letters he wrote to Europe during the remaining thirty-seven years of his life. There may have been more letters from him, but probably not many more; and the first of the three is only known in the version relayed to us by another man.

It is at Tabriz, in Persia, that his first letter takes up the story.[2] Written

2. All quotations from Montecorvino's letters, as well as from papal documents relating to him, come from A. C. Moule's translations in the *Journal of the Royal Asiatic Society*, London, 1914.

either in 1291 or in the following year, from India on the second of December, it says nothing of his feelings about leaving Rome and the intrigues of the Church. In Tabriz he was joined by "a great merchant," Peter of Lucolongo, who was to be his lifelong friend and companion. They set out in 1291. The first part of their route seems similar to that of the Polos, who had gone that way just twenty years before. Reaching Ormuz on the Persian Gulf, Friar John remarks, as Marco Polo had done, on the vessels used for the passage to India: "Flimsy and uncouth, without nails or iron of any sort, sewn together with twine like clothes, without caulking, with only one mast and one sail of matting, and ropes of husk." The "husk" was perhaps coconut fiber, still used for ropes in India and other Oriental lands. Unlike the Polos, it was by one such ship that Friar John and his friend continued their journey.[3]

From the gulf, missionary and merchant proceeded by coastal navigation to Malabar—the southwestern coast of India—and sailed round the southern tip of the land to the Coromandel or eastern coast at St. Thomé, near modern Madras. Here they stayed for thirteen months near the shrine of St. Thomas, the "doubting apostle." St. Thomas is traditionally supposed to have preached in India and to have been martyred at St. Thomas' Mount, a hill standing up from the plain about eight miles outside Madras. (Indians call it Faranghi Mahal—the Hill of the Franks.) It is mentioned by Marco Polo, and much later, in 1504, the Portuguese built a church in Madras itself over the tomb of the saint, which they had removed there. The whole story of St. Thomas is probably apocryphal, but at least from very early times there was a Nestorian Christian community in South India. Friar John remarks that the Christians, like the Jews who were there too, were not much thought of, and few.

Here, around the town of St. Thomé, Friar John made a hundred converts among the people, whether indigenous or expatriate he does not say. Here there was no winter and the land was lush beyond anything he had ever seen. He kept looking for the southern polestar, but the continual haze on the horizon foiled all his efforts to see it, as it does today. Here, he says, there were few flies and no fleas—quite the opposite of the present state of things. He found, moreover, none of the one-eyed and other human monstrosities, nor the people who live on the smell of wild apples, which former travelers

3. Of the ships at Ormuz, Marco Polo says: "The vessels . . . are of the worst kind, and dangerous for navigation, exposing the merchants and others who make use of them to great hazards. Their defects proceed from the circumstances of nails not being used . . . the wood being of too hard quality, and liable to split or crack like earthenware. . . . The planks are bored . . . and wooden pins . . . driven into them. . . . After this they are bound, or rather sewn together, with a kind of rope-yarn stripped from the husk of the Indian nuts. . . . Pitch is not used for preserving the bottoms of the vessels, but they are smeared with an oil made from the fat of fish. . . . The vessel has no more than one mast, one helm, and one deck."

had said existed in India. Not even the advent of many later Europeans brought them to life, though it brought the flies and the fleas—and of course the miraculous metal birds which nowadays swoop round the hill of St. Thomas, for the airfield is quite near its base. Friar John was interested to see reaping and sowing going on at all times, and fascinated to find that the aromatic spices so much sought after in the West were plentiful and cheap. And there were those "Indian nuts," mentioned by Marco Polo, which were as big as melons and as green as gourds—coconuts. Cinnamon, he says truthfully, came from an island close by, and this is the first mention of Ceylon in Western literature, a generation before the Arab traveler Ibn Battuta described it.

Friar John continues with an outline of the religion, the use of talipot leaves as writing surfaces, and the burning of the dead. He makes it plain that he liked the South Indian peasants, an agreeable people then as they are today, well formed and not black but olive. Near the end of his letter there is some description of the despotic monsoons—the southwest blowing from April to October, and the northeast from October to March. Friar John realized that it was by virtue of these winds that travel was possible in a regular pattern between the Persian Gulf, the coasts of India, and onward to the East.

This letter came in due course to Rome, providing the first description of southern India, giving the exact site, till then unknown, of St. Thomas's shrine, and opening up in Roman eyes the missionary possibilities of a new continent. At this time, little though Friar John suspected it, the three Polos were plowing through the seas off the South Indian coast with a bride for the dead Arghun. Friar John seems not to have heard of them, not even in Peking, and never to have suspected their long stay in the Mongol capital.

At the end of thirteen months at St. Thomé, Friar John set sail for China, his real destination. Before his departure, however, he saw the death of Nicholas of Pistoia, a Franciscan who may have accompanied him from Europe or may have been already in Madras; and surely he must have recalled, as he left the shores of India, the party of five Franciscan friars who had left for China more than ten years earlier, never to be heard of again. But he was not deterred. Perhaps Friar John thought the five were already in Peking and looked forward to meeting them, equipped as they were with a letter from the pope to "My dearest Son Kublai, Grand Khan, Emperor and Moderator of All Tartary."

Friar John himself bore a similar missive from the still-hopeful pope: "To Kublai the great Khan, famous prince of the Tartars, grace in the present time to lead to glory in the future. We rejoice in the Lord, noble Prince, and give Him devout and abundant thanks that He . . . has filled your inmost heart with such feelings that the desire of your mind is directed toward the enlarging of the boundaries of Christianity. For shortly after the beginning of our promotion we received in audience trusty messengers who had been sent by the Magnificent Prince Arghun . . . who told us very plainly that your Magnificence bears a feeling of great love toward our person and the Roman Church. . . ." (The messengers, of course, were Mark and Sauma.) The pope continues with a flood of verbiage calculated, one might feel, to raise little but a wry smile on the face of Kublai at the pontiff's apparent credulity. "We have chosen our beloved son Brother John of Montecorvino with his fellows of the Order of the Minors [none of whom reached Peking] . . . earnestly praying you to receive them . . . and to grant them the help of your Royal favor for the healthful work committed to them. . . ."

But if Pistoia died and one other friar never reached India, at least Friar John was not entirely alone; for the friend of his life, Peter of Lucolongo, was always with him. He seems to have supplied the finance in Peking for the missionary effort that, in Hudson's words, was "destined to give the Catholic Church half a century of life there, before the revolt which drove the Mongols outside the Great Wall."

The next news comes from Peking itself, in the second of Friar John's three letters. Perhaps it is natural that he says nothing of the voyage, nor of the port where he landed in China, nor of how he traveled to the great city; for the letter was not written until he had been eight years in Peking and the strangeness of his surroundings had worn off. The letter says simply: ". . . And I, proceeding on my further journey, made my way to Cathay. . . . To the emperor I presented the letter of the lord Pope, and invited him to adopt the Catholic faith. . . . But he had grown too old in idolatry."

The wording of his statement about the emperor has incited much learned argument. Was it to Kublai that he presented the letter—to Kublai who was by then an old man and died on the February 18, 1294? Or was it to Timur, the Ch'eng Tsung emperor, who succeeded Kublai on the tenth of May in the same year? The fact that we do not know the date (even the year) of Friar John's arrival in Peking, and also that he says nothing of the recently departed Polos, of whom, had he been received by Kublai, he might well have heard, makes it possible—and even probable—that Kublai was already dead when the Franciscan reached the capital. There is no mention in his letter of the death of an emperor, and this in itself also points to Friar John's having arrived after Kublai's death: had the great Kublai died while Friar John was actually in Peking, it is hard to believe that he would not have mentioned the fact. "Grown too old in idolatry" does not necessarily define the emperor of whom it was said as actually being old. Even a man of thirty or so, as was Timur when he was enthroned, might be so described if he was resistant to conversion.

Friar John, however, has the honor of being the first Catholic priest to set foot in Peking, indeed the first to step on Chinese soil, so far as we know.

As he reached Peking we could wish that his friend Peter the merchant had bestirred himself, as Marco the merchant was about to do at this time in his Genoese prison on the other side of the world, and given posterity a word or two of Friar John. But he seems rather to have busied himself in making money for the support of the mission that his Franciscan friend was to establish in that remote capital. Marco Polo had departed only two or three years previously, from this city of Peking, and the Yuan dynasty was still more or less in its heyday. The cracks in its facade had not yet appeared on the surface, although they were inbuilt in its fundamental structure. But excesses of the entirely idle Mongol nobility, who existed parasitically on the corpus of China, were beginning to affect the economy. Laws passed in 1298, not long after the arrival of Friar John, attempted to protect the Chinese, in the south especially, from the extortions of Moslem moneylenders, who were the tools of Mongol overlords. And a few years earlier there had been legislation attempting to ease the oppression of slaves, tenant farmers, and laborers. In the first year of the reign of the new emperor, Timur, there was even an edict forbidding Mongols to wreck crops by riding through them; and shortly after that another provided 107 strokes as punishment for a proprietor who beat a Chinese laborer to death.

On the politico-religious side (under the Mongols the two aspects were always fused) the preference shown by Kublai for Tibetan Lamaism, however much sentimental attachment he might also have shown to Christianity because of his mother's upbringing, agreed very well with the original shamanistic cults of his Mongol ancestors. And his successor, Timur, concurred with emphasis in this. At the same time Timur found it judicious to placate other sections of opinion, just as Kublai had done in honoring the direct descendants of Confucius—the K'ung family, who still lived in the philospher's old village of Ch'u-fu in Shantung. Timur, in fact, went further, issuing an edict that Mongols as well as Chinese should worship Confucius. This won him some sympathy from the Chinese literati. Kublai had favored the Nestorian Christians during his lifetime, partly—as we have seen—from attachment to the faith of his mother, the Kerait princess, but partly also for the political reason that many of his Asian and other subjects who were highly important to the administration and trade of his regime were of that faith. In 1275 he had created a Nestorian archbishopric in Peking. The strength of Nestorian claims on the Mongol hierarchy and emperors was therefore founded to some extent on establishment. It was this, doubtless, which made them bothersome to Friar John.

Continuing with his letter he says: "[The emperor], however, bestows many kindnesses on the Christians"—the Catholic converts he had made in Peking. "I have now been staying with him for eleven years." The letter is dated January 8, 1305, and eleven years would place his arrival within the life of Kublai; but Friar John's dates are often inaccurate. He says in this same letter that Brother Arnold of Cologne arrived in Peking the year before he was writing, and infers that this too was eleven years after he reached there, although it could mean eleven years since Nicholas of Pistoia died in India. If one does not have a regular calendar to consult, years are apt to slip by or, in the press of new events, to seem more than they actually are.

"The Nestorians," Friar John goes on sadly, "men who bear the Christian name but deviate very far from the Christian religion, have grown so powerful in these parts that they have not allowed any Christian of another ritual to have even a small chapel. . . . And so . . . both directly and through others they bribed, they have brought on me persecutions of the sharpest." The accusations brought against him—that he was a spy, a magician, an impostor—went on for five years. "After some while . . . they produced false witnesses who said that another messenger had been sent [to Peking] with presents of immense value for the emperor, and that I had murdered him in India and stolen what he was carrying. . . . Many a time I was dragged before the judgement seat with . . . threats of death. At last, by God's providence, through the confession of certain individuals, the emperor came to know my innocence and the malice of my rivals, and sent them with their wives and children into exile." These were Nestorians, jealous of a rival faith which also called itself Christian. We learn from a Dominican, John di Cora, who seems to have known a great deal about Friar John in Peking (though he himself never went there from Persia) that there were about thirty thousand Nestorians in China at the time. The figure is conjectural.

In his late forties when he arrived in Peking, Friar John must have found it a hard and exacting task to fight his way through the established Nestorians to a place of favor. He does not complain, simply stating the bare facts. But now the first obstacle was overcome, at least for the moment, and it seemed that fortune smiled on him.

"I . . . was alone on this pilgrimage . . . for eleven years until Brother Arnold . . . of Cologne came to me last year. I have built a church in the city of Peking. . . . And this I completed six years ago; and I also made a bell tower there, and put three bells on it. I have also baptized there, as I reckon, up to this time about six thousand persons." And, had the troubles with the Nestorians not been so long, "I should have baptized more than thirty thousand." What he does not tell us is that, while his church was being built, the Nestorians attempted sabotage every night on what had been accomplished during the day. For this information we are indebted to John di Cora. The courage needed to go on, quite alone in a strange land whose language he surely had only imperfectly learned by this time, and in the face of Nestorian and official obstacles, must have been considerable. But Friar John persevered.

"I have bought one after another forty boys, the sons of pagans [Chinese] aged between seven and eleven years, who up to that time had never learned any religion. I have baptized them and taught them Greek and Latin, and our ritual; and I have written out for them thirty psalters with hymnaries and two breviaries, with which eleven boys now know our office . . . and take their weekly turn of duty . . . whether I am present or not. . . . And the lord emperor is greatly delighted at their chanting. I strike the bells at the canonical hours, and with the congregation of babes and sucklings I perform divine service, but we sing by ear, because we have no service-book with music." Later in the letter he asks for the music to be sent to him in Peking.

The sound of chanting and the ringing of bells recall momentarily the lament of Po Chu-i (quoted in chapter 1). But times had changed, and to the Mongols—in contradistinction to the Chinese—the knowledge that they were being prayed for by all and sundry of every religion seems to have brought spiritual comfort. Such was the uncertainty of their philosophical background.

There were further triumphs for Friar John to recount in his letter. In his first year in Peking he was fortunate enough to capture the affections of a "certain king," a Nestorian, "who was of the family of the great king Prester John." This King George was probably ruler of the region called Tenduc by Marco Polo, who passed through it. Tenduc seems to have been located somewhere round the great loop of the Yellow River. It is apparent that Friar John had absorbed some of the European legends of Prester John, who was supposed to rule a vast Asian Christian kingdom. The legends seem to have had their origin about 1145 and to stem from the foundation in Central Asia of a kingdom called the Western Liao. This was formed by amalgamation of the Liao dynasty, which was driven from North China by the Jurchen Tartars (who established its successor, the Chin dynasty in China) with Uighur people. The Western Liao ruled around Kashgar until swept away by Genghis Khan in 1211. Nestorianism flourished in the kingdom, along with Buddhism, shamanism, and Islam. The cultural contacts of this brief kingdom with Russian princelings and their domains seem to account in part at least for the transmission of the legend of the mighty king to Europe. But the subject is still obscure, if fascinating. Just how King George was ruling from Peking a kingdom situated on the Yellow River, or how the original site of Prester John's Western Liao kingdom had shifted to the Yellow River area from Central Asia, has not been explained.[4] Nonetheless, King George "attached himself to me in the first year of my coming hither, and being converted by me to the truth of the true Catholic faith, took the

4. Moule has suggested (*Journal of the Royal Asiatic Society*, 1914) that King George has no connection with the figure named Prester John, and that both Marco Polo and Friar John passed through George's territory in the region of the Yellow River loop, the latter perhaps meeting the king there before he reached Peking.

lesser Orders. And when I celebrated mass he served, wearing his royal robes." Other Nestorians might accuse George of apostasy, but the convinced and converted king was apparently unshaken in his new faith, and brought over to Catholicism most of his own people.

In his own kingdom George "built a beautiful church on a scale of royal magnificence to the honor of God, of the Holy Trinity, of the lord Pope, and of my [Friar John's] name, calling it the Roman Church." This descendant of the fabled Prester John (if we allow ourselves to swallow the evocative legend) unfortunately died in 1299, leaving a son, whom he had called John after the friar, in the cradle and unable to prevent the Nestorian princes (brothers of the late George) from reconverting all the people back to that faith. "And because I was alone," Friar John says with infinite regret and sorrow, "and was unable to leave the emperor the Khan, I could not go to that church [which King George had built] which is twenty days journey distant. Yet if some good helpers and fellow-workers come, I trust in God that all may be retrieved. . . ."

This last sentence and the following ones echo with the long and agonizing efforts Friar John was making in Peking. His forbearance is large, for if ever a man needed, and failed to be given, support, it was he. It seemed that in Europe they had forgotten him, until Arnold of Cologne arrived. And it does not appear that Arnold was much help in the struggle with perfidious Nestorians and mistrustful mandarins, for little is said of him.

"I say again that if there had not been the aforesaid slanders [by the Nestorians] great fruit would have followed. If I had had two or three comrades to aid me, perhaps the emperor the Khan, too, would have been baptized. I ask for such brethren to come . . . as will studiously show themselves as an example." Friar John even details the best route for the friars to take. Because of the hazards of that route, he continues (with sublime forgiveness toward the unresponsive authorities in Europe), it is now twelve years since he has heard news of Italy and home. And "it is now two years ago that a certain Lombard physician and surgeon came and filled these parts with incredible blasphemies about the Court of Rome and our Order and the state of the West; and on this account I much desire to learn the truth." We do not know who this "Lombard leech" (the term is Beazley's) was, nor what took him to Peking. Although his frankness on the state of the Roman church and its priests in Europe was not entirely without justification, it was also, to say the least, tactless. But maybe he was bribed by the Nestorians.

Friar John's position suffered accordingly. But in due course this problem too was overcome, and "now I am in the act of building another church, with a view to distributing the boys in more places than one. I am now old and grey, more from toil and trouble than from age, for I am fifty-eight years old; I have a competent knowledge of the Tartar [Mongol] language and writing . . . and preach openly and in public. . . ." There is a wealth of understatement in those few words. Just how much "toil and trouble" had it taken him to get one church built and another started in the hostile environment of Peking? How much slow and painstaking labor had he spent in learning a language for which there was no grammar, and learning it well enough to translate the Scriptures? And over how much opposition had he slowly ridden to be able to preach his cherished religion "publicly"? Alas, we shall never know precisely.

But it is doing no violence to the facts to imagine Friar John, dressed in the plain habit of his order, going with a certain decision and dispatch through the unpaved streets of Peking—that city which Marco Polo was describing for us as he waited release from the prison at Genoa. In the cosmopolitan but hardly sophisticated streets of Kublai's city, Friar John would not have been very conspicuous: "There are many monks of different schools wearing different habits," he says. No picture of him exists which can pretend authenticity, but somehow his letters suggest a tallish man, thin, a longish face now drawn as well as ascetic. Whatever dislike he may have felt in his native Italy for the intrigues of the papal court, in Peking he had found that intrigue of exactly the same order was the blood of the imperial administration. Nothing could be achieved except by lobbying and scheming. And it is certain that, to judge only by those two churches he succeeded in building, Friar John had taken his share in such intrigues. Perhaps, when the supreme wish of his life was at stake, he found it less distasteful to curry what favor he could, and build a church out of it. But he is a man of scrupulous fairness, of a fine dedication to an impossible task—that of converting Peking people to Roman Christianity. That much, and more, shines out of his brief letters.

The postman who took his first letter from Peking, for forwarding to Rome by the Franciscan community in southern Russia to which it was addressed, was a member of an embassy from Persia, returning now to his own land. He left one copy with the friars in Russia and carried another on to Tabriz. And it is to the Franciscan friars in that Persian city, many of whom were undoubtedly known to Friar John from his sojourn there, that he writes his next and last letter. By this date, February 13, 1306, two years had elapsed, and in the interval he had been much heartened by a reply to the previous letter, by the knowledge that his earlier letters had actually been received. The desire to communicate at least something of the achievements and frustrations of those long years in Peking, together with a certain joy at hearing news from Europe, prompted Friar John to write again.

"The requirements of love demand that those who are separated far and widely, and especially those who travel for the law of Christ . . . shall at least comfort one another by words and letters," he wrote. It is a gentle rebuke from one cut off for more than a decade; if Friar John had been a less humble and straightforward man, one might suspect him of sarcasm. "I have thought that you may reasonably be surprised that living so many years in so distant a province you have never received a letter from me. But I have wondered no less that never until this year have I received letter of greeting from any Brother or friend, and it seems that no one remembered me; and most of all when I heard that rumors had reached you that I was dead." The stark isolation of his life in Peking cries out from those simple sentences.

But he turns at once to practical matters. The second church, mentioned in his last letter, is now complete, standing just outside the gate of the imperial palace, and he has had made six pictures of Old and New Testament subjects "for the instruction of the ignorant . . . the explanations written in Latin, Tarsic and Persian letters so that all tongues may be able to read." Not, we may note, in Chinese—another argument against his having converted any substantial number of them. Some of the Chinese boys he had bought have died, but he has baptized several thousand persons.

Friar John then gives a brief account of the difficulties that attended the building of his second church. "Master Peter of Lucolongo, a faithful Christian and a great merchant who was my companion from Tabriz, himself bought the site for the place . . . and gave it to me by the love of God. . . . I received the site in the beginning of August" and by the fourth of October it was finished, "with an enclosure wall and houses, complete offices, and an oratory which will hold two hundred persons. But on account of the winter I have not been able to finish the church [itself]. But I have the timber collected at the house, and by the mercy of God I shall finish in the summer. I tell you it seems a sort of marvel to all . . . when they see the place new built and the red Cross placed aloft at the top. . . . We in our oratory sing the office regularly by ear, because we have not the notes [music]. The lord Khan can hear our voices in his chamber; and this wonderful fact is published far and wide among the heathen, and will have great effect. . . ."

The distance between the two churches, Friar John states, is about two and a half miles across the city of Peking. So we may picture him passing daily through the city on foot, from one to the other. "I have divided the boys and placed part in the first church and part . . . in the second; and they perform the service by themselves. But I, as chaplain, celebrate in either church by weeks, for the boys are not priests."

To have achieved all this in little more than a decade was almost miraculous. Small wonder that Friar John felt his powers failing "more from toil and trouble than from age." But by now, he says, at least he had his rightful place at the court of the emperor, who "honors me above all other prelates. . . ." History does not bear out the implication that Timur really favored Roman Christianity, but it is very probably true that he honored this gray and upright man. Amid sycophants and self-seekers at court, he doubtless found in Friar John that persistent truthfulness an emperor needs. Friar John himself, on this subject, had warned in his first letter against the sending out to China of priests who would bring their "phylacteries"—their charms and amulets with holy writing in them—as a means to convert the people in Peking. Now in this second and last letter from the capital he states his doubts of European priests in general, saying that the monks of various persuasions in China are "of much greater austerity and obedience than Latin monks are." He had sized up the situation clearly, realizing that only by setting a wholly remarkable religious and moral example could Roman churchmen hope to draw to their faith the mass of people in the Middle Kingdom.

For the remaining eighteen years of his struggle in Peking, Friar John is silent. He writes no more letters to the West, or none have survived. For several reasons his silence is hardly surprising. Letters took upwards of a year, sometimes two or more, to reach Europe, and might never arrive at all. And in any case he received help in the years which followed his last letter. Much of it, in the shape of various friars, was not of outstanding quality. One suspects he simply gave up hope that letters would produce what he really wanted—a few men of his own caliber. Being a humble man, he just made do with what accident and the incomprehension of European popes provided. It is largely through the writings of these helpers who came to Peking that we learn, fragmentarily, of Friar John's later years there.

Word of Friar John's last letter was brought to the French pope, Clement V, at Avignon by a Franciscan called Thomas of Tolentino, later to be martyred in India. Thomas probably collected the actual letter, or a transcript, in Persia. He seems, however, to have had even more information about Friar John's work in Peking than the letter itself contained. The story of that heroic struggle in China, told by him in the papal court at Avignon, was sufficient to induce the opinion there that the East was ripe for Christian conquest—an idea which recurs throughout history and is always dashed. Clement at once named Friar John the first archbishop of Peking and dispatched seven bishops toward China to consecrate him. Later, probably in 1308, three of the bishops reached Peking. Of the rest, three perished on the way (they took the sea route via India) and one seems not to have set out at all. The three who reached Peking were Andrew of Perugia, Gerard Albuin, and Peregrine of Castello. They brought with them the papal bull appointing Friar John: ". . . Taking . . . into very careful consideration your conspicuous diligence in this holy work, we choose you . . . to be Archbishop in the great and honorable city of Khanbalig [Peking], in the realm of the magnificent prince the great king of the Tartars . . . committing to you the full charge and care of all the souls living in the whole dominion of the Tartars. . . ."

Friar John must have drawn a long reflective breath when he read those high words and wondered how—just how—a man could set about the mighty task they laid on his old shoulders. There were at that time at least as many souls in China as the papal authority—through thousands of churches and priests—commanded in all Europe and beyond.

But it is not hard to visualize the scenes of joy in the mission house at Peking when the three friars arrived from the fountainhead of the Roman church, empowered to consecrate Friar John in his archbishopric and empowering him in turn to make them his bishops. This must have felt like a divine recompense for those long lonely years of striving in the Eastern dark. The year was probably 1308, though the dates are confused by Andrew of Perugia, who, it turned out, was more concerned with comfort than with Christendom.

From Andrew we get the best, though slight, picture of Friar John as first archbishop of Peking.[5] Writing in 1326 (if we can trust his date this time) to the warden of the Franciscan house in Perugia, Bishop Andrew says he can "scarcely hope that the letters I have sent can reach your hands . . . on account of the vast distance of lands and seas that lies between you and me." He tells how he reached Peking after "much labor and weariness, fasting and hardship and anger . . . in which we were robbed of everything, even our tunics and habits, and came at last . . . to the city of Peking. . . . Here we consecrated the Archbishop . . . and stayed there almost five years."

During this time they received the grants of food and clothes that the emperor was accustomed to dole out to envoys of tributary princes. Conforming to that norm of Europeans writing about China to their friends at home, Andrew says he will "forbear to speak of the wealth and magnificence and glory of this great emperor, of the vastness of the empire and the number of its cities and their size . . . for it would be too long to write and would seem unbelievable to my hearers. For even I who am on the spot hear such things I can hardly believe them." Like Marco Polo, he did not dare tell the half of what he knew.

It is obvious that Bishop Andrew did not get along with the new archbishop, and apparently he went off to Zayton (the modern Ch'uan-chou-fu, near Amoy), at that time a great port on the coast in Fukien province in the south, opposite the island of Formosa, where John seems to have established a mission station a little earlier. "A rich Armenian lady had built a church [in Zayton], fair and large enough . . . with suitable endowment." She had given it to Gerard, who had been made bishop of the place by Archbishop John. When Gerard died, John wanted Andrew to take over in Zayton, but the latter would not consent. The third of the trio, Peregrine, did. He died there in 1332. Andrew was already at Zayton,

5. Information on Andrew of Perugia and the quotation that follows is in Dawson.

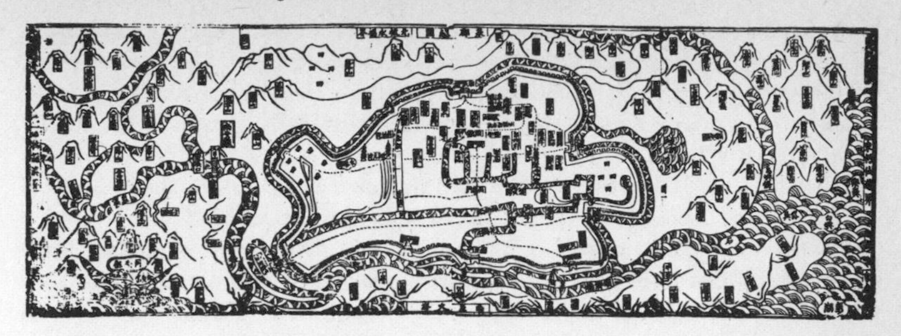

A CHINESE MAP OF ZAYTON, made in 1612, shows the legendary "carp" shape of the city. (From *Ch'uan-chou-fu chin,* 1612; British Museum, London)

"since for various reasons I was not contented in Peking . . . I got leave that the . . . *alafa* [imperial allowances] should be paid to me in the city of Zayton. . . . I travelled there very honorably with a train of eight horses allowed me by the emperor, and arrived while Brother Peregrine was still alive."

Andrew was nothing if not a successful schemer. Once there, he built a house for twenty brethren, "with four chambers, any of which is good enough for a bishop" and a church also, in a wood a quarter of a mile from the city. Here "I have taken up my abode, and I live on the bounty of the emperor," which he estimates at a very large sum. "And a great part of this alms I have spent in this house. . . . There is not a heritage among all those in our province to be compared with it for beauty and convenience."

In short, Bishop Andrew had opted out of the struggle, intriguing his way into imperial favor and establishing himself in considerable luxury in this residence outside the noisy city in a delightful grove. There, he goes on, he commuted to the town, and when Peregrine died he took over the cathedral of Zayton, which the expatriate Armenian lady had built. "And I am healthy in body and vigorous and active, so far as my age allows; in fact I have none of the natural defects and characteristics of old age except my white hairs." If ever there was a self-satisfied priest, it is Bishop Andrew.

Recalling poor John in Peking, one recalls too his warning on the quality of men he needed to work with him in China. The smug and otiose Andrew was hardly the stuff that saints are made of, nor missionaries either.

By the time he was writing, among those who had come to Peking all but himself and John were dead; but this fact does not greatly concern him. He is comfortable, well fed, well housed, supported by emperor and Armenian lady (among others of the large foreign community at Zayton). He is happily settled, with a ready-made, captive congregation of Europeans among the traders and their families, who doubtless provide him with company and conversation. Ironically, Andrew's is the only documentary evidence of the

Franciscans in Zayton, except for a fragment that came to light not long ago. During World War II a Chinese schoolmaster came upon a gravestone while the walls of ancient Zayton were being demolished. Among a large number of Christian and Moslem gravestones was a much defaced one which read: "HIC . . . SEPULTUS ANDREAS PERUSINUS ORDINIS . . . APOSTOLUS M. . .XII"—Here [lies] buried Andrew of Perugia of the Order [of Minor Friars] Apostle [of Jesus Christ] M[CCCXX]XII (1332).[6]

To that dogged saint John of Montecorvino there is no memorial, save those three short letters of his own and the few comments of his contemporaries. In 1328 he died at the age of eighty-one. The Dominican John di Cora gives a last glimpse of him: "This brother John the Archbishop converted there [in Peking] many men to the law of Jesus Christ. He was a man of very honest life, pleasing both to God and to the world, and stood high in the favor of the emperor. . . . To his funeral and his interment came a huge multitude of Christians and pagans, the pagans rending their mourning robes as the custom is. And these Christians and pagans devoutly took pieces of the Archbishop's funeral pall and revered them as relics. . . . They still visit his grave with extreme devotion. . . ."

Five years later, in 1333, news of John's death filtered through to the pope. A Franciscan friar named Nicholas, twenty other friars, and six laymen were sent out to Peking—but it seems unlikely that they ever reached the city in which John had lit the flame of Roman Christianity. In 1338 an embassy of sixteen men from the Mongol emperor himself (probably from Toghon Timur Ukhagatu, 1332–68, the last of the Mongol rulers in Peking) appeared before Pope Benedict XII bearing a letter from the khan. It was brief and quite genial. The khan asked for papal prayers (which his rotting Mongol administration sorely needed by this time), suggested diplomatic relations, stated his great longing for horses and other *mirabilia* of the "sunset lands."[7]

The khan also commended a letter from the Alans, which was brought by this embassy. The Alans were a Central Caucasian people, and a contingent of them was included in the Mongol troops in China. Originally Greek Orthodox, they had been converted by Friar John and had become some of his staunchest supporters. Their letter to the pope was distinctly tart both in comment and content:

". . . Let it be known to your holiness, that for a long time we have been conversant with the Catholic faith, and gently governed and richly consoled by your Legate, Friar John, a worthy, holy, and capable man who has been dead these eight years. During this time we have been without ruler, without spiritual consolation." They demanded a replacement at once, and no

6. The *Illustrated London News,* May 14, 1955, has an interesting article by Prof. John Foster on the subject, with photographs of some of the stones recovered.

7. Quoted in Yule's *Cathay.*

empty promises. "We ask your Wisdom to reply graciously to our emperor . . . for on several occasions . . . there have come . . . to the emperor our master, messengers who were graciously received by him, honored and loaded with gifts; and yet from then onward the emperor got no reply from you, although each promised to bring back such a reply." The Alans hoped for a reply this time, "for Christians are put much to shame in these parts when they are discovered to tell lies."

While it was not always the pope's fault that envoys did not reach Peking, it is abundantly clear that more consistent attention to the East on the part of the Holy See would have immeasurably benefited the state of its priests and converts there. The history of Western intransigence, though the Alans did not know it, was only beginning.

A LETTER FROM ASIA TO ROME. It is surprising how many letters managed to make their way across vast stretches of Asia to Rome, where they are still preserved. This example, dated 1304, is from the Chinese Nestorian patriarch Mar Yahballaha to Pope Benedict XI. The patriarch's seal is in the form of a stylized Nestorian cross within a border of what is popularly called the "Greek key" pattern, but which is a common and ancient Chinese design. (Vatican Archives, Rome)

But this time the pope responded with letters to the khan and to the Alans, and with a group of four Franciscan friars and other lesser missionaries whom we shall encounter in the following section of this volume. John of Marignolli, their chronicler, records that thirty-two of them reached Peking.

With that cry of the Alans echoing across the wilderness of Central Asia, we have all but the last note of the work and the life of Friar John of Montecorvino. Odoric, whom we meet next, has a small and charming tale of a man who is perhaps John, perhaps not.

The story of the Franciscan mission in Peking and in China at large had not long to run after the death of Archbishop John. It was principally a vision of *his,* and its vitality did not long outlast him; for those who came after were inferior men. His sincerity, the clean, clear arrow of his faith never wavered. His dedication to the strange purpose of converting the world's farthest and most alien country to his dearly held belief is a moving one.

It may be doubted if he made many Chinese converts besides those little boys he bought at a tender age, who (with what mental torture!) learned Greek and Latin and sang so sweetly in the choir. Almost certainly most of the souls he won—partly or temporarily or, in the case of the Alans, permanently—were Mongols and other expatriates in China. His remaining communicants were probably merchants, like Peter, who were born to the Roman faith. Assent to this assumption is perhaps given by the fact that as soon as the Mongols were driven out of China the Franciscan church was extinguished. Had the body of Friar John's converts been Chinese, it seems fair to conclude that there would have been some significant signs of their faith's continuing into the intensely Chinese air of the early Ming years. But there is none. The whole soul-racking effort of the noble Franciscan vanishes with a puff of Chinese dynastic breath; and, come the Ming in 1368, not even the wick of the lamp Friar John lit is there to trail a little Roman Christian smoke in the astringent Peking atmosphere.

CHAPTER FIVE

ODORIC OF PORDENONE

In a sense it is a relief to turn from the white intensity of Friar John the saint to Odoric of Pordenone. Odoric was a bit of a fraud, amiable, garrulous, quite energetic, and not profoundly wedded to his priestly calling—nor to anything else, it would appear, except traveling round and savoring the odd incidents that turned up in his life. In most ways he seems to have been a very human person, undisturbed during his foreign journeys by the call of higher things: a likeable vagrant in a Franciscan habit. We must not blame him for what eventually happened to him, for it happened when he was dead. He had no hand in it—or not much, anyway.

Odoric was born in the district of Pordenone, which lies on the plain of Friuli between Venice and Trieste, not very far from what is now the border between Italy and Yugoslavia. Local tradition says he came from a village called Villa Nova a mile and a half from Pordenone. And in the village today there is a two-story cottage said to be his birthplace. The date of Odoric's birth is said to be 1286, but there are reasons for putting it about ten or twelve years earlier. At an early age he joined the Franciscan community at nearby Udine, where his devotion to ascetic practices singled him out from his contemporaries. He is said to have lived on bread and water, gone barefoot, scourged himself severely, and worn haircloth and iron mail next to his skin. But all this was the usual regimen of self-mortification in the Middle Ages and, as a matter of course, would be tacked onto the story of a man who was later beatified. It is also said that, with the permission of his monastic superior, he retreated for years to the wilds (wherever they may have been in this plain of vineyards and mulberry trees) in order to pass a life of solitude and contemplation.

Now all these bits of information must be taken with a pinch of pious salt since after his death he became immediately the object of a cult. His biography drips with the holy water of putative sanctity. The only facts we know for sure about Odoric are: the area in which he was born; the incidents of his journeys in the East, which he narrates with a charm and frankness that single him out as the most readable, the most observant, and the most good humored of all the Franciscans who went to Peking; and the date of his death.[1] The rest is twopence-colored with pious invention.

Possibly Odoric did actually take to the wilds for a time. One may imagine him as a youth of lively imagination and inconsiderable intellect, but inspired by the temporal proximity and the example of St. Francis's life (as John of Montecorvino had been). And in that state of exaltation to which youth, seared by its passions, attains very easily, he may have seen for a time a future of communion with the God of St. Francis himself. Had his temperament and the monastic discipline led him that way, it would not be surprising to find him taking to the hills as a hermit of sorts. On the other hand he may simply have made this his excuse for wandering outside the gossip limits of the Friuli Plain and enjoying those very pleasures a priest renounces. Either explanation is possible in a man of the outlook displayed in his writings.

In due course he came back to the Franciscan fold, and we next hear of him sometime between 1316 and 1318, when he left Venice and sailed for Trebizond, on the southern shores of the Black Sea. From there he went overland to Tabriz and on to Ormuz, following much the same route as Marco Polo and John of Montecorvino before him. At Ormuz, in fact, he notices those unsuitable boats just as they had done, and, like Friar John, embarks on one for India.

No one knows whether he had in mind (or in his orders) to go to China or just to India, but it is in India that we find him about 1321, at Tana. Tana was a settlement on the landward side of an island now called Salsette, near Bombay. Was he perhaps sent to India in response to the description in John of Montecorvino's letters of the vast and virgin field for missionary activity there? Perhaps he was sent to succor the friars at Tana? These men had been taken there against their wishes by the crew of the boat they had hired to carry them much further south to Travancore; and no doubt they had managed to get word back to Italy of their predicament.

When Odoric arrived in Tana he discovered that four Franciscan friars had recently been martyred there. One of them was Thomas of Tolentino, whom we have already encountered bringing news of John of Montecorvino to the pope. For his pains in having reported the facts, Thomas, it would seem, was sent out with a party to southern India. With three other friars he had been living in the house of a certain Nestorian and his wife, and the husband one day beat his wife for some fault. The wife complained to the Moslem authorities of the town and put forward the four unfortunate friars as witnesses to her tale. Then, in a way not unknown in argument today

1. There are over seventy manuscript copies of Odoric's story still extant—testimony to the interest it aroused not long after his death. In this chapter we follow Yule's translation in *Cathay*.

between exponents of opposite ideologies, an entirely irrelevant quarrel sprang up between the friars and the Moslems about the identity of Christ. Friar Thomas (evidently spokesman on all matters, to popes and Moslems alike) had the better of the argument. This naturally upset the good Moslems, who then asked: "And what do you think of Mahomet anyway?"[2] Friar Thomas's reply was heated and unwise: "I tell you that Mahomet is the son of perdition and has his rightful place in hell with his father the devil—and not only he but everyone else who follows his law, false as it is, pestilent and accursed, hostile to God and the salvation of souls."

This, of course, was asking for trouble. As the story goes, the friars were seized and exposed in the furnace of Bombay's midday sun (praising God, Odoric says) while the *maidan* or public square was prepared for their immolation in the presence of a large crowd of local people. Friar Thomas (always the first) tried to throw himself into the flames but was prevented by the Moslems, who felt his willingness to do so might show he was proof against incineration. Instead, they threw in Friar James, who came from Padua. His voice could be heard calling on the Virgin Mary from the heart of the pyre; and when the blaze died down—there he was, unscathed. Undeterred, the Moslems stripped him, gave him a liberal coating of oil, made an even better fire, and cast him in a second time. But he emerged as before, to the consternation of the bystanders, who (according to Odoric) were now of two minds whether to change religions. At this juncture the authorities, judiciously, called the whole thing off and took the friars away. That night soldiers were detailed to kill them, which they succeeded in doing.

Friar Odoric, when he heard this saddening tale, did not tarry long in Tana. He gathered up the bones of the martyrs and set sail for Cathay. Or at least one suspects it must have been for Cathay this time, even if he had not at first started off with the intention of going to China; for India must have seemed at best an inhospitable country. It is quite obvious that martyrdom was not in Odoric's line.

He apparently left behind him at Tana a friar named Jordanus, for the abandoned friar says in a pathetic letter dated a few years later, in 1324: "Since then I have continued alone . . . for two years and a half, going in and out, but unworthy to partake of the crown of my happy comrades [the martyrs]. Alas me, my fathers! . . . thus left an orphan and a wayfarer in this pathless and weary wilderness!" Friar Jordanus sounds more the type that John of Montecorvino had been asking for in his letters from Peking. It is odd that Odoric never mentions Jordanus in his account of events at Tana; could it be that he did not wish to be accused of leaving him to the Moslem wolves?

2. Unless otherwise indicated, all quotations in this chapter are taken from Yule's *Cathay*.

ODORIC PREACHING TO THE HEATHEN. This bas-relief in the church at Udine, northern Italy, carved shortly after Odoric's death, is practically the only evidence we have that his travels in Asia might have been more than a pleasure jaunt. (Courtesy Ente Provinciale per il Turismo, Udine; photo by Brisighelli)

Odoric soon arrived by boat at St. Thomé round the tip of South India, but he did not stay very long there either. It is almost incredible in a Franciscan priest, which he was, but it is true that he says not one single word of any missionary activity there and confines himself to a description of the Jews, Nestorians, and "idol-worshippers," and some chatter on the local customs in which there is nothing very new. Then he was off again, much scared by India, one suspects, and shaking its dust from his sandals.

His voyage took him by a roundabout and interesting route through the great archipelago of what is now Indonesia. There is only one incident from this voyaging that is worth our attention in the present context—that intractable calm which set in at sea and fixed the ship in its motionless eye until Odoric bethought himself of the martyrs' bones he carried. Abstracting one of them (it could not have been the skull of Friar Thomas, for that was taken from India to Italy and in the seventeenth century was still to be seen in Tolentino), he handed it to a servant, who went boldly up forward to the bow of the ship and cast the relic into the sea. At once a fair wind rose and never failed them until they reached port.

The port Odoric eventually reached was Canton, where the most succulent of all the varieties of Chinese food is grown and cooked, where the people are volatile and garrulous, tireless in business and in love, and regard their northern brethren somewhat askance. Canton, Odoric found, was "one day's voyage from the sea" up the shallow estuary of the Pearl River. Today, and doubtless in his time too, the approach to land is heralded by a change in color of the sea from its ocean blue to a distinct mud-ochre as the Pearl carries thousands of tons of Chinese silt to the South China Sea each day. Guarding the estuary lies Macao, which had not yet seen the Portuguese who were later to settle there.

In Canton the general feeling and ambiance were pleasing to Odoric. He expatiates at length on the delights of Cantonese food, the wealth and variety of the poultry, and the "serpents" which were an indispensable part of all banquets. (Snakes, deliciously cooked in a dozen different ways, are still eaten regularly in season in the gastronomic province of Kwangtung, in which Canton lies.) With considerable emotion he writes that Canton was "a city as big as three Venices . . . and has shipping so great and vast in amount that . . . all Italy has not the amount of craft that this one city has." He is neither the first nor the last traveler from Europe to be astonished at the myriad boats of China which run like shoals of fish along her rivers and form towns of their own on the waters of every harbor and estuary.

When he blundered into South China a whole new world opened up to Odoric. His wonder and amazement are patent. There were two thousand towns in the area, he says, hardly believing it, of such size that Treviso and Vicenza in Italy, which he no doubt knew well, were not to be compared with any of them. And the throng of people in the Chinese cities was much greater than the crowds in Venice on Ascension Day. "The men, as to their bodily aspect, are comely enough, but colorless, having beards of long straggling hairs like mousers—cats, I mean. And as for the women, they are the most beautiful in the world!" Perhaps at that time in the fourteenth century the admixture of Arab blood in the South Chinese was stronger than it is today, which might explain the beards. For at that time, as during many a previous century, the Arabs were the great seafarers and traders between the Persian Gulf and the Indies and China. It was not until the coming of superior European ships sailed by the Portuguese that Arab supremacy dwindled and finally disappeared from the Chinese scene. As for the women of South China, if we may echo Odoric's words, there is no question that the Chinese at any rate think them the most beautiful in all the world. Many a saying to this effect is in current use even today, and the beauties of Soochow are considered to be the most luscious of all.

Coming to Zayton, farther north along the coast of China, Odoric makes one or two references to church matters. "I came to a certain noble city which is called Zayton, where we Minor Friars have two houses . . . and there I deposited the bones of our friars who had suffered martyrdom for the faith. . . . The city is twice as big as Bologna." Here he probably erred on the side of caution, for Zayton was at that time incontestably one of the largest ports in the world. (Ibn Battuta remarks this a little later in time.) It was certainly the largest of the China ports, and surely much more than twice as big as Bologna, which was a comparatively small town in the fourteenth century. The Chinese cities in general were much larger than any in Europe for a long time to come.

We hear nothing of Andrew of Perugia, who was in Zayton at the time; and as if to balance the slight, Andrew of Perugia tells us nothing of Odoric, although he does mention in his letter of January, 1326—a year after Odoric's visit—the sad story of the martyrs. At first sight it would not be surprising if the complacent Andrew had made friends with Odoric, whose interest in propagating the faith was as slight as his own. But perhaps he feared a rival in the sinecure of his own back yard.

On went Odoric, sampling the delights of the country and telling his story in very readable fashion, to Hangchow, which Marco Polo had documented with such admiration some time before. Odoric's description of the city confirms much of Marco's and succeeds in being less of a catalogue. "It's the greatest city in the whole world," he says, and this was probably literal truth, "so great that I should scarcely venture to tell of it, but that I have met at Venice people in plenty who have been there."

Writing as though he were talking to friends, amiably and in a completely worldly manner, with the appreciation that comes only of experience and not from mere pious observation, Odoric describes the wine, houses, people of Hangchow. Not a word is there of Christianity except when he tells us that four Franciscan friars in the town had converted a man in whose house he stayed, and who took him sightseeing. "And he said to me one day . . . wilt thou come and see the place? And when I said I would willingly go, we got into a boat, and went to a certain great monastery of the people of the country [perhaps the one on the shores of the West Lake described by Marco Polo]. And he called to him one of the monks, saying: Seest here this Franki Rabban [this European monk]? He cometh from where the sun sets, and goes now to Peking to pray for the life of the Great Khan. Show him therefore, prithee, something worth seeing, so that if he gets back to his own country he may be able to say, I have seen such and such strange things in Hangchow! And the monk replied that he would do so with pleasure. . . .

"So he took two great buckets full of scraps from the table, and opening the door of a certain shrubbery which was there, we went in. Now in this shrubbery there was a little hill covered with pleasant trees and full of grottoes. And as we stood there he took a gong and began to beat upon it, and at the sound a multitude of animals of diverse kinds began to come down from the hill, such as apes, monkeys, and many other animals having faces like men, to the number of some three thousand; and took up their places round him in regular ranks. And . . . he put down the vessels before them, and fed them as fast as he was able.

"Then, having fed them, the man beat the gong again and the beasts went off. So I, laughing heartily, began to say: 'Tell me . . . what this meaneth?' And he answered: 'These animals be the souls of gentlemen, which we feed in this fashion for the love of God.' "

Friar Odoric makes a slight protest at this point, saying that they are only animals. " 'No, forsooth, they be nought else but the souls of gentlemen,' . . . said the man. And say what I list against it, nought else would he believe."

Friar Odoric was not in the least shocked at this piece of rank heresy. He totally failed to understand it for what it was—the Buddhist transmigration of souls. And in this incident as in others of what he calls a "pagan" nature, he evinces little or nothing of the Christian orthodox attitude such as we would normally expect in his time and calling. He is a simple and frequently charming tourist, one who swallows most things, especially if they are wondrous enough. The feeding of the animal multitude is also attested by John of Marignolli a few years later.

Suitably diverted, Odoric left Hangchow and went on with several stops to Yangchow, where Marco Polo had held such high rank. And here we are favored with more of his excellent description "on that good quire of stationary," as he puts it, which he would have needed to tell all there was to tell of any single place in China. Here in Yangchow the Franciscans had another house. With great verve Odoric addresses himself to describing the life of the town, the well-patronized restaurants and the hospitality of its citizens—both of which amenities he obviously sampled to the full. It is all reminiscent of those volumes of letters written to fathers by young noblemen on the grand tour in Europe in the days when that cultural exercise was the fashion. In the pages of Odoric there is the same vivid interest in many aspects of polite society, though not quite the sophistication, the same comparisons of the size of foreign places with those familiar at home, the same notes on dress and curious local customs. Odoric was a splendidly responsive tourist, and for one of little education he did very well by his opportunities and by his public.

There is a little gem of a paragraph which relates to this southern part of China. "When I was in the province of Manzi I passed by the foot of the palace wall of a certain burgess. . . . He hath fifty damsels, virgins, who wait on him continually. And when he goeth to dinner . . . the dishes are brought by fives and fives, those virgins carrying them in with singing of songs and the music of many kinds of instruments. And they feed him as though he were a pet sparrow, putting the food into his mouth. . . . And thus," says Odoric, greatly titillated, "he leadeth his life daily until he shall have lived it out."

Replete with the experiences of the south, he takes a boat up the Grand Canal and at last reaches Peking, where John of Montecorvino was still laboring on into his old age. Never a mention of the fine old man passes the pen of Odoric, or at least not a direct one. Without a doubt Odoric stayed in one or other of the Franciscan establishments that Friar John had built with patient effort there, but Odoric dismisses it all in a phrase with truly tourist superficiality: "For we Minor Friars have a place assigned to us at the emperor's court and are always in duty bound to go and give him our benison."

Otherwise his picture of the city and the court of the khan is filled with interest and recounted in a lively fashion that outshines anything Marco Polo achieved in the way of entertaining writing, even if the account of Odoric is shorter and less accurate. His pleasure in the sights and sounds, the pomp and the ordinary bustle, the fine palaces and the verdant artificial hill in the imperial grounds is congenial to one who has visited the Peking of today.

"I, Friar Odoric, was full three years in that city. . . ." He made good use of the time and managed to add to our knowledge of ceremonial and other matters. Within the palaces he frequently saw great events such as the big feasts of the imperial year, and he tells about the mechanical "peacocks of gold. . . . When any of the Tartars wish to amuse their lord then they go out one after the other and clap their hands; upon which the peacocks flap their wings, and make as if they would dance. Now this must be done either by diabolic art," Odoric concludes with a sagacity unusual for him, "or by some engine underground." There must have been some latter-day William Boucher at the imperial court in Peking at this time, continuing the great tradition of the Parisian goldsmith who was friend to William of Rubruck. Besides the dancing peacocks there was a direct descendant of Boucher's magical fountain in the Peking palaces at the time of Odoric. Its design had come a long way from the complication of the Karakorum fountain, probably because in Peking it was style and luxury rather than magic that was needed to make an impression. The huge jade jar described by Odoric was an object of stylish simplicity and great cost. Apparently carved out of a single block of jade, it was hooped around with gold and had a gold dragon at each corner. The exterior was laced with a network of huge pearls, and the drink was conveyed into it by conduits from the courtyard outside. Many golden goblets were placed around it for the convenience of those who wished to drink from this well which was always full. The description makes it apparent that this was either the same buffet "in the form of a square coffer" which Marco Polo saw, or at least the next one after it.

But besides such marvels, Odoric supplies us with information we did not previously have from others—the Chinese habit of wearing very long fingernails was entirely new to the Western reader of the time. Marco Polo never mentions it. Nor does he remark on the binding of feet. But Odoric, sensitive to female allure, gives us a brief description of the custom, one which was intended to enhance the attractions of young Chinese women but was never followed by the Mongols. Odoric also paints an excellent picture of the high boot-shaped *bogtak* or headdress of Mongol matrons, which was supposed to remind the ladies that they had husbands even if these were not actually present. Much more importantly, Odoric describes the *kwei*, the ivory tablet held in the hand before the lower part of the face by every considerable court official since time immemorial.

To some information on the hunting parties of the khan, Odoric adds an intensely interesting and unique piece of detail on the security precautions surrounding the great man when he traveled. The khan was always at the exact center of a huge cross made of four detachments of soldiers, each one a day's march from him—one in front, one behind, and one on either hand. The khan's "two-wheeled carriage, in which is formed a goodly chamber all of aloe-wood and gold, and covered over with great and fine skins, and set with many precious stones . . . is drawn by four elephants . . . and also by four splendid horses. And alongside go four barons . . . keeping watch . . . over the chariot so that no hurt came to the king. Moreover he carrieth with him in the chariot twelve gerfalcons; so that even as he sits therein upon the chair of state if he sees any birds pass he lets fly his hawks at them. . . ." If proof of authenticity is needed, we need only refer to Marco Polo's setting of the same scene (see page 73).

The Mongol postal system, as revolutionary then, in its improvements over previous systems, as automation is today, impressed every traveler in China. Odoric was no exception, and he gives a good description of the system, including details of how the foot-runners have bells attached to their belts to warn the post station of their approach so the relay may be ready to snatch the pouch and leave on the instant.

But there is one story of his years in Peking which Odoric does not tell in his book. Significantly, it concerns Franciscans and was told by him to a friend in Italy, Marchesimus of Bassano, who added it to the Odoric manuscript. In it we have what may be our last glimpse of the saintly John of Montecorvino. "When the Khan was on his journey from Shangtu to Peking, Friar Odoric with four other friars was sitting under the shade of a tree by the side of the road along which the Khan was to pass. And one of the brethren was a bishop [Montecorvino?]. . . . The bishop put on his . . . robes [as the khan drew near] and took the cross and fastened it to the end of a staff. . . .[The friars] began to chant, *Veni Creator Spiritus*. . . ." The khan called them over to him, wanting to know what the words meant, and the bishop presented him with the cross to kiss. "Now at the time he was lying down, but as soon as he saw the cross he sat up, and doffing the cap that he wore, kissed the cross in the most reverent and humble manner." The rule was that none should come empty-handed to the khan. "So Friar Odoric, having with him a small dish full of apples, presented that as their offering to the great khan. And he took two of the apples and ate a piece of one of them, whilst he kept the other in his hand. And so he went his way."

Without gaining any more information on the Franciscans in Peking we leave the city with Odoric and journey through Tenduc—the country of Montecorvino's converts who so sadly lapsed into Nestorian heresy when their king died and Friar John had no chance to go to their rescue. Next, passing through what we may perhaps reconstruct as some gorge with colossal statues of the Buddha (not a rarity on the route Odoric took), the traveler treats his readers to a full-blown marvel-story. Traversing this valley "upon the river of delights" and seeing countless bodies of the dead, hearing all kinds of music but especially that of drums "marvellously beaten," he shudders with apprehension. "And so great was the noise that extreme terror fell upon me," for the valley was said by the Moslems to be death to an unbeliever. But, "going in, that I might see for good and all what this matter was, I saw in a rock upon one side a man's face, so terrible that my spirit seemed to die within me utterly. . . . Close up to that face I never dared to go. . . ." Once out of the vale he could see nothing, though the drums continued to mutter. But on the hilltop, he tells us without further explanation, was silver heaped up like the scales of fish, which he gathered—only to cast it from him. And he remarks that it was only by the grace of God that he emerged from the valley, and much to the surprise of the Saracens, who had not expected him to survive. Bunyan could have written this tale. But in fact it was that renowned and mendacious fabulist Sir John Mandeville who cribbed most of Odoric and rehashed it as his own. Odoric's stories are always well told, and enlarge considerably our knowledge of the times when he was in China.

Before leaving the Eastern segment of the miraculous world of his day and following him home to Italy, we have to notice briefly one other traveler, another Franciscan, named John of Marignolli. He was born about 1290 of a noble Florentine family (there used to be a street in Florence named after them). Taking holy orders, he joined the monastery of the great church of Santa Croce, which still stands like a long rock among the surrounding Florentine roofs. He appears to have had pretensions to erudition and is said to have lectured at the University of Bologna. On his return to Europe from his China trip (1338-53) he was made bishop of Bisignano in the southern province of Calabria. But somehow or other he came to the notice of Emperor Charles IV, who took him to Bohemia and set him to work—now a very old man—to write the chronicles of that state. In the midst of those rambling histories he inserted his own recollections of his Oriental journey, together with a funny little note in which he hoped that the emperor would allow this section to stand. It had no possible relevance to the history of Bohemia, but apparently Charles, if he ever read those confused and worthless histories he had commissioned, did not strike it out.

Marignolli's journey was the result of that embassy from the khan and the Alans to the pope described in the chapter on John of Montecorvino. A party of thirty-two set out from Europe for Peking, reaching their destination in 1342 (a date about which for once the Franciscan and Yuan-dynasty histories agree). The Yuan history records that the party brought with it a large horse eleven and a half feet long and six feet eight inches high—exactly what the khan had requested in his letter to the pope. The horse must have created something of a stir in the palace on account of its size, for there are several contemporary Chinese poems about it—one called "Ode to the Supernatural Horse."

The khan lodged the party of Franciscans in an imperial apartment for three years: Marignolli tells us how he led a Christian procession before the emperor, carrying candles and crosses and chanting the Nicene Creed.

The emperor apparently wanted at least one of the party to stay on and take over the work and position of Montecorvino, now long dead. But none of the Franciscans had been empowered to do so, and none was a bishop. Their refusal must have seemed to the Mongols and others in Peking (especially the Alans) yet another example of that Western intransigence to which, as the Alans said, they were used. For the Alans' letter had specifically stated their need of a bishop. The envoys eventually left Peking and returned home by sea.

Marignolli's tale is certainly among the most confused and confusing of all the writings by Europeans on China. What he writes and his manner of writing it reveal an odd, vacillating character, quite sharp in some ways, birdlike in his observation and inconstant in his opinions, at times even feebleminded. Vanity is one characteristic most obvious and least endearing in him.

His epitaph was written (all unconsciously) by the bishop of Armagh in Ireland, Richard Fitz Ralph, who had, at some time after Marignolli's return from the East, been offered his services. "Let him come on then (say we), that old beggar of a Bisignano Bishop! . . . We'll take the measure of him, though he does paycock about the Kaiser's Court and call himself (save the mark) Apostle of the East! . . .'Twill be a pity if I . . . find it a hard matter to put a collar on a poor old wheezing tyke, who has scarcely a bark left in him, and never had the least repute for brains!"

Odoric, whom we left somewhere on the way back to Italy, at last reached his native soil in 1329 or 1330. When his story was read there he was not much more believed than was Marco Polo. Some of his compatriots even seem to have thought the East had turned his brain a little. But there is no reason to suppose any such thing. His are the memoirs of a nice enough man, a bit of a wanderer, a little self-centered, entirely without Christian fires burning in the heart or in the head, who made a marvelous trip to Peking and lived to recount a delightful tale.

Among Christians, however, his fame rests on other foundations (if we may so dignify them with that term). This fame descended on him, very much by accident, the moment he was dead. After a visit to the papal court at Avignon he returned to Udine, took to bed, and expired on January 14, 1331. The friars were about to bury him quietly, as they would have done anyone else, on that same day, but they were prevented by the chief magistrate of Udine, who was a great friend of Odoric's. He intervened and ordered a "solemn funeral" for the following day, and it was "attended by all the dignitaries and created public excitement. The people began to push forward to kiss the hands and feet of the dead friar, or to snatch a morsel of his clothing. Rumors of miracles rose and spread like wildfire. A noble dame . . . who had long suffered from a shrunken arm, declared aloud that she had received instant relief [of what precise sort we are left to imagine] on touching the body." So then the whole town flocked to the church. "Lucky were those who could but put a finger on the friar's gown," whilst those who had the chance grasped his hair and beard. The learned Yule remarks at this point: "Just as I have seen Bengalis snatch at the whiskers of a dead tiger, and from like motives," thus putting the scene in perspective. "One virago," he continues, "made a desperate attempt to snip off the saint's ear with her scissors, but miraculously the scissors would not close."

In deference to public demand the burial was delayed another two days, and when it finally did take place the body was dug up again on the same day by popular demand. Now the county aristocracy came crowding into

THE TOMB OF ODORIC in the Chiesa del Carmine, Udine. (Courtesy Ente Provinciale per il Turismo, Udine; photo by Brisighelli)

Udine with wives and families; and finally came the nobles. One great lady —Beatrice of Bavaria, Countess Dowager and Regent of Goritz—arrived with a large retinue. So now, patronized by royalty no less, poor Odoric was removed from his coffin and put in a much better one. But this was only the beginning of a comical round of *viaggi postumi* or "journeys after death," to use the words of the local guidebook to the Udine church where his remains now lie.[3] The somewhat ghoulish farce of these frequent translations from place to place—there were six of them—was not complete until 1931, when the last occurred. But in the meantime his beatification, tardily, took place in 1755.

We can only wonder with Yule, whose treatment of the story is one of restrained and scholarly wit, "what odd chance picked out Odoric as the wanderer to be accredited with such exceptional sanctity. . . . Had . . . that zealous patriarch John of Montecorvino, striving for faith at the world's end to the age of fourscore years, been made a saint of, one could have understood it better."

But there in Udine still lies the blessed Odoric in his tomb, his body occasionally exposed to pilgrims until at least last century, when Yule just missed seeing it one day. We may perhaps wonder how much of the hair and beard escaped the clutching hands of those female devotees on his original burial day. What does remain, however, is his attractive and too

3. *Cammino ad Oriente,* by F. Cargnelutti (Edizioni F.A.C.E., Udine, 1964).

little known book of travels, the last worthwhile memento in a European language of the brief Mongol sway in China.

PART THREE

VISITORS TO THE GREAT MING

Pirates, Casuists, Scientists, Diplomats

CHAPTER SIX

THE WAY FROM PORTUGAL

NOT LONG after the comedy of Odoric's burials and exhumations in Italy, and almost at the same time as foolish old Marignolli was scribbling his thoughts on China, the Mongols were driven out of Peking. They had come a long way in their century in China, absorbing something of its civilization. The descendants of the nomad chieftain Genghis Khan had achieved an all but Chinese nobility in the person of Kublai Khan. But after him the Mongol rulers degenerated into a succession of vacillating nincompoops who sat on the Dragon Throne but could not rule. By their ineptitude, the later Yuan emperors permitted a situation to arise in China which, in 1368, spelled their own doom at the hands of the peasant usurper Chu Yuan-chang, who chased them out of the capital and set himself up as first emperor of the Ming—or Brilliant—dynasty.

With the departure of the Mongols, the thousands of Westerners they employed disappear from the pages of Chinese history, just as did the Nestorians for different reasons on the fall of the T'ang. And when the West returned again, more than a hundred years afterward, it was with the Chinese Ming dynasty they had to deal. In Ming China they discovered an entirely different set of circumstances from that experienced by their predecessors in the Mongol dynasty.

Events in China, in the century before the Westerners' return, showed that vigor which always characterized a new administration determined to seat itself firmly in the imperial saddle and to reinstate the ancient good while suppressing the bad old ways of the preceding dynasty. The Ming were extremely thorough in this. For one thing they were hyperconscious of their Chinese-ness in contrast to the foreign-ness of the Mongols. They took as their model the glories of the T'ang era. It was partly this which forced on the Ming their extremely conservative outlook. Unlike the T'ang, however, they closed China's frontiers to foreigners. And, equally at variance with the T'ang, which had invited the world to China, the new dynasty embarked in its early days on an ambitious program of overseas voyages.

Under the great eunuch-admiral Cheng Ho a newly-built fleet of huge Chinese ships sailed out into Southeast Asian and Indian waters, collecting vassals like souvenirs, replacing recalcitrant kings with more compliant rulers willing to pay tribute to the august majesty of the emperor in Peking. Thus Malacca, on the west coast of Malaya, became a tributary of the Ming and a Chinese trading station—an important development, as we shall see. This maritime adventure, remarkable as it was, lasted only a few years before it ceased as suddenly as it had begun.

At home, the capital of China, first at Nanking in the south, was transferred to Peking in the north after Yung-lo, fourth son of Chu Yuan-chang, had seized the throne in 1402. The Peking that Yung-lo built left almost nothing of the former Mongol city, which Marco Polo and the Franciscans had known, to remind him or the Chinese of that century of subjection. It is this Peking of Yung-lo, with embellishments at the hands of the later Ch'ing, or Manchu, dynasty, which remains today one of the most astounding monuments to Chinese traditional thought, architecture, and planning genius and one of the most charming and genial of its outward expressions. It is in this Ming Peking that the flux of much later Chinese contact with the Westerners took place. And it is here that much which was to mold the future attitudes of Westerners and Chinese to each other originated. Here the Chinese watched with curiosity, with irony, with occasional amusement, the unsolicited arrival and the often abrupt departure of many a European traveler.

The defeat of the Mongols in China tended progressively to isolate the country from land contact with the West, for the old trade routes now passed through hostile territory. Once more, China was intellectually isolated. The twin planets of East and West, tenuously linked in the T'ang by merchants and Nestorians, and quite strongly linked in the Yuan dynasty by a continuous flow of Westerners into China, now moved abruptly away from each other in the universe of the late fourteenth century. The inhabitants of China and Europe might just as well have been living on two actual planets in space for all the communication they now had with each other.

The cumulative effect of Western culture on China proved to have been insignificant.[1] Nestorian Christianity had all but vanished, and the Christianity of the Franciscans vanished too. Whatever the Chinese had learned of the affairs of the West was soon forgotten. China of the Ming enclosed herself hermetically in her geographical confines, bounded to south and east by

1. It would be wrong, of course, to suppose that nothing at all from the West, or from outside China, had been accepted by the Chinese up to this date. Readers of Hirth, Goodrich, and Needham will find much fascinating discussion on the non-Chinese origins of various techniques and materials in use at this time. But it can be said that virtually no important substance, idea, or technique that had any appreciable influence on future Chinese life had by this time come from Europe.

A CHINESE IN FOURTEENTH-CENTURY EUROPE. The man in the conical hat, said to be a Chinese, is one of a crowd in the Calvary of an altar cloth called the Parement de Narbonne, woven between 1364 and 1380. (Musées Nationaux, Versailles)

mountains, and on other frontiers by hostile desert and by the sea of which at the moment she had the mastery.

Meanwhile, in that other planet called Europe, the mysterious coincidences of history were brewing up something entirely different from the Ming Chinese wine. Just as Odoric was returning from China, Marco Polo, living in Venice, was releasing his "Tartar" servant Peter from bondage. In fact, at that time in the thirteenth century and for another hundred years or so, there were numerous Chinese and Mongol slaves in Venice, Florence, and other Italian cities. Most were women, but there were a few men. Several Chinese commodities were much prized in the households of rich Italians and others at this time. One was Chinese silk, which Europe had been importing sporadically in the raw state for over a thousand years, but which now came in the form of cloth. At Perugia (where Friar Andrew, lately bishop of Zayton in South China, was born), Pope Benedict XI was buried in 1304, wrapped in the faith and also in a shroud of Chinese brocade. And about the same time a lord of Verona who was wealthy enough to indulge his fancy for Chinese silks took the title Can Grande (Grand Khan, like a Mongol emperor) della Scala. This Italian khan also was buried in Chinese

PEGOLOTTI'S BOOK OF DESCRIPTIONS. The first page of an exquisitely illuminated manuscript copy of the book given to the Riccardian Library of Florence in 1765.

silk. And even the great poet Dante, who was a guest at his house in 1304, refers in the *Inferno* (XVII, 17) to the prevailing delight in *drappi tartareschi*—Tartar cloth.

It is surprising to discover that, considering the lively interest in Italy in things Chinese, no one collected together the numerous writings of the travelers who came back. Perhaps significantly, the most important book on the general subject of the Orient at this time dealt not with what had been gleaned about Oriental ways of life, but with trade. Francis Balducci Pegolotti, its author, was agent for the great Florentine trading house of Bardi. His *Book of Descriptions of Countries and of Measures Employed in Business and of Other Things Needful to Be Known by Merchants* must have been a boon to traders setting out on journeys across Asia on the trade routes.[2] Pegolotti had not personally been further than northern Persia, but he assiduously gathered all the information he could on the routes to China and wrote it all down in a volume of profound unreadability and execrable style. There is a sentence embedded in Ralph Fox's book on Genghis Khan which perhaps sums up not only Pegolotti but the temper of the times as well: "The Mongol

2. The book apparently circulated only in manuscript copies, one of which is preserved in the Riccardian Library at Florence.

horseman, who spoke always in images of the purest epic poetry, taken from the life of his steppe, master of the four fierce hounds who drank dew and rode on the wind . . . could never have imagined that his life and conquests would have ended in the adventures of the cautious Master Francis Balducci Pegolotti, riding in his comfortable wagon, with his woman, his dragoman, and his little convoy of merchandise. . . ."

So much, then, for the effects of China on Europe as the Mongols were driven from China. In the same peninsula of Italy more important things were afoot very soon after. By the middle of the century after Odoric died, the Italian Renaissance was in full swing. We have only to recall a few names from the period to realize how the civilization of the West was beginning to take a giant stride forward, while that of China was backpedaling. Leonardo da Vinci (1452–1519) was looking with a scientific and critical eye into the workings of every natural phenomenon he saw, while his contemporary on the Chinese planet, a man called Wang Yang-ming (1472–1528) was emphasizing a traditional and ultraconservative way of Chinese life.

But it was not the Renaissance that caused Westerners to set out for China again—or at least not directly. The reason was in fact a culinary one. One of the perennial problems of European life was how to make meat palatable in the winter. Successful winter feeding of cattle was little evolved and a large part of the beef herds had to be killed off each autumn, the meat salted and stored. To improve its taste, vast quantities of spices were necessary. During the Mongol domination of Asia the spice routes from India and Southeast Asia had been open, but on the fragmentation of the Mongol empire the spice trade had fallen into the hands of the Turkish Mamluks, who controlled Egypt and the Red Sea. Spices carried by Arab and Indian boats were now bought by the Mamluks, sold at a big profit to Italian traders, who shipped them to Europe and sold the cargoes at an even greater profit. This state of affairs pleased no one in Europe except the Italian merchants.

By this time the Portuguese were the foremost maritime power in Europe. Between 1493 and 1519 they had discovered no less than three routes to the East: across the Atlantic and over Mexico or Panama to the Pacific; round the southern tip of South America; round the Cape of Good Hope to India and points east. This last is the route that concerns us. In 1497, Vasco da Gama reached South India at Calicut with the help of an Indian pilot.[3] At Calicut he was not exactly welcomed. Fired by the jealousy of the Arab

3. Da Gama had set out bearing a letter to Prester John, whom popular fancy had now relocated, since the time of the Franciscans in China, in an equally unlikely place—Ethiopia. And in fact Da Gama found that with the sole exception of himself no one on the African coast had ever heard the name. But there is a tenuous connection between the rumored sites of Prester John's domain: in one of his letters from Peking, John of Montecorvino remarks that an embassy of Ethiopians had come to see him in the Chinese capital. Possibly they were Coptic Christians.

THE BEGINNING AND THE END OF THE VOYAGE. Lisbon and Macao in the late sixteenth century. The busy port of Lisbon on the Tagus River is shown with its crowded street plan, in sharp contrast to the tiny isle of Macao at the eastern end of the long sea journey. Portuguese merchants and Chinese mandarins are seen walking in the shade of parasols held by attendants. (Lisbon plan from Schutte; Macao drawing from De Bry)

traders there, who feared they would lose their trade to the Portuguese, the local ruler opposed him. The undignified exit of Da Gama from the port when the situation turned into actual violence is well known. Returning to Portugal, da Gama told his story of India. The imagination and cupidity of the court and merchants were aroused, and before long Portuguese ships were establishing trading stations in India at the point of the sword. Armed enclaves such as Goa were set up. Alfonso de Albuquerque became viceroy of India with his capital at Goa in 1507, and proceeded to sail for Malacca. In 1511, Albuquerque besieged and captured the town, which was a Chinese trading station.

The temperamental violence displayed by the Portuguese in their exploits in the East was a product of several factors, among which two might be mentioned. First, they had an inbred hatred of the Moslems—the Moors, who had long occupied the Iberian Peninsula—and in Malacca they found a native population of Moslems. Second, the papal bull dividing the non-European world between Spain and Portugal had given the latter the East to exploit. Having subdued Malacca, Albuquerque, with a typically Portuguese sense of crusading mission, and also with the usual high hopes of trading and gain, looked still farther afield. He sent out a party to explore the possibilities of China.

The irony of the Portuguese movement toward China consists in the fact that, without the despised Arab pilots and the deceptively simple Chinese invention of the compass, Western ships would almost certainly not have reached Chinese waters until much later in history. The first compasses, or at least some description of the device, probably arrived in Europe (again ironically, via Arab seamen) in the twelfth century, and shortly after the year 1300 there appeared the first portolan charts, whose making depends in part on the use of the compass.[4] These new charts, showing chiefly the coastline, were much more accurate and practical than any in previous use. Thus the Chinese invention of the compass hastened what its inventors never dreamed it might accomplish—the conquest of the East by Europeans. It was perhaps as well, from the Chinese point of view, that their invention of watertight compartments in ships did not also reach the Portuguese, or many a Western man-of-war would have survived the fury of the typhoons of the Eastern seas, and the depredations of Europe on the China coast would have been even more severe.

While the consciously anachronistic grandeur of the Great Ming in China was settling in for what turned out to be a three-hundred-year rule, the history of European man was just beginning one of its most remarkable phases—the Renaissance. The logic and science of ancient Greece, the arrogance, territorial lust, and legalistic heritage of Rome, the perversion

4. The story of the Chinese compass and the Western portolan charts is told in great detail in Needham.

of the originally peaceful Christian credo, were all combining in a new fire, at once divine and hellish. To the discoverers of the Eastern world the least interesting aspect of that world was the human life it contained. In general they treated Orientals as local mammals, subhuman material for slaves, and producers of desirable exotica.

With the arrival at the port of Canton in South China of Albuquerque's advance party, led by Fernão Peres d'Andrade and Tomaso Pirès, in late September of 1517, a momentous, tragic, but fascinating new page of world history was begun.

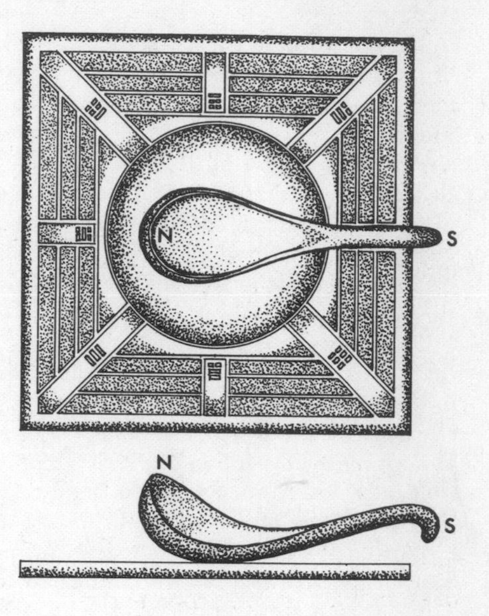

AN EARLY FORM OF THE CHINESE COMPASS. This conjectural drawing shows how the magnetic lodestone spoon rotated on its dish in response to magnetic forces. (Redrawn from Wang Chen-to in the *Chinese Journal of Archaeology*, 1948)

CHAPTER SEVEN

POOR PIRÈS

SOMEONE had to be the first European in Peking under the august Chinese Ming. Fate chose poor Pirès, a victim of his own countrymen, a martyr to an unworthy cause. Arriving at Peking, he stayed there for a short time as a virtual prisoner. Then he was sent back in ignominy to Canton, having made almost no impression at all on the capital. The misfortune of poor Pirès was that he represented the Portuguese. And they, first of all the Europeans, were responsible for setting an example which ever afterward was to sour relations between China and the West.

As the sixteenth century dawned on China, the Ming rulers had all but closed the doors of the country to foreigners from the West. In Asia's northern steppes the old foe—nomad power—was growing strong again; up and down the long scroll of China's coast Japanese pirates rampaged hungrily; and—a new threat—European sea power in Southeast Asian waters was on the increase, an unknown and so far unconsidered factor in the long history of China. To all those circumstances the reaction of China was to draw in her antennae, to man slightly contracted frontiers, to promulgate decrees forbidding Chinese to emigrate, and, metaphorically, to draw herself up to the immense grandeur of her traditional stature, gazing abstractedly out from the Flowery Kingdom with a kind of measured dignity.

It was a fine, Chinese, obscurantist dream. The hostile world was more or less successfully excluded for the time being as China prolonged that intensely national reverie, a sleeper reluctant to wake to the morning of the world's reality. She neither knew nor, had she known, would she have cared that a pope and his church in their infallible wisdom had assigned the whole continent of China to the attentions of the Portuguese and that those fierce venturers from across the world were already showing interest in China. Certainly the Chinese court and administration at Peking had heard that its trading port of Malacca had been seized by the Portuguese in 1511. But, viewed from the remote enclave of Peking, or even from more commercially minded Canton, Malacca was very far away in the Nan Yang, the Southern

PORTUGUESE GALLEONS, a painting attributed to Cornelis Anthoniszoon (about 1500–55). (National Maritime Museum, Greenwich)

Seas; and the great, tightly organized central empire of the Ming thought it could afford to ignore the city's capture.

Even before Pirès, the earliest appearance of a lone Portuguese trading ship of hitherto unseen design on the southern coast of China had apparently not disturbed this tranquility, had not injected the tiniest drop of disquiet into the settled calm of China at this time. Its captain, Rafael Perestrello, a relative of Christopher Columbus, had probably been the first Portuguese to set foot on the soil of Ming China. Some historians say that for his pains he and his crew were imprisoned; but this is most doubtful, since he returned to Malacca (if a little tardily), having managed to exchange his goods for a highly profitable cargo of Chinese wares. Likely enough, the tale of his woes is a later fabrication of Portuguese traders and historians, who by that time felt they had a lot to complain of.

A Florentine, Andrew Corsalis, writing in 1515 to Duke Lorenzo de' Medici, remarks with interest and some satisfaction on this novel adventure of Perestrello's: "During this last year some of our Portuguese made a voyage to China. They were not permitted to land, for they [the Chinese] say it is against their custom to let foreigners enter their dwellings. But they sold their goods at a great gain, and they say there is as great profit in taking spices to China as in taking them to Portugal, for it is a cold country and they make great use of them. It will be five hundred leagues from Malacca to China, sailing north."[1] Corsalis's ideas of geography were a little vague, as a glance at the map will reveal. But in an earlier section of his letter he makes one of the first observations on the Ming Chinese with a frankness and lack of condescension very rare in the annals of European comment on that people. He says: "The merchants of the land of China also make voyages to Malacca across the great Gulf to get . . . spices, and bring from their own country musk, rhubarb, pearls, tin, porcelain, and silk . . . of all kinds . . . of extraordinary richness. For they are a people of great skill, and on a par with ourselves, but of an uglier aspect, with little bits of eyes. . . ." His phrase *di nostra qualità,* "on a par with ourselves," is the comment of a cultured European—an unusual one at that time and, as we shall see, in total opposition to that of Portuguese merchants with whom the Chinese mostly had to deal.

It would be interesting to know what Tomaso Pirès (Thomas or Tomé Pirès, as some writers call him) thought of China and its inhabitants when he sailed up the Pearl River in 1517 aboard one of a flotilla of Portuguese vessels in command of Fernão Peres d'Andrade. Pirès had been sent from India. An apothecary by profession, he was brought to Malacca and taken onward to China, perhaps as the flotilla's doctor. As it turned out, however, fate had a different role in store for poor Pirès. Whatever may be thought of the shiploads of Portuguese pirate-traders who were his companions on the voyage to South China, it is quite certain that Pirès himself was a man of some culture, in many ways a worthy representative of Western civilization. While in the East he wrote a book called, in its English translation, *The Suma Oriental,* which deals mostly with the economics of trade at the time. Perhaps we should not expect more than occasional moments of illumination from it. His description of India, the East Indies, and other parts of the Far East is as important as that of Marco Polo dealing with the terrain through which he passed. Moreover, Pirès is more trustworthy than Marco Polo.

On the Portuguese trading station at Malacca, Pirès is enthusiastic and not without humor. There, he says, "you find what you want, and sometimes more than you are looking for." The town is the "end of Monsoons and beginning of others. . . . No trading post as large as Malacca is known [in the East] nor any where they deal in such fine and highly prized merchandise. Goods from all over the East are found there; goods from all over the West are sold there."

In this account of the East, Pirès includes the first descriptions of the Chinese that reached Europe after the passing of the Mongol regime, hav-

1. Quoted by Cordier.

ing doubtless gathered his information from Chinese traders at Malacca. "According to what the nations here in the East say, things of China are made out to be great; riches, pomp and state in both the land and the people, and other tales which it would be easier to believe as true of our Portugal than of China. . . . All Chinese eat pigs, cows and other animals. They drink a fair amount of all sorts of beverages. They praise our wine greatly. They get pretty drunk. They are weak people of small account." Continuing with the earliest known description of eating with chopsticks, he then says: "No Chinese may set out in the direction of Siam, Java, Malacca . . . without permission from the Governors of Canton . . . where the whole kingdom of China unloads all of its merchandise, great quantities from inland as well as from the sea." Correctly estimating the situation, Pirès's account of the East and of the prospects for trade adds up to a well-calculated eulogy on the subject of commerce centered on Malacca. For this reason, and certainly on account of his cultural achievements, there can be no doubt that Pirès was a good choice for the role to which he was soon to be assigned.

The flotilla under the command of d'Andrade reached the coast of China at the island of Tamão[2] in the Pearl River Delta, perhaps near Macao. Here, according to Mendoza,[3] all foreign ships were required to remain while they traded their cargoes with the Chinese, the goods being taken by boat to Canton. D'Andrade notified the Chinese authorities, in the person of the commander of a Chinese antipirate fleet, that he had come in peace for commercial reasons, and that he had also brought an embassy from the king of Portugal with letters and presents to the emperor of China. R. M. Major, in his introduction to the English translation of Mendoza's history, remarks blandly that due to "various delays and difficulties occasioned by the numerous gradations of rank among the Chinese authorities, their ceremoniousness, and the mistrust, imperfectly veiled by civility, of the Chinese toward strangers, d'Andrade reached Canton at the close of September, 1517, and ran into the harbor with all the nautical ceremonies." And it is elsewhere recorded that these ceremonies included the flying of the Portuguese flag from his ships and a discharge of cannon. "When surprise was expressed at this," the introduction goes on with its usual equanimity, "he

2. The identity of this island is disputed. One authority believes it to be Lintin (T'un-men), some miles northeast of Macao. Some historians call the island Tamang, which is nearer the Portuguese name used by Pirès, which in turn is said to be a corruption of the Chinese Ta Ngao, Big Bay. The island of St. John or Sanscian (Portuguese rendering of Chinese San Shan or Three Hills), which is fifty miles southwest of modern Macao, was later used by the Portuguese, who also called it Beneaga, probably a corruption of a Malay word, picked up in Malacca, meaning "market."

3. Juan Gonzalez de Mendoza, author of the earliest account of China ever printed in English (1588). Together with a book by Ortelius, it was also the first book printed in Europe to contain Chinese characters—although these are hardly decipherable.

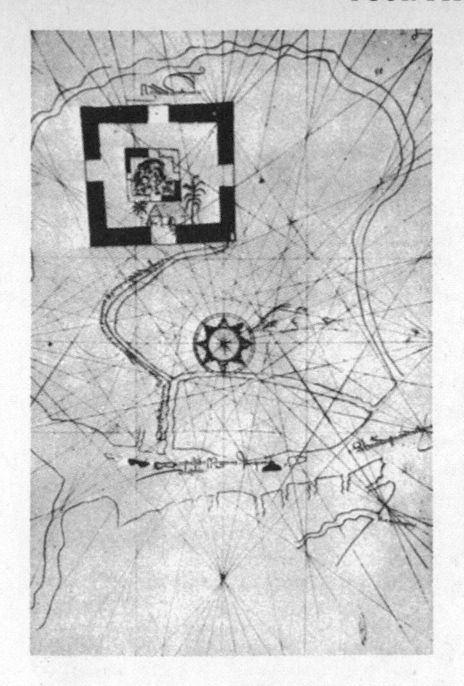

SKETCH OF THE RIVER AT CANTON, an early Portuguese chart. The dark square probably represents Peking, and the waterway leading from the river northward to it may have been the result of stories about the Grand Canal. (From Pirès)

justified himself by referring to the practice of the Chinese in this particular when their ships came to Portuguese Malacca"—which must have seemed to the Chinese nothing short of a slap in the face, since Malacca had been theirs until the Portuguese forcibly took it from them a few years before d'Andrade sailed so blithely into Canton. After this first howler, however, d'Andrade appears to have done better.

To the Chinese of Canton in 1517, whether mandarin[4] or man in the street, the sight of those strangely shaped Portuguese ships with their proud standards flying in the breeze of an autumn day in defiance of immemorial law, with their squat, malevolent cannon poking through apertures in the sides, and their curious, big-nosed seamen in oddly fitted clothes (men whose mere presence was also in defiance of Chinese law), must have been both outlandish and unwelcome. Never before had the like been seen in the crowded river. And to the Chinese way of thinking, any such gross departure from the settled normality of occurrences was to be regarded as a portent

4. "Mandarin" probably comes from Portuguese *mandar,* to command, although some authorities trace it to a Malay word. It was generally used for any Chinese official having the right to wear a button in his cap indicating to which of nine ranks he belonged. The word came later to be applied also to the official pronunciation of Chinese as spoken at Peking.

of calamity. More than that. The foreigners had arrived and fired their guns in the very midst of the city—as if hostilities had broken out. The "harbor" to which Mendoza refers is in fact the Pearl River as it passes the heart of Canton, running yellow with South China mud beside the flat expanse of what even then was a great city of low, black-roofed, wooden houses.

To Pirès, if we may endow him with a little poetry of vision, this spreading mass of buildings, entirely unfamiliar in shape, crowded with unfamiliar Chinese busy about their own incomprehensible affairs, must have been as strange as was the sight of Pirès's own white face to the Chinese. Sailing up from the sea along that sprawling delta, seeing for the first time the intensely paddied and irrigated and nurtured land, Pirès must have experienced something of that sensation which so many later travelers have remarked—the feeling that here was something different, something ageless, tightly organized, industrious, close-knit, and utterly un-European. Here indeed were the Chinese in their immemorial land and at their immemorial occupations, as unconscious of the other side of the world as Pirès had been of them not so many years before.

The river itself must have been a revelation to him. Never before in any part of the world—neither in Europe nor in the harbors of India or South-

A Portuguese Man-of-war in an engraving by Breughel, about 1560. The ship is flying a Portuguese flag. (National Maritime Museum, Greenwich)

east Asia—could he have seen such a multitudinous fleet of boats: boats of all shapes and sizes, boats whose dusky sails were set and cut to patterns unfamiliar in the West, boats laden with merchandise to which he could not put a name, boats containing the home and the lives of whole families of three generations, boats princely and decorated with a richness and panache reminiscent of the barges of the greatest European kings (but which apparently belonged to mere officials of this incomprehensible realm). Even their rudders, he may have remarked, were of a design unknown in the West.

On that short trip up the river Pirès must have realized, too, how vast in numbers were the Chinese, and he must have conceived of the vast extent of their agricultural lands in which the villages huddled gray and close to economize on ground of which every inch, every Chinese foot, was cultivated. And coming in sight of the great capital of Kwangtung province—the city of Canton—he must have seen above the low clutter of dwellings, the taller, grander roofs of temples and palaces of officials; and here and there the low hills with their pagodas and their curly roofed pavilions against the brilliant skies. Almost certainly it was then that he heard for the first time from the throats of great multitudes of people the musical speech, the sensuous range of tones of the Cantonese language. As the ships hove to in the river, Pirès was surrounded by this spreading capital of the southern marches of China, its streets with their massed shop signs, their banners emblazoned with written characters. For miles along the bank, wharves swarmed with gangs of coolies shifting tremendous weights with an ease dependent on a method quite strange to Europeans—the distribution of the weight among many men by means of interlaced bamboo poles. Now and then some merchant would appear, carried on his palanquin by servants; give his orders without the trouble of descending to the street; and be carried off like some great lord. And meanwhile the river would be alive with the fleets of small boats, sampans and many others, often sculled by women with babies slung on their backs. We may wonder what Pirès thought. Nothing that he may have written on the subject has been preserved.

By this time Pirès very likely knew that he had been chosen to represent the king of Portugal as ambassador to the emperor of China, who resided in the uncharted distances north, in Peking—wherever that might be. And he must have been troubled. Perhaps he too wondered why he had been chosen over men of so much higher rank. Could it be true, as some historians have conjectured, that d'Andrade was to have been the envoy but that when he realized he would have to perform the kowtow before the emperor at court, his vanity and Portuguese pride rebelled and he delegated the job to humble Pirès?

The affair of the Portuguese fusillade of cannon was satisfactorily explained to the Chinese authorities as a mark of respect, and apparently the Chinese in turn winked at the insulting explanation given of the flying of

Portuguese flags on the ships in a Chinese port. Pirès was invited to come ashore and be treated as tributary ambassadors were normally treated while they waited for word to be returned from Peking as to their disposal. D'Andrade agreed to the disembarkation but refused to allow Pirès and his party to be supported at Chinese expense (as was the custom). He also refused an invitation to go ashore himself, sending instead his factor and some others, who arranged for a godown to be put at their disposal. The authorities, says Major, "allowed the merchandise to be landed by degrees, and an interchange of traffic commenced." It must have appeared to the Chinese that the Portuguese were a good deal more interested in immediate trade than in pursuing their scheme of sending their ambassador to Peking. And in this they were doubtless very near the mark.

"Matters were in this prosperous condition" when d'Andrade had to leave Canton. Malaria (and perhaps scurvy, which the visitors could have averted by eating any one of the dozens of varieties of Cantonese cabbage being cooked in boats all round them) had taken its toll. The factor and nine others were dead. Another probable reason for d'Andrade's departure at the end of September was that the remainder of the Portuguese fleet at Tamão had just been attacked by pirates.

Before he left, d'Andrade made it plain to the Chinese that anyone who had demands on the foreign traders should apply and have his affairs settled to his satisfaction before they left Canton. This action must have enhanced the prestige of the foreigners and indicates that d'Andrade was an essentially reasonable man. Mendoza is of the opinion that by the time of the fleet's departure the situation had become one of peaceable and amicable trade. And d'Andrade, he says, returned to Malacca "loaded with riches and renown." So the trade must have been profitable for the visitors, a state which was to be the norm of East-West commerce for the remainder of Chinese-European relations, right down to 1949, when China succeeded in closing her doors again to all but a regulated trickle of foreign merchandise.

The situation in Canton, however, was far from being so simple. With the decline that had taken place in Chinese maritime trade carried on in Chinese ships, it is possible that the attitude of the Chinese toward the Portuguese capture of Malacca was less hostile that it might otherwise have been. Probably the sultan of Malacca, a Chinese vassal, failed in his representations to Peking to get help against the Portuguese because the administration there saw in the foreigners a means of bringing the goods of Southeast Asia to its doors without the trouble of keeping and superintending a fleet of its own merchantmen. If this is so, it would explain the comparatively benign attitude shown to Perestrello, and later to Fernão d'Andrade when he first came to trade at Tamão, even when he high-handedly sailed upriver to Canton without Chinese permission.

TRIBUTE. A late-Ch'ing drawing of barbarian envoys presenting their gifts as tribute to the officials at Peking whose task it was to list the presents and other details regarding the persons and countries of the bearers. (From Medhurst)

Once there, d'Andrade's actions conformed more or less to what the Chinese expected of traders. The Cantonese merchants in whose capable hands trade lay, once it had been authorized, were doubtless overjoyed at the money which accrued to them. (Cantonese even today are perhaps the most skillful and assiduous of all Chinese traders.) And there is the final fact that it lay entirely within the competence of the local authorities to capture, put to death, or otherwise dispose of the Portuguese at Canton, and that they made absolutely no move to do so.

Trade was certainly not unwanted by the Chinese. For centuries they had been peacefully trading with the ships and merchants of India, Malaya, and other Oriental areas. Even the touchy nature of the national situation at the time, with Japanese pirates looting and burning along the coastal reaches, does not seem to have blinded the Chinese to the need for commerce when it could be peaceably conducted on their own terms. Nor was the desire for gain on the part of the merchants and authorities diminished by a certain risk. The mandarins doubtless got their cut of the proceeds and were disposed to turn a blind eye to the letter of the law.

But an embassy from a European power—a power increasingly troublesome to the Chinese—this was a different matter. But Pirès was lodged at Canton and determined to take his embassy to Peking. In the end, the Chinese simply fell back on precedent. In their eyes Pirès was just another tribute-bearer to the Emperor of All the World, and despite the peculiarities of his particular situation, as such they decided to treat him. Until the nineteenth century, with insignificant exceptions, all ambassadors were so considered and so treated. As tribute-bearers they were lodged at Chinese expense while written tidings of their arrival and their country of origin were sent to Peking. In due course the embassy was accepted by Peking, and word was sent ordering it to proceed, by an exactly specified route, to the capital. Arriving there, all embassies were quartered in a compound together, irrespective of whether they came from some tribal nomad chieftain or from the most powerful king in Europe. Here, further and more detailed information was sought about them and committed to writing, together with a list and valuation of the presents they had brought for the emperor. Then the ambassadors and their chief aides were minutely instructed in the observance of ceremonies at court when audience was given them (including the correct performance of the kowtow). Finally they were summoned in the dead of night, generally without warning, to attend on their knees the dawn audience at the palace. Many never in fact even then saw the emperor, who was not invariably present. Sometimes only his awesome but vacant throne stood with baffling symbolism in the open doors of that great gilded and lacquered hall, embowered in mists of incense from a hundred burners, and in the deep hush of a thousand stilled human breaths.

Although, as we shall shortly see, Pirès himself did not even have the chance of considering whether or not he would kowtow, this vexing question was to face all the many envoys that followed him, and we might pause a moment to consider its significance. The Chinese invariably demanded the performance of the kowtow; some Westerners refused and as a consequence suffered agonies of frustration; others performed it, and suffered agonies of humiliation.

The Chinese term transliterated as kowtow means simply "head knocking." We have to seek the origins of the custom to understand its meaning. From the time when Confucianism became accepted as the orthodox morality and philosophy of Chinese life, China had regarded herself as the natural leader and ruler of mankind. And there was some justification for this view, for at the time China was surrounded by peoples of demonstrably lower attainments in most human fields. Pursuing this line of thought, it was natural for the Chinese to feel that, in the occasional visits of non-Chinese people to the capital, they were conferring a privilege by receiving the outlanders, loading them with presents, and granting them audience with the emperor. The "barbarians" were thought by the Chinese to have gained spiritually by being thus enabled to partake of the great things, and to see the great ones, of China. Thus did the Chinese, fortunate possessors of a great culture, deal benevolently with those less fortunate ones who, by the accident of their birth, were non-Chinese.

While the envoys were at court they were instructed in the performance of the kowtow—a perfectly normal Chinese ceremony performed by every Chinese before his parents on certain days of the year, by parents and everyone else to their ancestral shrines, by mandarins and everyone else to the emperor, and even by the emperor himself to his parents and also to Heaven twice a year. In short, it was nothing more that a social custom quite similar in intent to the low bows and curtsies with which Europeans of the time greeted their social superiors.

The ceremony in itself varied from time to time. But its basis remained three kneelings, each accompanied by touching the forehead to the floor three times. The performer fell to his knees and literally knocked his head on the floor. There can be no doubt that the physical actions themselves left little room in the mind of the performer about who was ruler and who was ruled. Chinese accepted this relationship among themselves and did not find it in the least irksome. To them the natural accompaniment of achieved power was to be deferred to in this manner. There was no feeling in the Chinese mind that to kowtow was to place oneself in the position of a slave: that was something quite different.

But for Westerners, most of whom could not appreciate these facts in the Chinese light, to kowtow seemed tantamount to declaring slavish obedience and subjugation. (The reader who wishes to experience something of their sensations has but to perform the kowtow described above to some friend and see what his mental sensations are when he does it with a certain attempt at respect.) For the Chinese, however, those Western feelings and fears

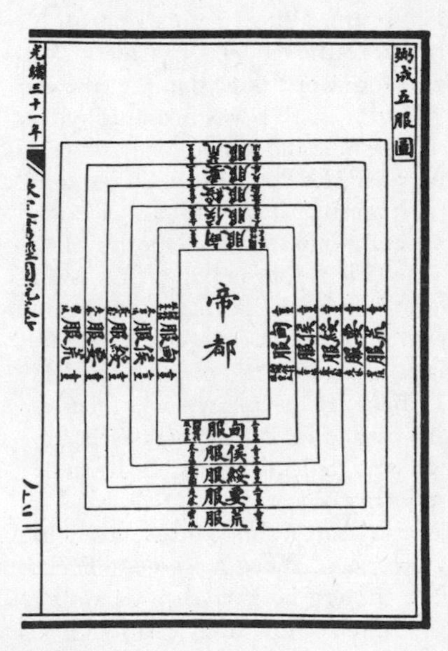

"The Middle Kingdom." A graphic representation of the manner in which the Chinese considered their culture to radiate from the capital outward, as given in the *Ch'in-ting shu ching t'u shuo* (Imperial Illustrated Edition of the Historical Classics). In the central rectangle are the characters meaning Imperial Capital. This area is surrounded by five bands: Imperial Domains (i.e., China itself); regions of feudal princes tributary to the emperor; the "zone of pacification" (the borders where Chinese civilization was supposed always to be winning adherents); the zone of "barbarians" allied to China; and the Chinese equivalent of our Biblical Outer Darkness, where total savagery reigned supreme. (From Needham)

were all but totally incomprehensible. In reverse, we in the West might be equally nonplused were all Chinese visitors to refuse to shake hands with us and to insist, for example, on always entering and leaving rooms in front of rather than behind us.

The Westerners in Peking mostly attempted to avoid the kowtow. In doing so they either spoiled their chances of success in the capital or were summarily dismissed. But in fact even those who performed the ceremony with tolerably good grace were ensured only of a better hearing, not necessarily of the success of their mission. Reading their accounts of how they did or did not kowtow, one is driven to the conclusion that for the most insular of reasons most of them refused in order not to compromise (as they saw it) their personal and national dignity in China. This, as we have seen, was not understood as an idea by Chinese, who had no such *hauteur*. The Chinese who visited abroad in the past did in fact quite willingly conform to other nations' ceremonies without feeling they had lost anything. Similarly, those Westerners who *did* kowtow, it is equally evident, sacrificed in the ceremony no dignity or honor or privilege in the eyes of the Chinese.

But the kowtow was only a small part of the elaborate ritual of an imperial audience. Once the envoy's presents had been accepted, gifts were allotted in return (in strict conformity with the supposed importance of the ambassador's own ruler), and the embassy was instructed exactly when and how it must depart from Peking, and the precise route it must take out of China toward its own barbarous domains.

Of most of this Pirès was doubtless ignorant. With his small party, his presents, his letter from the king of Portugal to the emperor of China, Pirès waited at Canton for well over two years before word came that the embassy should come north. Even by Chinese standards, this was an inordinately long time. Perhaps the delay was due to Chinese indecision, or perhaps to the suppression of a rebellion then in progress. The Portuguese did not quite fit into any of the accepted categories. A lingering doubt probably clouded the official mind of the administration, compounded of memories of the Portuguese sack of Malacca (however convenient that may have proved in some ways) and of the anomalous position whereby the Portuguese appeared to treat Tamão as their rightful place in Chinese waters (whereas it was merely a place set aside for *all* foreign ships to trade). And possibly the delay also resulted from internal jealousies, bribery by the merchants of Canton through the governor. The wily administrators in Peking may well have formed their own opinions of a governor who delicately pressed the suit of the foreigners (as the Canton governor possibly did).

At last on January 23, 1520, Pirès set out from Canton, reaching Nanking four months later (a very slow progress). When he came to Peking toward the end of the year his fate, little though he knew it, was already compounded. For by now the sultan of Bintang in Malaya, through his envoy, had most effectively poured out his tale of woe, of deprivation at the hands of the Portuguese in Malacca; and he had backed up the tale with others concerning the reprehensible Portuguese methods in the Moluccas, making the case (quite truthfully) that European trading visits were no more than the prelude to annexation of territory. With the tiny sea power at this time available to the Chinese, and that occupied in repelling Japanese raiders, such an argument was telling. While the administration was making up its mind about the Portuguese, events in Tamão came to the official ear. They were hardly such as to endear poor Pirès to the court at Peking.

Fernão d'Andrade's successful trading venture at Canton "excited the cupidity of the Portuguese" in Malacca, as Cordier puts it, "and Simão d'Andrade obtained . . . permission to be sent as a replacement for his brother." He arrived at Tamão with three junks in August, 1519, took the island, and fortified it with stone and wooden battlements on the pretense that these were needed for defense against pirates. More, he committed what was from the Chinese viewpoint nothing short of *lèse majesté* when, taking the law into in own hands on Chinese soil, he put up a gallows and hanged a sailor and also raised a *padrão* (a stone pillar with the arms of Portugal on it) to mark his arrival on the island.[5] In the wake of this, native marauders took the opportunity of assaulting various mainland towns in the vicinity, in the name of the Portuguese.

At the same time a crop of horror stories bloomed like poisonous flowers throughout South China.[6] The best of those which were recorded in Chinese at the time has a strangely modern air about it: "These men [Portuguese] love to eat little children. . . . Their method is to boil water in a great jar, and when it bubbles, to put a little child in an iron cage, placing this over the jar and steaming the child in the vapor until the sweat comes out. When it has all come out they take the child from the cage and with a brush of iron bristles remove the softened skin. The child is still living. Then they kill it, slitting the abdomen and removing the intestines and stomach; the body is then further steamed before it is eaten." Irresistibly one is reminded of those World War I stories of the Kaiser, who allegedly caused French and Belgian children to be rendered down for soap and lubricating oil.

Other tales tell of the Portuguese ravishing children under ten years old and eating the corpses roasted, a commerce in which the local gangsters

5. Estimates of Simão d'Andrade's character are all but unanimously unfavorable. The Portuguese historian Barros says: "He was a great gentleman in his own estimation, very pompous, vainglorious, and of an adventurous disposition. . . ." Major concurs: "Simão d'Andrade himself gave frequent occasion for complaint by inconsiderate or unjust regulations, contrary both to the laws and to the received opinions of the country, and provoked the Chinese against the Portuguese: and even his personal behaviour seems to have been calculated to provoke animosity."

6. Some of these Chinese stories, like those recounted here, will be found in *T'oung Pao*, No. 38, 1949, in an article by Paul Pelliot.

made great profit. "And all the people of Kwangtung province were thus afflicted so that none could be sure of his safety." However much the culinary details may be in error, from the numerous references in sober Chinese contemporary works it is certain that a foundation for the stories existed. The Portuguese were far from having endeared themselves to the populace. Even today one of the common Chinese expressions meaning "foreigner" is *yang kuei tzu,* "ocean devil."

Antiforeign feeling was further incited by Simão d'Andrade's assumption of control of all trade and shipping at Tamão, his refusal to pay customs duty, and (to the Chinese) the terrible crime of severely beating an important Chinese official. In their eyes this all added up to Portuguese assumption of sovereignty, not over some far-away Malacca but over an island that was actually part of China.

Because of these events and suspicions, in Peking the innocent Pirès was branded as a spy, and all his countrymen as mere pirates. The final blow came when news arrived of a series of raids which d'Andrade had made along the China coast. Poor Pirès had no luck. And the ill omen of the emperor Ming Cheng-te's death, which took place while Pirès was in Peking, did nothing to quiet the stormy atmosphere surrounding him. He was escorted back to Canton in ignominy, arriving on August 22, 1521, there to await the outcome of Chinese investigation of the whole Portuguese affair. That outcome was a foregone conclusion for which the Chinese cannot be blamed. Their patience had, to put it mildly, been sorely tried.

In a letter written from a Canton jail, one of Pirès's companions, who was still languishing there in 1534, tells the remainder of the story.[7] As soon as the party arrived from Peking they were arraigned before the authorities and then taken "to certain jail-houses that are in the storehouses of foodstuffs; and . . . Pirès did not wish to enter them, and the jailers put us in certain houses in which we were thirty and three days. . . ." Then Pirès was taken to another prison with six of the party, and the remainder of the unhappy band were taken to another place, there to remain for ten months. "All the goods [by which he seems to mean the presents which Pirès had taken to Peking for the emperor, which had been returned with him to Canton; but he may also mean the remaining Portuguese trade goods in the warehouses] remained in the power of Tomaso Pirès. They treated us like free people; we were closely watched in places separate from the prisoners." The separate parties were eventually brought together again and confronted with the "Malays"—presumably the envoys of the sultan of Bintang, whose accusations about Portuguese deeds in Malacca had had such an adverse effect on the fortunes of Pirès in Peking. The magistrate pronounced that the emperor had decreed that the king of Portugal "should

7. The letter is quoted by Ferguson.

deliver up to the Malays the country of Malacca which he had taken from them. Pirès replied that he had not come for that purpose, nor was it meet for him to discuss such a question; that it would be evident from the letter he had brought [addressed to the emperor by the king of Portugal] that he knew nothing of anything else." This doubtless sounded very like the defense usually put up by spies when caught.

The Portuguese were kept on their knees for four hours at this hearing. "And when [the examining magistrate] had tired himself out he sent each one back to the prison. . . ." On August 14, 1522, fetters were attached on the wrists of Pirès, while his companions were fettered at both wrists and ankles. (By this time Pirès must have been revising his opinion that the Chinese were a "weak people of no account.") The Chinese then "took from us all the property we had. Thus with chains on our necks and through the midst of the city," they were taken to another prison, where their fetters were replaced with even stronger ones. At the gate of this prison one of the company died from the rigor of the experience, and the arms and legs of all were swollen and cut by the tight chains. "This, with a decision that two days afterward they would kill us. Before it was night they . . . [conducted] Pirès . . . alone, barefoot, and without a cap amid the hootings of the boys to the prison of Cancheufu [the Canton city jail] in order to see the goods that they had taken from us which had to be described; and the mandarin clerks who were present, wrote down ten and stole three hundred." It was decided in the presence of the prisoners to write to the emperor and ask what was his intention as to the ultimate fate of the Portuguese. "On the following day they struck off our fetters, which if we had borne a day longer we should all have died, and they brought Pirès back once more to his prison."

Two years later, in May, 1524, Tomaso Pirès died in that prison of some sickness whose nature we do not know. And so in time, one may reasonably suspect, did all those of his compatriots who survived him. Presumably the throne preferred not to make any edict as to their eventual disposal and to allow them gently to languish.

Such, then, is the sad little story of poor Pirès and his trip to Peking. Among the many Portuguese involved in the events of this time and of later years until the establishment of the Portuguese colony of Macao around the year 1557, it seems that only Fernão d'Andrade and Tomaso Pirès showed any intelligence in dealings with the Chinese. About the rest, including Fernão's brother Simão, all the information that has come down to us (even from Portuguese historians) is almost uniformly grim. Distinguished for neither their cultural attainments nor their charm (which might have compensated for the lack of the former), nor even for common sense, they invited animosity and obstructionist reaction to their actions in South China.

For their ill fortune "the Portuguese had chiefly themselves to thank. Truculent and lawless, regarding all Eastern peoples as legitimate prey, they were little if any better than the contemporary Japanese pirates who pillaged the Chinese coasts. The Ming authorities can scarcely be censured for treating them as free-booters."[8]

The moral of poor Pirès and the Portuguese pirates in their confrontation with early-sixteenth-century China has of course elements of the obvious, at least when we look at the story in the light of twentieth-century conditions. There is no point in dismissing Portuguese acts as mere piracy, for piracy was an accepted aspect of Christian Portugal's (and of other countries') world outlook at that time. There are much more complex, and extremely important, aspects at which we must look briefly if we are to understand the future of European-Chinese relations in the years that came after—during the remainder of the declining Ming dynasty, during the ensuing Manchu, or Ch'ing, times, and right up to the present.

Even the most general reading in the history of relations between the Chinese and the foreign peoples coming to their country reveals an almost black-and-white contrast in the reception accorded, and the treatment meted out, by China to different people at different times. In the T'ang, Sung, and Yuan all comers were welcomed. With the coming of the Ming dynasty, things changed radically. Due to their desire to tighten the contours of China, maritime trade (after a first exploratory flutter in Chinese ships) became the main source of contact between Chinese and foreigners. Seafaring Indians, Malays, and Arabs were still welcomed. Even the first recorded Portuguese, Perestrello, appears to have traded peacefully. Foreigners, though none of them Westerners as far as we know, were residing in many seaboard and inland Chinese cities, apparently in peace with their Chinese hosts. Yet under the Ming and later the Ch'ing the attempt was continuously made to restrict contact with Europeans (and Europeans only) to the one port at Canton, via one or other of the islands of the Pearl River Delta, or to the peninsula of Macao. Europeans, moreover, were forbidden to penetrate into China, and it was made an offense for a Chinese to assist them to do so.

"This contrast," as Fitzgerald remarks, "must have an explanation, and it is only too plain from the records, both Chinese and European, that the unfavorable treatment of the Europeans was the consequence of their own violent and barbarous behavior."[9]

Here we reach the crux, the radical, the fundamental disaster, whose effects and consequences have afflicted the whole course of relations between the West and China ever since. This is not to suggest that only the *Portuguese*

8. From Latourette's *The Chinese*. For the interesting story of Portuguese piracy and settlements on the South China coast a little later on, and how they were driven from the ancient port of Zayton, see Chang T'ien-tse, who has collected many of the Chinese sources.

9. *The Chinese View*.

follies in China were operative from then until now. But it is quite definite that it was just those follies of the time when the Ming Chinese were being initiated into trading contact with the nations of Europe that shaped the Chinese attitude to later ventures, whether Portuguese or other. The difference between the Portuguese and the Dutch, who came later, or, still later, the English was not apparent to the Chinese. In all that occurred during the later years of the sixteenth century and until the fall of the Ming, and then throughout the centuries of Manchu rule, it is those aggressive and barbarous attacks of the early Iberian traders, which were the first Chinese experience of the maritime West, that bedevil the scene. And in the same breath the emphatic point must be made again that China had for centuries been trading on the whole quite peacefully with a large variety of foreigners of Oriental nationality in her own ports, without the necessity, as far as is known, of antiforeign decrees against such trade or contacts. By way of elucidation something must be said on the Chinese attitude to foreigners in general. However much the educated Chinese, who decided these matters, looked upon all foreigners as lacking the ineffable blessings and benefits of Chinese civilization, as, therefore, existing in that outer dark of the spirit which gloomed around the whole circuit of the shining and perfect world of their Central Kingdom, they never evinced the slightest desire to thrust their own way of life, their philosophies, their forms of religious belief, on the rest of the world. The militant conviction that others are in dire need of one's own religion seems to be confined almost exclusively to Christians (and has been shown by Moslems too, at some periods of their history), and is often accompanied by such extreme measures as foreign conquest, forcible trading, and commercial rapine on the part of these propagators of their faith. The Chinese, with sagacity and a certain haughtiness and aloofness inherent in Confucian teachings, made few serious attempts to conquer foreign lands unless attacked. Even those great expeditions sent overseas in the early Ming under the Chinese admiral Cheng Ho, while they returned with various petty kings, seem to have had as their aim a desire for knowledge of the outer world and a wish to found trading stations in its ports. The phase was brief. It succeeded in establishing groups of Chinese merchants in some Southeast Asian ports, men whose activities were basically peaceable and whose lives took on much of the color of their new homes. There is no example of the Chinese sailing out to some remote land, conquering it and subjugating its inhabitants by military means, and thereafter establishing a Chinese realm there, governed by the Chinese from the home country, from Peking.

The Iberians—Portuguese and Spaniards alike—on the other hand, did exactly that. Schooled for several centuries in all but continual warfare with the Moors, they had come to regard all foreigners as despicable, inferior, worthy only to be put to the sword if they would not be converted to that searing brand of Christianity which the inhabitants of Spain and Portugal had come to accept as the only true faith. The towering arrogance of men such as those ignorant Portuguese who came first to China's coasts, their clear belief that the only thing to do with non-Christian foreigners was to kill them if they were militarily weak enough—this fierce and baseless pride of the Iberians seems now nothing less than monstrous. Only when the heathen—of China, of America, and elsewhere—displayed some semblance of strength, some desire to defend their way of life against these depredations, did those shiploads of pirates calm their desires and make do with trade for the time being. But as the Chinese were not slow to realize, the ultimate aim was always to annex, to own, to cast into slavery. For thus was profit swelled, and, as an afterthought, thus was paganism conquered for the Christian faith and for the Christian monarchs of Spain and Portugal.

They came, then, to China, those Portuguese, and those many other Westerners after them, mostly with piratical intent well- or ill-concealed, sometimes with the approval of kings or of men of God, sometimes without, almost always with desire for gain. The few exceptions to the rule of cupidity were mostly missionaries. Many of them were sincere, and of those many were stupid, uncultured—dogmatists, plodders, reaping a small and volatile harvest of Chinese, but of few Chinese souls. But all, or almost all, came to China with a conviction which must nowadays seem singular to the point of extravagance: that Western culture and all that it entailed was a way of life infinitely superior to that of China. That this opinion was, at least until the sixteenth century, in almost every particular totally untrue, and later only true in certain respects, escaped almost all of those men and women. It must not escape us now or we shall fail altogether to understand the story of Europe and China.

So the story of Pirès, the luckless ambassador, with its moral which could even then have been understood by right-thinking men in Europe, escaped interpretation. The pattern was outlined. The Chinese had no reason for believing that in the West there actually existed a civilization and great and peaceful men, for the representatives of that civilization were uniformly barbarous by any standards. Not until the coming of the Jesuits, and then only with extreme difficulty, did the West manage to show it contained other beings. It is with a few of those who reached Peking and touched the nerve center and the fount of Chinese tradition and opinion there that we are now about to make the acquaintance. Theirs, also, is a curious tale.

CHAPTER EIGHT

MATTEO RICCI AND THE RELUCTANT DRAGON

As THE sixteenth century was drawing toward its close and the august dynasty of the Chinese Ming had run more than two-thirds of its course, there came to China one of the most remarkable men of the age. His name was Matteo Ricci and he was a Jesuit priest. Of all the Europeans who attempted the task of understanding the Chinese and their civilization he was the most talented, the most important. Among all the Westerners who sojourned in China, he was the only one to whom the Chinese accorded unreservedly their respect as a scholar in their own language and literature. To achieve that position, Matteo Ricci had to become in all relevant ways at least one-half Chinese himself.

Humility, which seemed the best approach to the persevering Friar William of Rubruck, is not what we need to follow the single-minded but subtle Ricci. He was not saintly, he was intellectual. While he was dedicated to his faith, he was at the same time a politician. The most European of Europeans, passionate but controlled in his passions, he realized that to come to terms with the Chinese he had first to understand them.

Something between admiration and awe, or a mixture of both, is the natural result of reading the story of his career in China and his own voluminous writings from there.[1] During the latter part of his stay in Peking it was

1. We shall have occasion to quote often from Ricci's writings. Rather than burden the reader with an interruption in each case, the following information should serve in a general way. The major works on Ricci's life and writings are those by d'Elia and Venturi. Since neither has been translated in full from the original Italian, I have relied on scholarly translations of parts of them by several persons, especially Cary-Elwes and Needham. Another important book is Trigault's compilation of the Ricci journals, which Trigault brought back to Rome from China in 1614, after Ricci's death. Written originally in Italian, the journals were translated into Latin by Trigault, who added much valuable information of his own, and from other Jesuit sources where Ricci had been too modest to recount his own share of achievement. I shall quote often from Trigault's book, in the translation by Gallagher. A few other quotations come from Dunne, Bartoli (in Cary-Elwes's translation), and Bernard.

with the same kind of respect that the Chinese themselves, literati and ministers of state, seem to have looked on this unique phenomenon in their midst. The earning of their respect had taken Ricci the last twenty-seven years of his life.

Ricci was an Italian born in 1552 in Macerata, a town near the Adriatic coast on the far side of the Apennines from Rome. His family were minor nobility, their coat of arms bearing a blue hedgehog (*riccio* in Italian) on a red ground. When Matteo was seven years old his tutor joined the Jesuit order, and shortly afterward a Jesuit school was founded in Macerata, which the young Ricci attended. Here, at the age of sixteen, a youth of sturdy physique, strong black hair, and penetrating blue eyes, Matteo finished his course, and was sent by his father to Rome to study law. The son had other ideas, or dreams perhaps they were at the time. Without a doubt by this time he had heard from the Jesuits in Macerata of one of the most important Jesuit activities of the age—foreign missions. The pioneering voyages made by St. Francis Xavier to India, Japan, and other exotic lands had doubtless been recounted to the pupils, and Ricci probably knew that the saint had died within sight of China in the same year as he himself was born. Once in Rome, many of his friends were students from foreign places. It was perfectly natural that very soon he should be absorbed in the dazzling ideal of converting those almost unimaginable tracts of the world on which the light of the Christian faith had not yet fallen. Many another youth was so inspired. Piety and a zest for adventure combined to form the intellectual climate of a thoughtful young man in those days, and inevitably led Ricci on a course opposed to his father's wishes. He joined the Jesuit College in Rome, and from that moment his future was decided. As he opened the door of the college for the first time he was met by a priest called Alessandro Valignano, master of the novices, who himself was soon to be sent to the East as Jesuit visitor and vicar-general. By coincidence it was this man's future decisions which were in years to come to fulfill the precise desires of his new pupil.

Matteo the young man was no fool. Besides natural piety, he had a passion for knowledge of all kinds, and for study. Soon the exact direction of his life was clear to him—he intended to go to China. Perhaps he had read in the college the letters of St. Francis Xavier from the East: "Opposite to Japan lies China, an immense empire, enjoying profound peace, and which, as the Portuguese merchants tell us, is superior to all Christian states in the practice of justice and equity. The Chinese whom I have seen . . . and whom I got to know, are white in color . . . are acute and eager to learn. In intellect they are superior even to the Japanese. . . . Nothing leads me to suppose that there are Christians there."[2] To Ricci, words such as these must have read like an open invitation.

2. Quoted by Coleridge.

His education at the college could hardly have been more suitable. All of it was to prove more helpful than he could have known at the time. To the study of rhetoric he brought a fresh intelligence, learning also the philosophy of Aristotle and the mathematics of Euclid under Clavius, the foremost mathematician of the day. He absorbed not only the principles of Ptolemaic astronomy and some knowledge of geography, but also more practical crafts such as the construction of clocks, sundials, and various celestial measuring instruments—quadrants and the like. Even as a student the exceptional quality of his memory was remarked.

With his mind made up, the conviction that his vocation lay in the Orient strengthened day by day as he studied under the Jesuits. But it was quite another matter to be *chosen* for the Eastern mission. His time did not come until he was twenty-five. And then, after an audience with the aging pope Gregory XIII, Ricci set out at last. Not east but, paradoxically, west to Genoa and thence to Lisbon. For the East, inclusive of China but not the Philippines, had been assigned by papal bull to the Portuguese. All missionaries bound for the Indies and farther were obliged to sail from the Tagus in Portuguese ships.

From Lisbon, Ricci sailed away from Europe for the first and last time. There is not a word of regret at leaving and hardly an instant of nostalgia for Europe in his writings. Throughout the rest of his life, whatever he may have felt of those two emotions, he never committed them to paper. While his affectionate nature always remembered in letters the thinning ranks of his friends in Rome, that astonishing singleness of purpose which first drew him East tenaciously held him there, uncomplaining, until death.

With him in the nine-hundred-ton carrack sailed his friend and fellow Jesuit Michele Ruggieri, twelve other missionaries, and a preposterously crowded company of about five hundred adventurers, traders, and army personnel. So high was the fever for mercantile riches to be wrested from the East, so keen the competition for appointments to the trading stations established in India, Malacca, and elsewhere by the Portuguese, that the infrequent ships were stuffed with men who accepted animal living conditions for the privilege of going there. As the burdened little ship crept down through tropical African seas, its human cargo diminished with disease, so that as they rounded the Cape of Good Hope and reached Mozambique the master had room to take on not only the fresh water and fresh food they needed, but a goodly number of Negro slaves for sale in the markets of the Indian possessions.

At Goa, a filthy, overcrowded, colonial town on an estuary halfway down the west coast of India, a place where the debauchery of its European inhabitants was only matched by the number of churches and mission stations, Ricci and his friends disembarked thankfully. It is not hard to imagine the shock that the young idealist must have felt at this first sight of the Orient.

Even in the twentieth century to set foot in the teeming lanes of certain Oriental ports, to wander innocently outside the fairly seemly areas of the main streets, tends to banish the glamor and romance that still cling in the Western mind to the picture of life in the East. For the young Ricci, with his passionate and still idealistic longing to make a start on converting the Orient to his own innocent flame of Christian belief, Goa was a sobering event. In sight of the cathedral of his faith and a stone's throw from the seat of the Portuguese viceroy, who was the upholder of that Christian faith, lay the slave market. Here—amid ordure, perfidy, and degradation—men, women, and children from Africa and India and other lands were offered for sale. Here Portuguese *fidalgos,* rigged out in splendid silks, paraded with retinues like little kings, returning to their houses, where harems of slaves waited. To the earnest Ricci, Goa must have been a shock.

Like most European colonies before and since, those of Portugal in India had been achieved by sword and flame, consolidated by economic subjection, and held at gun point. The Portuguese had massacred hundreds of thousands of Indians, whose only fault was that they resisted the taking of their native earth and their traditional means of livelihood. The history of Portuguese struggle in their own country against the Moors, who had done the same thing to Portugal and Spain, made such an outlook normal, and doubtless on this account its application to the "infidels" of all foreign parts seemed right and natural to most of the colonizers in India at that time. Only perhaps the Jesuits, particularly Valignano when he was in India a year or so previous to Ricci's arrival at Goa, put forward a doctrine of conversion by peaceful means. This basic technique was to be the mainspring of the Jesuit approach to the Orient.

In Goa, Ricci resumed his studies in theology. After a year he found himself a schoolteacher instructing the more promising of the local children at St. Paul's College in Latin and Greek. His health broke down. Probably the frustration of his hopes of missionary work and the unaccustomed heat of Goa had lowered his vitality. He took to bed with a high fever. Recovering, he was sent south to Cochin, where, in another Portuguese settlement, he took up teaching again, if unwillingly. But at last in Cochin he was consecrated priest; and soon he went back to Goa to continue theological study. In the spring of 1582 the opportunity he had so long awaited at last arrived. Valignano sent for him. He boarded the next ship for Macao, the remote Portuguese settlement and mission on the southern Chinese coast. He was now thirty years of age. The story of his life was just beginning.

Once a year the Portuguese carracks sailed from Goa, bound on favoring monsoon winds for Malacca, Macao, and onward to Japan, where a thriving Jesuit mission already existed. To all those stations they brought mail, personnel (priests and civilians), money, stores. The early-Ming days of Chinese ocean voyaging were long over, and Portuguese vessels in the Orient were largely responsible for the transport of Chinese goods destined for Japan, Southeast Asia, and elsewhere. In every way these voyages were profitable, for the Portuguese were in the happy position of being the sole middlemen, the sole carriers. They were not slow to exploit their monopoly.

Still weak from another fever on board, Ricci disembarked at Macao in early August, 1582, and was greeted by his old friend Michele Ruggieri, who had been there for some time already. Just up the hill from the quay he met the man who, years ago and on the other side of the strange world, had admitted him to his novitiate in the Roman College—Alessandro Valignano. It must have seemed the moment of culmination, the moment of beginning on the real tasks of his life. But he little knew that day on the sunlit shores of Macao how different and how incredibly more difficult, how much more secular in many ways, his future work was to be from the usual missionary activity about which he had already learned a little.

Macao at that time in the late sixteenth century must have been a charming enough place, a little lost, perhaps, but restful after the stews of Goa. A peninsula, virtually an island, hanging from the coast of China in the Pearl River Delta where the yellow water reached the sea from China's deep south, Macao has a subtropical climate. Summer is warm but not stifling, although the humidity is high. Spring is a recognizable season and so is autumn, both gay with their particular flowers and blossoming trees, both summery as days in a European June. And winter is a long dry spell when skies are more often than not pale and cloudless blue with days something like a good European spring. At that time there were only about ten thousand people in Macao, a thousand of them Westerners. Ricci says little enough about it, but this must indeed have been a change from his fever-stricken Indian days. It was also an exciting place. Here, at last, he stood on the doorstep of the vast empire of China with its unnumbered peoples, with its immense and singular challenge.

Quite soon Ricci was to discover that Macao, far from being the doorstep of the house of China, was rather the prison to which the Chinese, with considerable subtlety, had relegated those few "barbarians" who were in some ways useful and in others a continual nuisance to the Celestial Empire. In fact the historian-priest Semedo (whom we met in connection with the Nestorian stone) tells a story of Valignano gazing one day out of the window of the Jesuit College in Macao toward the continent of China, whose hills formed the northern horizon. It seemed to Valignano in his despondency that the prospect of its conversion to the true faith was so dark that he "called out in a loud voice and the most intimate affection of his heart, speaking to China: 'Oh, Rock, Rock, when wilt thou open, Rock?' "[3]

When, indeed? This was Ricci's question too. His response to it was

3. Quoted from Semedo's history in Latourette's *History of Christian Missions*.

ardent, vigorous, characteristic. He set about the task for which his whole thirty years had prepared him. Quite contrary to the shock tactics of Xavier, and of several parties of Dominicans and Franciscans who had recently attempted to enter China (mostly from the Philippines, which that blissfully unreal papal bull had given in the same parcel as the Americas to Spain's tender mercies), Valignano's ideas on converting the Chinese were much more subtle. "The only possible method of penetration will be completely different from that which has been adopted until now in conducting all missions in [the East]," he wrote.[4] What he proposed instead was a kind of "cultural accommodation." And it was in applying this idea that Riccci made his greatest contribution to Sino-Western relations and understanding. But it was as a result of an earlier Jesuit attempt to enter the country that Ricci's first chance came.

Not long before Ricci's arrival, Michele Ruggieri had been sent up the river to Canton with traders, bribing Chinese officials to allow his passage. Ruggieri made no attempt to preach the faith, yet he had nevertheless been sent back promptly by the Chinese; but not before he had reached a town called Chao-ch'ing, some distance west of Canton. It was this trip which, in the summer of 1583, bore surprising fruit in the shape of a letter from the governor of Chao-ch'ing. Having heard of Ruggieri's mathematical skills, he invited him back. He even hinted that a plot of land might be made available where Ruggieri could build a house. The letter is so remarkable, so un-Chinese for its period and in view of its author's position in the Chinese bureaucratic machine, that it would be interesting to know why it was ever written. We can only speculate that the curiosity value of the Jesuits overcame the governor's more sober judgment of the temper of the times and made him forget or ignore the imperial edict forbidding the entry of foreigners to the countryside, which had been promulgated in early-Ming times and recently reinforced.

The Jesuits at Macao were in a quandary. They had no money to finance Ruggieri and Ricci—who was to accompany him and, in fact, to be the dominant force—on a long stay in China. The situation was only saved by a Portuguese merchant of Macao who donated the funds. So, early in September that year, Matteo Ricci and his old friend set out for China up the muddy river to Canton. Neither of them had much reason for the hope they ardently felt; but both were set on grasping this providential chance of a footing in the baffling, closed, hostile land of China. The "Rock" of Valignano seemed to have opened a fraction, just enough to let them pass inside. Before leaving Macao they shaved their heads and beards and put on the rough, hooded cloaks of Chinese monks.

In Canton, as if to underline the insecurity of their position, they came

4. Quoted by Dunne.

face to face with some Spaniards who had been wrecked off the coast and brought captive to the city to explain themselves. And there was also a party of ten Franciscans in town. Their ship had run aground in a storm off Hainan Island and they were being held as pirates. The delicate situation of these groups was explained by the fact that piracy by Japanese and others, including the Portuguese, was very common at the time and the Ming authorities had little enough power to quell it. Ricci pleaded for the prisoners, a brave thing to do and a very risky one. Somehow he managed to have them all sent back to Macao without compromising his own position. Within a week of leaving their base, Ricci and Ruggieri arrived safely in Chao-ch'ing and were led inside its walls, through its crowded and prosperous streets to the hall of the governor. It was their first real contact with official China—a surprisingly mild one by later standards.

There sat the governor at the end of the ceremonial hall, surrounded by his aides and minor officials, all dressed in that gaudy anachronism, the ceremonial dress of their kind which had hardly altered for hundreds of years and which was to persist for hundreds more. The governor wore his black-silk robes embroidered with a plaque on the breast showing a wild goose, the insignia of an official of the fourth grade.[5] Before him was a table covered with brocade, on which lay writing brushes, seals, ink sticks of red and black, and his fan. Before him, too, were the Jesuits, heads and beards shaved, clothed in the coarse gray cloth of indigenous priests, prostrate on the floor in attitudes of abject submission. Only thus was it possible for them to enter his presence at all.

As yet they knew little enough Chinese. Ruggieri had spent a year or so learning the spoken language of South China (Cantonese) only to discover on his first trip that his labor was wasted as far as communication with officials went; for the official speech of all China was the spoken form of the North (called Mandarin by non-Chinese). They presented their written petition to the governor (the only way of communicating with him, despite the fact that he had invited them) through an interpreter—a Chinese Christian whom they had brought along from Macao. It was simple enough: "We belong to an order of religious men who adore the King of Heaven as the one true God. We come from the uttermost reaches of the West, and it has taken us three years to reach the Kingdom of China, to which we were attracted by the renown and glory of its name. . . ." They sought permission to build a house and a church for worship, and to reside there for the

5. Chinese officials—"mandarins" to the Europeans—were divided into nine grades, each grade having a principal and a subordinate rank. Each mandarin wore a button at the top of his dress hat, its substance varying according to grade from ruby (first class), through sapphire (third), to worked gold (ninth). They also wore on the breast of their robes a plaque of embroidery (called *p'u tzu*) that differed from grade to grade; the first was a white crane, second a golden pheasant, third a peacock, eighth a quail, etc.

rest of their lives, promising to live within the law. Nothing at all about converting the Chinese to Christianity. Nothing about Christ or the Virgin Mary. In Macao when the form of declaration had been drafted, Ricci had protested. But Ruggieri and his fellows had informed him of the state of affairs in China, where all foreigners were automatically classed as barbarians and any suggestion that a foreign cult could be accepted by the Chinese was entirely offensive.

The governor of Chao-ch'ing looked favorably on the two foreign priests and suggested they take a piece of land near a pagoda which was being built by public subscription near the town. Habited as Chinese priests (bonzes, the older books call them) and about to be saddled with a site tending to underline the idea that they were merely representatives of some curious variant of Buddhism, or of Taoism (of which they had not at that time heard), Ricci and Ruggieri were unhappy; but they accepted, and began to build a combined mission house and chapel. Their appearance, together with the strange objects they had with them, both intrigued and aroused doubts among the populace. All were fascinated by the large glass prism that threw the light of the mundane sun in a spectrum band to its side; and everyone was overcome by the strange beauty of the paintings of Christ and of the Virgin Mary (the realistic Western style in painting being quite foreign to Chinese art). But the committee in charge of the new pagoda project was enraged at the mere presence of the foreigners, let alone their intent to build a mission house nearby. They regarded Ricci and his possible building as inauspicious omens in a climate of rigidly calculable *feng-sui*. (The word, meaning "wind-and-water," is used by Chinese to denote auspices dependent on such matters as the lie of the land, color of buildings, free flow of benevolent influences from the south, and many another arcane influence.) But the committee could not openly oppose the governor.

The governor meanwhile had obtained permission, from the viceroy of the whole province, for the Jesuits to live in Chao-ch'ing. Hard to make up one's mind about this governor. Very curiously for a Chinese of his rank and delicate placing in the hierarchy of the civil service, he seems from the first to have courted difficulties. It was expressly forbidden to introduce foreigners inland; yet he had invited them. Then he found a loophole which provided that all foreigners must be self-supporting, must wear Chinese dress, must marry Chinese if at all. "You will become," he said, "in all but physical appearance, men of the Middle Kingdom, subject to the emperor." It was a prophetic statement, little though he knew it. Ricci and Ruggieri hesitated. As yet they knew little of Chinese laws. But once more they accepted. The governor was pleased. Possibly he hoped that by fostering the phenomenon of the two foreigners, men who were skilled in mathematics and other strange subjects, he would gain in the official or public eye. Pos-

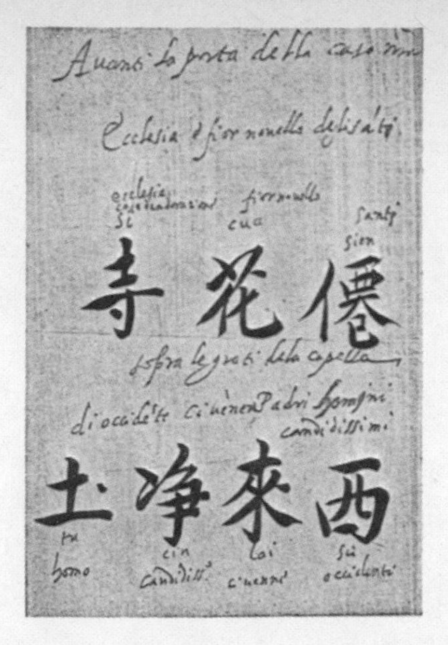

THE GOVERNOR'S PLAQUE FOR RICCI'S MISSION. The governor of Chao-ch'ing presented this calligraphy to Ricci for displaying on the door of his mission house. The characters mean "The Sacred and Beautiful Hall of the Westerners of the Pure Land." The latter words almost certainly refer to the popular form of Buddhism called Pure Land, and very probably the governor thought that Ricci, dressed in his Chinese priest's robes, was expounding some form of the Pure Land doctrine. Ricci's knowledge of Chinese was still scanty and he probably missed the allusion. Above the characters, and perhaps at a later date, Ricci has written a literal translation into Italian and also an attempt at phonetic transcription. At the top he has written "In front of the door of my house." (From Venturi)

sibly he hoped for some private or personal delectation by virtue of these skills. The explanations are all unconvincing; and as it turned out, he misjudged the temper of his own people.

The pagoda committee soon began to pinprick the Jesuits, objecting that the site for the mission house had been intended for a pleasure ground. Open confrontation ensued, and in the end Ricci accepted defeat. He agreed to transfer his building outside the pagoda limits, and in compensation received a quantity of building materials from the committee, who gave it so as not to appear to have vetoed the governor's scheme.

With remarkable speed the mission house was completed. Great crowds arrived to inspect it—a curious example of Western-style building—and to gaze at the wonderful glass prism. The picture of the Virgin that hung over the altar, however, proved to be an unfortunate choice, since the Chinese at once decided the Jesuits worshiped a female god. In China's male-centered society this was inconceivable, and Mary had to be demoted in favor of a painting of Christ the Saviour. Adroitly styled by Ricci "Lord of Heaven," Christ met with more general approval.[6] Already, in this re-

6. "Lord of Heaven" is *t'ien-chu* in Chinese, a term used throughout the Chinese Classics. Ricci also allowed the term *shang-ti*, "supreme ruler," for God. The use of these terms came under fire later during the Rites Controversy of the seventeenth and eighteenth centuries on the grounds that, to Chinese ears, the terms simply did not mean the God of Christianity.

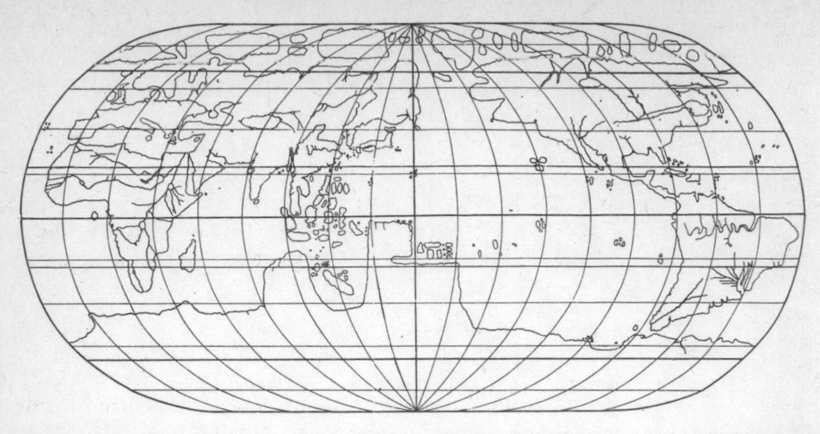

OUTLINE OF RICCI'S WORLD MAP. This outline of Ricci's great map of 1602 (see photograph on page 192–93) is doubtless similar to the earlier map Ricci made while at Chao-ch'ing. Although China is near the center of the projection, it does not dwarf the rest of the world.

naming, an adaptation had been made to suit the Chinese. The process of cultural accommodation was beginning.

One of the most significant actions of Ricci in Chao-ch'ing was the making of a world map. He describes what happened in his own words: "The Fathers had hung up in their hall a map of the whole world. . . . When the Chinese understood what it was, never having seen or imagined such a thing before, all the more serious-minded of them wanted to see it printed with Chinese characters, so as to understand its contents better. So the Father [Ricci refers to himself in the third person] who knew something of mathematics, having been a disciple of Clavius when in Rome, set about the task, helped by one of the literati, a friend of his: and before long they had made a map of the world bigger than the one in the house. . . . It was the best and most useful work that could be done at that time to dispose the Chinese to give credit to the things of the Faith. For up to then the Chinese had printed many maps of the world with titles such as 'Description of the Whole World' in which China was all, occupying the field with its fifteen provinces; and round the edge they depicted a little sea where a few islets were dotted about, on which they wrote the names of all the kingdoms of which they had ever heard; and these [kingdoms] all put together would not have equalled the size of one of the provinces of China. . . . When they saw the world so large [on Ricci's map] and China in a corner of it . . . the more ignorant began to make fun of such a description, but the more intelligent, seeing such an orderly arrangement of parallel lines of latitude and longitude . . . could not resist believing the whole thing true.... It was printed again and again and all China was flooded with copies."

But in fact the printed copies appear to have shown China, in something like its true proportion indeed, occupying the middle of the projection chosen, an example of Ricci's consummate tact. We must believe him when he tells us what an eye-opener this map of his was. Although it is true that the Chinese maps in popular currency at the time were as he described them, the Chinese possessed others of a much more accurate kind. Chinese cartography was not so primitive as Ricci thought. A tactful gesture placed China in the center of the land masses, but his real contributions to Chinese knowledge in this field were the demonstration that the world was round and that the great area of China occupied quite a small part of the world's surface, with other countries bulking much larger than the Chinese had seriously thought. When, later at Chao-ch'ing, the astute father built world globes for the Chinese, those points were even more clear to them, and the local educated classes seem to have accepted the explanations given.

Ricci and his successors in China, however, never seem to have appreciated the real achievements of indigenous Chinese science. This is hardly surprising since the Chinese themselves, delighting in the traditional mists of their own literary elegance and conditioned to a nonscientific attitude, had forgotten much of their own discoveries and successes in the domain of science. It was for this reason, and because the era of Chinese voyaging, already referred to, was long past, that the current Chinese maps placed the country in the position which it *ought* to occupy—the center of the known world. This placing accorded with that concept of exclusivity which said (and had said for a long time) that China was the only civilized country in the world—the Middle Kingdom in name and in fact. The detailed maps and sailing instructions, with their descriptive notes, that had resulted (for example) from the long voyages of the Chinese admiral Cheng Ho a couple of centuries before the time of Ricci were conveniently forgotten, if indeed the literati had ever known about them.

In Nanking some years later, Ricci was shown an observatory whose instruments filled him with the utmost astonishment, for they were finer than any in the Europe he had recently left. The instruments had been made in China in the Yuan (Mongol) dynasty. The Chinese had forgotten how to use them—so completely forgotten, in fact, that they had moved some of the instruments from another town and failed to regulate the calibrations for the latitude of the new site. But still, while admiring these delicately and scientifically made tools for research in astronomy, Ricci does not appear to have tumbled to the fact that the Chinese must have known something of science ever to have made or used them. Perhaps we cannot unduly blame him. We ourselves have had to wait until the middle of the twentieth

禹跡圖
每方折地百里
禹貢山川名
古今州郡名
古今山水地名
阜昌七年四月刻石

◀ A CHINESE MAP OF CHINA, made before 1100, showing with great accuracy the courses of the Yellow and Yangtze rivers. The scale of the grid is one hundred *li* to each division. The original, incised on stone, is in the Lei Lin Museum, Sian. This map has been described by Needham as "the most remarkable cartographic work of its age in any culture." (From *Bulletin de l'Ecole Francaise de l'Extrême Orient,* III, 1903)

century for a full appreciation of the remarkable history of Chinese science.[7]

While he stayed at Chao-ch'ing, Ricci came to be regarded among the Chinese scholars of a wide region as a remarkable genius. The assessment was near enough correct, but the reasons they put forward were not. Ricci wrote later to a friend on this topic: "Once I was with some literati at a party, and I made them write a great number of letters [he means Chinese characters], and reading them once through I recited them to the assembly from beginning to end by memory. They were all so dumbfounded that it got about that I could remember a book by heart on reading it once, and had therefore no need to read it again." As to his mathematical powers, there was "no one in the world my equal, so it was said. Indeed if China were the whole world undoubtedly I could call myself the greatest mathematician and scientist. . . . It is amazing how little they know, for they are concerned entirely with moralizing and the elegance with which they . . . write it. . . . They think me a monstrosity of learning and that nothing like me has ever left our shores. All this makes me laugh. . . ."

In fact it did more than make Ricci laugh. Amused though he was at this adulation of his powers, he had already realized that in just those powers he held his trump card. He used it with determination and a flexible skill to further the cause of Christianity through the devious paths of applied science. This was in complete accord with the Jesuit plan—the plan of the far-sighted Valignano, amplified and applied over the next twenty years by Ricci.

This stategy was the crux of the Jesuit approach to the East and it was well adapted to Chinese attitudes. Expound and exhibit the marvels and attainments of Western science and technology, it went, make that sector of Western achievement respected with the ruling classes. Then, like the pill in the spoonful of jam, slip in morsels of Christian faith, and they will respect these too. Where Dominican fire and Portuguese swords had signally failed to do anything but arouse contempt and expulsion, Jesuit science and Jesuit casuistry (if we may use the word) were to be the weapons of Ricci, wielded, he deeply believed, for his God.

It is both tempting and perhaps justifiable to look with a certain skepti-

7. Needham is the best guide on the subject. It was almost a hundred years before the Jesuits realized how useful the records made in the past by Chinese of celestial observations could be to them, and how accurate these were in several respects.

cism now on the methods of this long Jesuit endeavor. The means were not ignoble in themselves, although slippery, and the aim (for Christians of the time) unimpeachable. We need not argue the point, perhaps, since we are dealing with a subject that for most of us today has lost its sharp bite. Few now believe that Christianity is likely ever to sweep the civilization of China with its Western flame, and few would now crave that it do so. The Jesuits, however, did; and so did many others of different sects who followed them. The mental and spiritual contortions they performed with more or less straight faces in order to make acceptable their various forms of Christian belief range from cunning to pious self-deception. Ricci, the first of the Jesuits of significance in China, was cunning but extremely scrupulous.

The odd thing was that at first the Chinese merely suspected the Jesuits of being spies for Portuguese or other foreign armies who would later invade. And the reason for this probably lies in the fact that educated Chinese hardly ever professed any religion. Religions—Buddhism and Taoism—were largely discredited among the literati, who despised their priests. Those priests were by now for the most part degenerate tricksters preying on the credulity and superstitions of the illiterate. So it was hard for a literate Chinese to credit that a man as learned as Ricci obviously was could at the same time believe in any religion. Religious systems, from the Chinese point of view, were fitted only for the unschooled masses.

The mission at Chao-ch'ing prospered only moderately. Ricci wrote in a letter: "As to what you ask, hoping to hear news from China of some great conversion, I tell you that I and all the others who are here, dream of nothing else night and day; and for this purpose we are dressed and shod in Chinese fashion, we neither speak nor eat nor drink, nor live in our house except in the Chinese manner. . . . The time in which we live in China is not one of harvest."

Converts at Chao-ch'ing were naturally few since much jam and few pills were the order of the day. Slowly Ricci and Ruggieri were learning some of the ways of Chinese life. But they felt extremely insecure most of the time. It was obvious that they had no sure tenure of their place in South China and that, despite the admiration of all for their novelties (which included, from the Chinese point of view, the glass prism and the pictures of Christ and the Virgin as equal attractions), they were themselves by way of being little more than free entertainment for the populace at large. If they were religious as well as masters of scholarly things, then, the local literati thought, it must simply be put down the fact that they were foreigners—all foreigners were strange, one way or the other. Not for a moment, yet, did the Chinese attempt to consider that the two Jesuits might be representatives of another *civilization,* one which was different but perhaps equal and coexistent with their own. There was still a long way to go before Ricci and his companions could even suggest such an idea.

Meanwhile other wonders from the West were arriving from Macao. "Many were drawn by the big clock [made for the governor at his request but later returned by him to the mission when he could find no one able to regulate it for him] . . . others by the various mathematical instruments. . . . The books also made them all marvel on account of their different bindings with much gold and other ornamentation, besides the books on geography and architecture in which they could see so many countries and provinces all over the earth, the beautiful and celebrated cities of Europe and elsewhere, the great buildings, palaces, towers, theatres, bridges and churches. Later, musical instruments [which were to be one of the stock attractions proffered by the Christians in China] arrived, which were much to their taste. . . . On such occasions [Ricci] began to speak of our holy faith. Consequently the house was full all day of grave personages and the street was full of their litters, the river bank in front of our house full of boats belonging to the mandarins."

But money was running short. Ruggieri returned to Macao to raise funds, while Ricci went on with his study of Chinese, using grammars and dictionaries constructed with enormous labor by himself and his companion. Very soon he was remarkably proficient. Part of the difficulty of learning Chinese (even for the Chinese themselves) consists in memorizing very large numbers of characters. Only by virtue of an astonishing memory could Ricci have acquired Chinese so rapidly.

Reading some of the fundamental classics of Chinese literature, he began to have some understanding of the Confucian basis of thought and outlook in China. And he began to search for a means of welding that system to the beliefs of Christianity. He began, too, to search his own conscience. How much, he asked himself, of the basic Chinese beliefs and ceremonies was compatible with Christianity? How little would he have to proscribe in converting a Chinese? The answers to these queries lay ahead. The whole future of the Jesuit mission in China, had Ricci known it, was to revolve around those questions and the varying answers to them.

The remarkable thing about his stay in Chao-ch'ing was that Ricci became to some extent *persona grata* with the intelligentsia. This was doubly remarkable in that their minds were all but closed to serious consideration of "barbarian" non-Chinese; and in that the Jesuits were clad in the robes of the despised bonzes, of whose lack of intellectual attainments no one had any doubt. These literati never showed the slightest inkling that they could bend their own Confucian philosophy a little to accommodate the elements of Christian dogma as tentatively expounded by Ricci. Even the terms which Ricci invented for God, and for other concepts in Christian religion, were in themselves a demonstration of Ricci's conclusion that, to be considered by the Chinese, Christianity must adapt itself to Confucius. There must have been moments when Ricci wondered who was being converted to what.

Complications of a much less esoteric nature soon arose at Chao-ch'ing. However much respect Ricci might command among the educated few, during the prolonged absence of Ruggieri the attitude of the ordinary townspeople gradually turned to one of hostility toward the foreigners. The people of Kwangtung province were perhaps the most antiforeign of all Chinese—for one simple reason. In the centuries before Ricci's advent their coasts and their ships had been peculiarly liable to the attacks of all kinds of pirates. Raids that wiped out coastal villages were commonplace. Their coasting junks were frequently harassed (not least by Portuguese) and intercepted, their cargoes transferred to pirate boats, and their crews slaughtered. Moreover, with the installation of the Portuguese at Macao (with the privilege of

A COMPARISON OF THE JESUIT AND CHINESE CURRICULUMS

Jesuit Roman College	*Required Reading for the Chinese Civil Service Examinations*
First year:	The Five Classics (traditionally supposed to have been written or compiled by Confucius):
Theology and other Catholic subjects	Book of History
Euclid	Book of Rites
Arithmetic	Book of Changes
Ptolemaic Astronomy	Spring and Autumn Annals
	Book of Songs
Second year:	
Theology and other Catholic subjects	The Four Commentaries (containing the opinions of Confucius):
Theory of Music	The Analects
Theory of Optics	The Great Learning
	The Doctrine of the Mean
Third year:	The Book of Mencius
Theology and other Catholic subjects	
Advanced Astronomy	Official Dynastic Histories
Advanced Mathematics	Ancient Book on Arithmetic

Chinese candidates for the examinations were required to show ability to write essays on anything contained in the books listed and had to have memorized the texts of most of them. With the exception of the book on arithmetic, none of the books contain anything of a directly scientific nature, although several are of great importance in reconstructing today Chinese scientific thought in early times. The arithmetic book dated from before 200 B.C. and remained unaltered, nor was any candidate ever failed in this subject.

Note that Matteo Ricci not only graduated with distinction from the Jesuit Roman College, but also came to know the Chinese Classics as well as most Chinese scholars, thus performing the prodigious feat of mastering both curriculums.

trading at Canton once a year) Chinese transport fleets had lost much trade. Foreigners of all kinds had a bad name. They stole your goods on the high seas, they raided your villages, raping or carrying off into slavery your women, they burned your houses and then made off to sea. There is no doubt also that Portuguese trade around South China had begun to alter the economy of the region for the worse, to affect the people adversely in terms of hard cash.

So the people of Chao-ch'ing, quite wrongly but understandably, began to murmur that the mission was nothing but a fortress built under false pretenses and later to be used by invading Portuguese whom the long-absent Ruggieri had gone to summon and escort to their town. Youths began to climb the nearby pagoda and pelt the mission at night with big stones. Exasperated at this, an Indian servant belonging to the mission

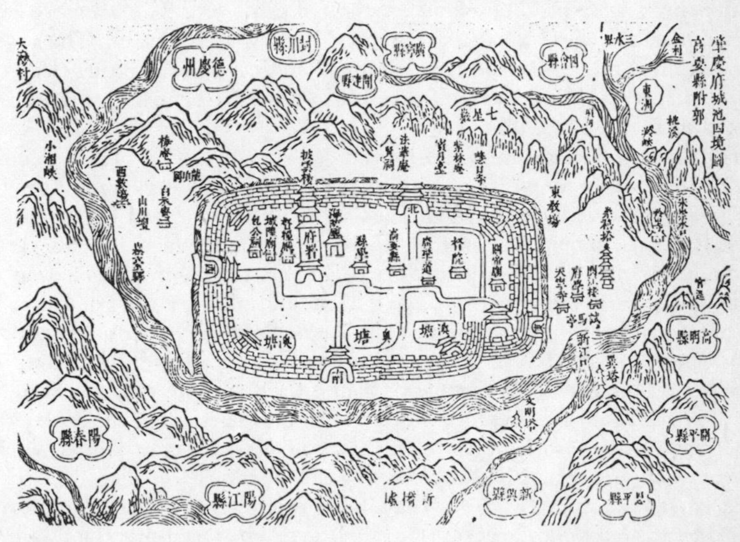

Plan of Chao-ch'ing. Taken from the annals of the town for the year 1673, this plan shows the city as it was long after Ricci's departure, and the surroundings of his mission have altered somewhat. Outside the walled town with its river on three sides, we see mountains and villages, the names of the latter enclosed in lozenges. The largest building within the walls is the office of the governor. Compass directions are the same as in Western maps. Outside the eastern gate, between the walls and the river, is a group of four buildings; beginning from the wall they are: the Temple of Heavenly Peace, the Government School, the Tower for Viewing the River, and the Pagoda for Venerating the Gods. Ricci's mission was on the site of the Government School, and the showers of stones were probably thrown from the roof of the Temple of Heavenly Peace. (From d'Elia's *Fonti Ricciane,* retouched)

caught one of the stone-throwers and brought him inside. But the youth was soon released on Ricci's advice. Other people took the incident up, and Ricci was accused of detaining and drugging the young man so as to send him to slavery in Macao. And the youth himself decided to make his reputation as a hero on whose account the *yang kuei tzu,* "ocean devils," had been run out of town. With wild hair and mad cries, he ran hither and thither inflaming the townspeople against the Jesuits.

The case was heard by the governor before a crowd of partisan citizens. The false witnesses were harkened to and the governor was within an ace of pronouncing sentence. But at this juncture the Indian servant, who acted as Ricci's interpreter, suddenly let go the ends of his long sleeves. A cascade of stones (those thrown at the mission) tumbled noisily to the floor of the governor's hall—a well-timed gesture whose effect impelled the governor to further investigations. The whole story then came out. The accusers were routed, the false witnesses slunk away, and Ricci was vindicated. The governor furnished the mission with a document stating that the fathers had the permission of the viceroy himself to reside in peace in the town; and this was posted on the doors for all to read.

The governor acted justly, but his verdict was unpopular. The xenophobia of the Chinese, of which this is an early instance, has generally proved in the end to have been justified. It could have been proved by Ricci that he had no evil intent, but in fact his intention to convert the Chinese to Christianity in itself constituted a threat to the settled traditional way of Chinese life as it was lived in Chao-ch'ing and all over the country. We cannot credit the illiterate populace of Chao-ch'ing with the understanding of these exact terms, but in their intuitive way they knew very well that the presence of foreigners, whose actions round the coasts were blatant piracy, was a poor augury for present and future contacts between themselves and the foreign bonzes from Macao. The story of Westerners in China, which we are examining in this book, tends to bear them out.

Ricci was one, and powerless; the Chinese were many. A combination of events finally led to his eviction from the town. The viceroy of the province was replaced by another who—perhaps because he had recently been in charge of suppressing pirates—was much less favorably disposed toward the Jesuits. Ricci was accused of knowing how to transmute cinnabar into silver and of refusing to divulge the information. Since the currency of China was silver, this caused deep anger. Then, taking advantage of a flood as the river broke its banks, the townspeople entered the mission on the pretext of collecting wood to repair the breach. Once inside, they partially destroyed the establishment. And the governor, at last sensing the tide of opinion against him, grew more and more cold and distant. Ricci and Almeida, a priest who had joined him after Ruggieri left, were eventually expelled from Chao-ch'ing. Ricci was offered sixty gold pieces in compensation for the property, which he refused. With his usual forethought, he obtained a document stating he had not been sent away for any crime, little knowing how useless such papers were when presented to some other official at a distance in order to vindicate or to prove his rights.

Thus, seven years after he came to Chao-ch'ing he left it and sailed reluctantly under military guard downstream to Canton. He was in great distress of mind at this development, which he saw as a failure; and the sight of a boat drawing alongside the craft on which he was living did little to relieve him. For this boat came from the viceroy who had recently expelled him, and whose envoys now again offered him the gold pieces. Again he refused. At once he was taken back to Chao-ch'ing. To refuse what a viceroy offered was a punishable offense, and Ricci had every reason for believing he might lose his head. But in fact, the viceroy, for political reasons, did not wish to take the mission ground without payment, since he had plans of his own for it. When he and the Jesuit met, an angry scene broke out. Ricci conducted his case coolly, and to such effect that he eventually persuaded the official to send him elsewhere in order to establish another mission. Then he accepted the gold, and he was also given (a subtle viceregal warning, perhaps) a volume recounting the official's success in suppressing the pirates off Hainan Island. He was packed off to Shao-chou, a town more than one hundred miles up the Pearl River from Canton.

A lesser man than Ricci, one less convinced of the spiritual absolutes of Christianity, less courageous, would have gone back to Macao defeated. It was a bitter thing for him to see seven years of incredibly difficult labor in Chao-ch'ing lost in an instant, and to contemplate starting all over again in an unknown, probably hostile place, to begin to rebuild even the foundations of respect—far less to begin to save souls for his God. But Matteo Ricci was in no single respect less than a great man. He set off once more, with the frail Almeida, and began again.

In a brief chapter on Matteo Ricci it would be foolish to attempt to detail all the events which occurred in the long haul before he reached Peking, the ultimate goal of all who go to China. At least one learned Jesuit scholar, Pasquale M. d'Elia, has spent a lifetime elucidating the subject, and many others have contributed to a vast documentation. We are more concerned with the broad pattern of events, of struggle and study, of expulsion and new endeavor, which had been established at Chao-ch'ing and was to be repeated elsewhere; and with their significance in the context of Chinese-Western understanding. During the years that followed, Ricci was slowly accumulating the experience in many fields which was to be his mainstay in Peking. Nothing but determination matched by intelligence—his most remarkable qualities—could have succeeded, as they did in the end.

Ricci and Almeida went slowly upriver to Shao-chou, hiding their un-Chinese faces and still clad in the robes of bonzes. At Shao-chou, a city twice the size of Chao-ch'ing, the authorities installed them in the pagoda of a local temple. It seemed to be the right place for priests. At once Ricci set about removing the "idols" and replacing them with his own Christian images. (The irony of simply substituting Christian images for Buddhist ones can hardly fail to strike the reader today, but was not apparent to Ricci.) Protracted negotiations eventually got the Jesuits a field near the pagoda, and largely with their own hands they built a house and church—this one in Chinese style for fear of being accused of building another Portuguese fortress. In a year it was done. For another six years it was to be their home, and the site of exhausting work. Despite the natural beauty of the

Ricci's Route to Peking. Names in parentheses indicate former or later designations of places. Otherwise the map shows China as it was at the time of Ricci. With the exceptions of a few small sections of the southeast coastline, the whole of this vast area was completely unknown to Europeans.

country around, the situation of the city seemed to Ricci and Almeida much less salubrious than that of Chao-ch'ing. Clouds of mosquitoes made malaria endemic, and both the fathers were gravely ill from it. Characteristically, Ricci recovered from his first fever when he heard they had been granted the land. Almeida recovered, only to fall ill again. Soon he died, despite the administration of the best drugs in the Chinese pharmacopoeia and the the prayers of Ricci. Once again, nearing forty by now, Ricci was alone. Ruggieri never returned to China, having been sent to Rome to ask for a papal ambassador to the emperor—a request which came to nothing.

The mission at Shao-chou had made only seventy-five converts by the time Ricci left it in 1595. Even counting fifteen at Chao-ch'ing, whose lapsed state he saw for himself on a visit, he had less than one hundred converts to show for twelve years of labor in the reluctant fields of China.

But at Shao-chou matters other than conversions were as important in the long run. There Ricci had time for study (after being joined by a fellow Italian, Father Cattaneo, in 1593). His Chinese improved and by now he had mastered about five thousand characters, enough to enable him to read with more fluency. Delving into Chinese literature, he discovered "in the Canonical books [the Five Classics] many passages which are favorable to things of the faith, such as the unity of God, the immortality of the soul, the glory of the blessed." This had the force of a revelation for Ricci. Now, he argued, he had a respectable and ancient Chinese basis to use in discussion with educated men on the subject of Christian faith. Whatever their beliefs, all educated Chinese were imbued from childhood with profound respect for the Classics, much as nineteenth-century education in England inculcated respect for ancient Greek and Latin writers. In China the tradition was stronger than that in Europe, partly because the Chinese Classics were written in the characters familiar to the educated and not in a foreign language, and partly because the content of those Classics was the foundation on which state, society, and personal conduct had been based for nearly two thousand years. The strength of Chinese tradition was both the strength and the weakness of China; it supported and moderated society and government, but at the same time tended to discourage evolution within those areas.

We must pause here a moment to accord Ricci his due. In a very precise sense his was the first sortie by a European in the field of Chinese scholarship. His was the first translation of any Chinese classical text, and he was the first Westerner to master the Chinese language. Not only to master it, but to become a scholar whose only equals were the Chinese literati themselves. This last achievement has seldom been equaled by a European since then, and never surpassed. And Ricci had to make his own vocabularies, to disentangle absolutely single-handed the gentle bafflements, the maddening allusions of Chinese language, in which so often meaning hangs on knowledge of the use of the character by some ancient author, much more than on its precise modern significance. The achievement of translating the Five Classics strikes one now as the work of a lifetime; but in fact it was but a part-time occupation for Ricci.

The Chinese sage K'ung Fu-tzu was unknown in the West before Ricci translated his works and gave him the name that has become a household word—Confucius. We may wonder if, as Ricci studied the works of Master K'ung, he pondered a moment the last recorded words of that ancient sage:

> The great mountain must crumble;
> The strong beam breaks;
> The wise man must wither away like a plant.

The mountain must have seemed to him like the Chinese people; the beam, their obdurate clinging to tradition; and the wise man, himself.

Meanwhile he was trying to fit Christianity into the gap left by Confucianism in the realm of the supernatural. It is difficult in a word or two to define the teaching of Confucius. But perhaps its essence may be called rationalist and humanist. In the works of the sage, and those attributed to him, there is none of the mysticism of other ancient teachers such as Buddha and Mohammed. His outlook on life, on people and the way of ordering

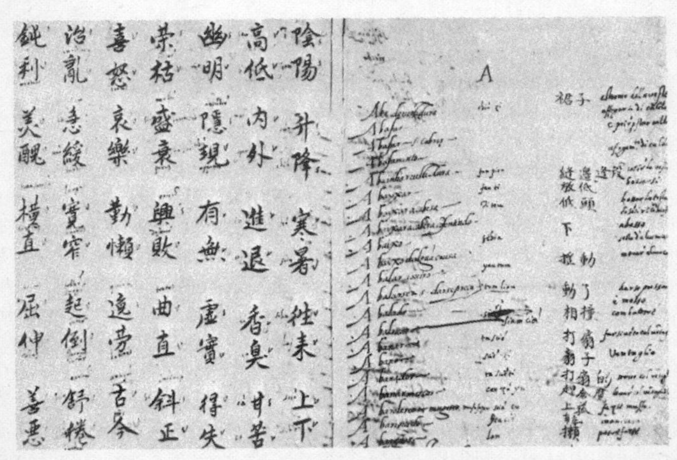

RICCI'S CHINESE DICTIONARY. The left-hand page consists of pairs of Chinese characters of opposite meaning—"blunt" and "sharp," "beautiful" and "ugly," etc. The right-hand page is the first of the dictionary proper. The first column contains Italian equivalents. The fine script of the Portuguese was written by Ricci himself, while the Italian is in the hand of his friend Ruggieri. But Ruggieri's translation ceases after ten pages or so. (Istituto Storico della Compagnia di Gesù, Rome)

affairs, is one of reason and compromise. Holding to the ancient Chinese beliefs, he was convinced that the laws of Heaven and of the other world were not susceptible to man's explanation, and better left alone. His doctrine, which was to capture and hold the imagination and affection of the Chinese for twenty-five centuries, to serve as the foundation of their society and government, was one of love and respect for one's fellow men and of obedience to lawfully constituted authority. Coupled with this went respect for ancient precedent, and the abandonment of force for the sweetness of reasoned discourse in the context of ancient examples.

The idea of divine presence in the souls of men was entirely absent in this structure; and it was into this lacuna in Chinese thinking that Ricci saw he might insert the tenets of Christian faith. Logically, he had every justification. But in practice he was to find it incredibly hard to get any Chinese to follow suit.

During his work in Shao-chou, Ricci met a man called Ch'u T'ai-su who came of a rich family but who had squandered his inheritance and was now in search of the philosopher's stone that would turn cinnabar to silver, a substance he desperately needed. The Jesuit reputation in that alchemical chimera occasioned their meeting. With his immense patience Ricci soon weaned the younger man's brilliant mind from such fancies and filled it instead with the mathematics, geometry, and astronomy he had learned in Rome. As this new sun dawned on his traditional world, Ch'u's enthusiasm was akin to, but more acute than, that of a Western youth when the order and logic of science first grip his budding mind. Setting aside his alchemy and astrology, Ch'u embraced Western science with nothing short of rapture.

It would seem that the Chinese are like any other people, some having a mathematical and some a literary bent; but in the cage of traditional society only literary activity was respectable or profitable. Ch'u proceeded to write elegant treatises in Chinese on the captivating knowledge he had just acquired, and his work in Chinese ought by rights to have been the point of departure from which China learned Western science and the Western scientific discipline. We cannot blame Ch'u that no such thing occurred: we must blame Ricci. For, apart from being a scientist, he was fundamentally a priest. The time came when Ch'u T'ai-su (who had already heard much of Christianity) stopped his science classes "and wished to speak of this subject, and he did so very much to the point. For in order to do so to the best advantage he had made a book of the difficulties he had found in the Catholic faith in order to put underneath the answer and solution the Fathers would give to each." Ch'u was soon ready to be received into the Church. There was just one impediment. His wife was dead and he had a concubine "of low birth," whom he refused to marry. Worse, he had no son by his wife or by his concubine, and it looked as though he would have to take other concubines until one of them bore him an heir to worship at his shrine when he died. For such was the venerable and inescapable necessity of Chinese custom, hallowed by Confucian texts. It was some years before even Ricci could overcome these problems.

When, eventually, Ch'u married his concubine and was baptized, there occurred an event as shocking to us of the present as it was pleasing to Ricci then. Ch'u T'ai-su, the learned Chinese, sent his whole collection of rare books and manuscripts dealing with every aspect of geomancy and other related subjects "to our house to be burned, [and] certain plates . . . which they use for printing books, together with three or four porters' loads of books of the doctrines of the schools . . . part manuscript awaiting printing, all of great value." How much was lost in this merry little Christian fire to the historians of Chinese science, we shall never know. "How tragic it was," Needham remarks in this context, referring to the origins of the Chinese compass, "and how paradoxical in view of the exceptionally learned nature of the Jesuit mission, that the ideal of 'holy ignorance' should thus have closed, perhaps for ever, some of the doors of knowledge on the origins of one of the greatest of all Chinese contributions to science." The same thing was to happen again in Peking when, at Ricci's behest, Li Ying-shih, a distinguished scholar who "had rather a good library," took three days (in Ricci's own words) "to purge it of books on subjects prohibited by our [Catholic] laws. . . ." Again the loss to the story of Chinese genius in the sciences and protosciences was as irreparable as it is incalculable. "It will not do to say that the Jesuits were combatting superstition in the capacity of rationalists," Needham observes. "They did not disbelieve in the geomantic art: they considered it diabolical."

Ricci's thoughts while in Shao-chou turned increasingly on how to reach Peking and the emperor, the center of power. He realized, correctly, that only *there* could Christianity smell the flowers of success in China. What seemed an excellent opportunity finally presented itself in the person of a high-ranking official of the regional military commission who had been summoned from his home near Shao-chou to attend to affairs in Peking. Ricci went with him as far as Nanking, but there his protector left him, for even that exalted official feared to introduce the Jesuit foreigner into Peking. His reasons were the same as those for which Ricci was shortly to be expelled from Nanking—the war between Japan, whose armies had invaded the Chinese tributary state of Korea, and the Peking government. Nanking (the Southern Capital) had formerly been the Ming capital and still retained the privileges of a royal city even though the center of power had been moved long ago by the emperor Yung-lo to Peking. No one at Nanking would countenance Ricci's stay there, and he retired discomfited to Nan-ch'ang, more than a hundred miles south. It was now midsummer of 1595. He had already been twelve years in China.

But even in retreat one significant advance had been made. On the journey north Ricci and Cattaneo, the priest who had replaced Almeida, had put off the habit of bonzes and, growing their hair and beards, had donned the robes of scholars or graduates. The change was as dramatic in its social effect as in the difference it made in their appearance. No longer were they associated in the Chinese mind with the cringing, illiterate priests despised by educated society. Suddenly Ricci found himself received on terms of equality. No more falling on the knees in great men's presence, no more prostration before officials. In his new gown, ankle-length, purple silk with generous sleeves, its neck bordered with light blue, and wearing the tall black hat that was not unlike a Christian bishop's miter, Ricci felt a new being. This was the dress of the Chinese leisured, and therefore educated, class (even the embroidered silk slippers showed that the wearer did not go far without his palanquin), not dissimilar in style and luxury to that of church dignitaries in Rome.

With the robes went the appurtenances of the class—the visiting books of fine white paper in red covers, inscribed with the characters Li Ma-tou (Li for Ricci, there being no *r* sound in Chinese, and Ma-tou standing for Matteo). When he paid a call the book was handed to the gatekeeper of the house by Ricci's white-clad servant, and he was then admitted and welcomed by the host. Host and guest played the game of attempting to make each other enter first, the guest always losing in the end. Inside, the host offered a chair, placing it to the north and flipping nonexistent dust from the seat with his sleeve; the guest did likewise for his host, placing the chair to the south. Tea was brought, and some expensive little delicacy to eat. Polite sentences were exchanged, each with its ordained and proper response. Only after that might the real subject be broached. And, the visit over, the guest rose and bowed, hands clasped in his sleeves. He then resisted the attempts of the host to go with him beyond the threshold of the house. As the guest climbed into his conveyance, the host, who had withdrawn for the instant, again appeared in the doorway to bid farewell. The elaborate ritual was not yet done, for no sooner had the guest reached his own home than a servant came from his host to bid him a final goodbye; and the guest was required by civility to send back one of his own servants with his compliments. Ricci accepted and performed those social rituals for the rest of his life. But his final judgment on them is revealing: "This desert of gentility," he observed.

He was still far from Peking. But at least his name and his amazing intellectual capacities, his sincerity and his honesty, were beginning to be known to many powerful men. The story of his travels before he settled in the capital, however, was still to be crossed, like love, by the hesitations, doubts, and vacillations of the Chinese with whom he had to deal. There was always a reason why they could not altogether accept him. Instinctive Chinese

MATTEO RICCI wearing the gown and cap of a Chinese scholar. This portrait was painted in Peking in 1610, shortly before Ricci died, by Yu Wen-hui, a Chinese Christian baptized Emmanuel Pereira. After Ricci's death the painting was taken to Rome, where it still hangs with those of the founder of the Jesuit order and Francis Xavier in the Jesuit House. The picture's legend is mistaken as to Ricci's age. (Istituto Storico della Compagnia di Gesù, Rome; photo by Brian Brake)

isolationism, instinctive belief in their possession of the sole valid culture, instinctive distrust of foreign influences creeping in with the gifted stranger and threatening the immaculate tower of their equanimity and traditional ways—these were the obstructions that rose again and again in the Jesuit's path.

Ricci went on doggedly entrenching, converting where he could, making friends and influencing people. In Nan-ch'ang, in the company of great scholars, great hereditary princes who were dilettantes and often also debauchees, he dazzled the circle. One evening he recited a whole book of Chinese poetry after reading it once; and to refute the feeling around the table that he already knew the book, he repeated his old parlor trick of having those present write down a random list of unrelated Chinese characters—this time five hundred of them—and was able to repeat the list from memory after a single reading. Not content with this prodigious feat, his audience asked for more—and Ricci recited the same list of gibberish backwards with equal accuracy.

Out of this unexampled capacity to memorize, Ricci continued to make capital. Such a feat as he had just performed impressed the Chinese in a particularly important way since so much of their education required extensive memory of the classical books and commentaries, and the mental acrobatics of committing to memory the many thousands of characters essential in order to read these books in the first place. Coveting his attainments, they were deeply impressed—not least because here was an "ocean devil" capable (whoever would have thought it possible?) of meeting them on scholarly ground, among the hallowed thickets of their own literature and tradition. It was not long before a trip to Peking was arranged. And thither Ricci sailed, with ardent hopes, up the Grand Canal in the fall of 1598.

But, subtle and learned as he now was in the ways of China, Ricci still did not find a way through the miasma of officialdom and official perfidy which was the climate of affairs in those closing decades of the degenerating Ming. En route he all but lost his liberty, such as it was, at the hands of a powerful eunuch; and when at last he reached the capital all efforts at audience with the Wan-li emperor were effectively blocked by the jealousies and suspicions of a legion of functionaries. After a brief stay, perceiving he was getting nowhere, he abandoned the attempt and returned south, downcast, to Nanking.

This time he managed to remain there. And in Nanking, city of past splendor, home of many a sophisticated man of letters, he established another mission, which in time prospered well. Many converts were made. More important for the future, numbers of the city's most influential men were attracted to the philosophical and even to the spiritual content of Christianity, which Ricci never ceased to expound in varying watered down forms. Most of his upper-class converts were caught on the hook of Ricci's science and painlessly beached in a sort of trance on the shores of Christianity's more rational areas. Ricci's writings in Chinese at this time included a little work on friendship in the form of a dialogue between himself and a Chinese scholar, and this became a minor classic, much reprinted by the Chinese themselves. It was the first original writing in Chinese by a European ever to be printed in China for public sale. The outlook seemed bright for the mission and its lay and priestly brethren who from time to time were sent from Macao to join Ricci at Nanking.

But through the writings of Ricci that deal with this period there is woven an insistent thread of discontent. More than ever was he certain that the final answer for Christianity in China lay in Peking. The setback of his recent trip there had done nothing but sharpen his desire and strengthen his will to return, to grapple with what he saw as the central problem of his work.

With the death of the Japanese shogun Hideyoshi in 1598 the catastrophic war between Japan and China in Korea ceased. But the Chinese had not won it in any real sense. Despite hundreds of thousands of men poured in, millions of ounces of silver expended, their armies had been no match for the Japanese, and only the demise of the shogun saved the day, leaving China with a depleted treasury, and Korea with wounds from which she never fully recovered. Now it was the turn of the eunuchs to take advantage of the situation. Thousands of them, wallowing in corruption and in unlicensed debauchery, roamed the provinces of China, with and without imperial sanction, extorting money on any pretext they could invent. Little of the proceeds reached the vaults of Peking. The power in China was by now only nominally in the emperor's hands and only sometimes in the hands of the great ministers of state. Eunuchs ruled, everywhere. Wan-li, who had ascended the throne twenty-six years previously at the age of nine, had been tutored exclusively by eunuchs. Under their malign influence he remained, and in the last twenty years of his reign he gave up any pretense of governing through proper official channels, allowing all the business of state to be dealt with by the eunuchs and by his favorite concubine.

Ricci knew something of all this, but naturally he did not know how close the Ming dynasty was to its end. On his second venture to Peking he was to have a further taste of eunuch strength.

By the first year of the seventeenth century he had gathered together the credentials and the indispensable presents for the emperor in the shape of Western objects unfamiliar to Chinese; and he had arranged what seemed safe transportation for himself and the complete furnishing of a mission station from Nanking to Peking. On May 18, 1600, he set off on the now familiar route, northward on the Grand Canal. "As far as Nanking [the Yangtze] flows north, then turns somewhat south and flows rapidly toward the sea, forty miles beyond Nanking. In order to go by water from Nanking to the royal city of Peking, the Chinese emperors had a long canal constructed from this river to another, called the Yellow River because of the color its turbulent waters. . . . This Yellow River has no respect at all for Chinese law and order. It comes from a barbarous region and, as it were, seeking vengeance for the hatred the Chinese have for outsiders, it frequently ravages whole districts of the realm when it fills up with sand and changes its course at will." Ricci notes the propitiatory rites performed by boatmen on the river, comments correctly on its content of one-third silt. "Entrance to the city of Peking . . . is made by the canal constructed for boats bringing cargoes. . . . They say there are ten thousand boats engaged on this commerce." Private boats are not allowed on the canal "to prevent the multitude of boats from clogging the traffic and cargoes destined for the royal city from being spoiled. And yet so vast is the number of boats that frequently many days are lost in transit." Then, shrewdly, correctly, he says: "All this may seem rather strange to Europeans who may judge from maps that one could take a shorter and less expensive route to Peking by sea. This may be true enough, but the fear of the sea and the pirates who infest the seacoast has so penetrated the Chinese mind that they believe the sea route would be far more hazardous. . . .

"Besides the cities, there are along the banks so many towns, villages and scattered homes that one might say the entire route is inhabited. Nowhere . . . is there any lack of provisions, such as rice, wheat, meat, fish, fruit, vegetables, and wine. . . . Through the canal . . . they bring great quantities of wood for royal buildings; beams, columns . . . especially after a royal palace has been burned down. . . . Each year the southern provinces provide the emperor with everything needed . . . to live well in the unfertile province of Peking; fruit, fish, rice, silk cloth for garments, and six hundred other things, all of which must arrive on a fixed day, otherwise those who . . . transport them are subject to a heavy fine. . . . During the hot summer season much of the foodstuffs, which are perhaps a month or two in transportation, would spoil before reaching Peking; so they are kept on ice to preserve them. The ice gradually melts, and so great stores of it are kept at certain stops. . . . Hence it is said that nothing grows in Peking but there is nothing lacking there."

By the Grand Canal, Ricci sailed up to Linch'ing, a town on the border of Shantung and Hopei provinces. And here he was imprisoned. The story is too complicated to retell in full; but briefly the cause was eunuch rapacity. The eunuch who conducted him by boat sold out to another, Ma T'ang, more powerful than himself, partly in case he should be late in Peking with his imperial cargoes and so incur a fine, and partly for private gain.

Ma T'ang, whose evil reputation had already reached Ricci's ears, drew alongside Ricci's boat in the most sumptuous craft the Jesuit had ever seen —"a large and very elegant boat, suitable even for the emperor to travel in, with saloons, rooms and numerous cabins, all very wonderful and commodious. The galleries and the window casings were made of an incorruptible wood carved in various designs, shining with a coat of Chinese sandarac [a transparent varnish] and resplendent with gold." Ma T'ang made a great show of deference to the statues and pictures Ricci had with him for the emperor, prostrating himself with a mockery of reverence before the Virgin and promising her he would find a place for her in the Peking palaces. After this nauseating exhibition, he proceeded to make Ricci and his companions virtual prisoners, detaining them while he thought how best to milk them of those treasures. But meanwhile he entertained the Jesuits to feasts and theatrical shows the like of which Ricci had never seen. The eunuch had a houseful of "tight-rope walkers, sleight of hand artists, jugglers of goblets and other such parasites . . . whom he supported for his entertainment. It was thus that he passed his days."

After some time the little party and its stock of presents went onward precariously, under Ma T'ang's guards, northward to Tientsin. It was the beginning of August when they reached the port near the mouth of the Pei River—a mere eighty-four miles from Peking, but, in their plight, as far away as Rome. There, with a further mockery (this time of procedure normally carried out in Peking whereby imperial tribute was counted and detailed in writing before being taken to the palace), Ma T'ang personally took possession of Ricci's gifts and removed them to his house. Then he left for the south. Ricci and his small band were alone and apparently without hope of escape. For months they languished in captivity in Tientsin.

Ricci never discovered what happened—but suddenly orders came from Peking for Ma T'ang to forward the Jesuits to the capital. Later, Ricci attributed this turn in his fortunes to the intervention of providence. But we may suspect that someone in Peking recalled one day that there was an odd foreigner held up in Tientsin with his cache of gifts, and issued the order to expedite a possibly profitable matter. Ma T'ang was exceedingly angry but, since the order came from sources more powerful than himself, could do nothing but obey, taking care, however, to prepare his henchmen in Peking to keep Ricci under observation. The presents were returned, and after dark Ricci removed the relics (among which were, it seems, some pieces of the "true cross").

These Ricci concealed in his personal baggage for safety. The party set out with an escort provided by imperial rescript, traveling with the dignity required of those who come with tribute. And so, on January 28, 1601, in his forty-ninth year and in the eighteenth of his stay in China, Matteo Ricci came again to Peking, this time to remain for the remaining nine years of his life.

He tells us nothing of his feelings on that momentous day. Already middle-aged, with the better part of two arduous decades in China behind him, haunted by the malaria he had caught at Shao-chou—physically and mentally he must have been deathly tired. Before him lay the magical Chinese beauty of the capital, the measured statement of its massive walls, the fascination of its wide, straight, teeming streets, the shut palaces hiding he knew not what opportunity, what check to a lifetime's hopes. The autocratic smile of Ming splendor still dignified a face corrupted by moral decay. Before him, too, lay other great imponderables—the hierarchy of ministers and scholars divided into endless mutually antagonistic categories, and the further, even less knowable, cabals of eunuchs, their functions and influence undefined, whose faces were turned with enmity and cunning against those scholar-ministers. On the interactions and schemes of both, and at the whim of an unknown emperor, hung the lives of that myriad harvest of a hundred and fifty million souls—the Chinese people.[8] Before he could reach those souls, however, Ricci knew he had first to conquer the nation's rulers. He could hardly have been blamed if, as he neared Peking under a cold yellow winter sun in the first year of the new century, his courage had failed him.

In fact, it did nothing of the sort. He seems to have revived like a weary horse which sees the water and the hay trough at the end of a long day. At this as at other times one senses the real quality of this remarkable man.

THE EUNUCH MA T'ANG.
(From Favier)

8. Estimates of the population of China are always dubious, but this seems a fair figure. The total population of Europe in 1860 was about 187 million, and probably much less than 100 million in the time of Ricci. Hence, since he judged in this matter by European conditions, his frequent surprise at the enormous numbers of Chinese.

Everything about him added up to being a "whole man," the European ideal of his times. Sad to reflect that Ricci was one of the very few of that typically European breed that the West ever sent to represent it and its ways of life to the great men of China. On account of these qualities, one can never feel sorry for Ricci. Unlike Rubruck, unlike the saintly Montecorvino, and very unlike poor Pirès, he never seems the plaything of the greater forces he encountered in China. In his own way he was in command of every situation, because his intelligence, as much as his faith, was impregnable. He came now to Peking, a great man with the essence of germinating Europe within him.

Within three days after his arrival the presents for Wan-li had been despatched to the Great Within through the eunuchs who were Ricci's sponsors and also the satraps of the *éminence grise,* Ma T'ang. The gun in the palace was fired, informing all Peking that tribute had come for the emperor. And around Ricci a heavy silence fell, full of rumor, empty of news. He did not know the Chinese chroniclers were writing in their old, elegant brush-strokes: "In the second month [of 1601] the eunuch Ma T'ang of Tientsin sent to the Court Li Ma-tou, a man from the Western Ocean, who had some rare gifts for the emperor. The emperor sent the eunuch's memorial to the Board of Rites who replied: 'The Western Ocean countries have no relations with us and do not accept our laws. The images and paintings of the Lord of Heaven and of a virgin which Li Ma-tou offers as tribute are of no great value. He offers a purse in which he says there are bones of immortals, as if immortals when they ascend to heaven did not take their bones with them. On a similar occasion Han Yu [a scholar and anti-Buddhist of T'ang times who advised an emperor on the matter of a reputed finger of the Buddha offered as a "relic"] said that one should not allow such novelties to be introduced into the palace for fear of bringing misfortune. We advise, therefore, that his gifts should not be received and he should not be permitted to remain in the capital. He should be sent back to his own country.' "[9]

For a week he waited, wondering which of the presents would take the fancy of Wan-li. Half were of a religious nature: a breviary with a gold-thread binding, a cross adorned with precious stones and containing relics of the saints, the four Gospels. The remainder were calculated to interest by their novelty—a big clock worked by weights, a small spring-wound clock in gold, two of the crystal prisms that had been such a success elsewhere, a clavichord, two hourglasses, a rhinoceros tusk said to cure diseases, European cloth. He knew nothing of their fate, and viewing the vast imperial palaces, the reports of ten thousand peculating eunuchs inside, and the great luxury surrounding the emperor, Ricci had grave doubts of his paltry gifts.

9. Quoted by Fitzgerald in *China.*

Suddenly he was summoned to the palace. The command came from Wan-li himself. The larger of the two clocks had ceased to strike the hours. Ricci may well have given a sardonic chuckle. After all those years of study which had made of him a first-class Chinese scholar, all those years of patient struggle to reach the emperor's ear with his message, it was the farcical necessity of winding a clock that at last gained him entry to the Great Within, the Forbidden City. Maybe he smiled. But more likely, with perfect sincerity, he put his good fortune down to providence. He usually did.

Escorted by a posse of eunuchs, the palanquins of Ricci and Pantoja (a priest who had joined him earlier) passed through the swirling dust clouds of the Peking streets. The occupants were decanted outside one of the side gates of the great yellow-roofed rhomboid of T'ien An Men, the front entry to the palace, whose name means Gate of Heavenly Peace. As they entered, history turned a small but significant Sino-European corner. They were the first Europeans ever to pass through the outer walls of the Ming Great Within. Inside, they saw an immense coffin-shaped courtyard stretching north with Chinese-compass precision and blocked at the far end by another enormous gate structure, Wumen, the Meridian Gate, locked, forbidding as it was intended to be, its double roof a hundred feet high. On either hand the walls of the courtyard rose smooth and stuccoed, colored dusky red and topped by a trim of imperial-yellow tiles in ridges, as if in imitation of bamboos. At two places the walls gave way to wooden buildings lacquered red, their windows filled with paper. Ricci did not know it at the time, but north beyond the barrier of Wumen further vast courtyards stretched one after the other, punctuated by pavilions and gates of greater and greater luxury and intricacy, all standing on white marble podiums above seas of dark paving stones; until half a mile away from where he stood the final barrier rose, beyond which were the inmost places penetrated only by the imperial family and their eunuchs and concubines. There the mysterious emperor lived his unknown life and ruled with a Chinese writing brush dipped in vermilion ink. Even the unromantic Ricci was impressed.

But on a deeper plane he was excited. To set foot here was perhaps the beginning. With a mad sort of rightness, the care of the clocks had been given to the imperial college of mathematicians in the palace, a band of eunuchs into whose heads it took Ricci three whole days to instill the elements of mechanical sense. For every part of the clocks he had to invent terms in Chinese. To demonstrate how the timepieces worked he took them apart and put them together again, showing his robed pupils how the weights gave motive power and how the performance of the hands could be regulated. And of course he explained the elementary necessity (which they had overlooked) of winding the weights up at regular intervals. During his three-day stay in the palace he was the target of an almost continuous salvo of questions sent by the emperor. Relays of servants came running from

THE EMPEROR WAN-LI, an official portrait (Imperial Palace Museum, Peking)

Wan-li demanding answers to a selection of queries on everything European that entered the emperor's head—customs, land fertility, clothes, architecture, gems, marriage and funeral ceremonies. The questions of a birdlike, pampered mind. And when Wan-li sent for his clocks somewhat before the expiry of the three days he had set for their regulation, the eunuchs fell into a panic, swearing they might lose their heads if they could not return the clocks and also keep them going correctly.

As it turned out, Wan-li was delighted to see his timepieces going again and chiming the hours melodiously. At thirty-eight, a life of idleness and debauchery, together with the unbridled exercise of his arbitrary powers, had made of the emperor a pathetic gilded monster so gross and so afraid of assassination that even his ministers had seldom seen him in years. His twice-yearly journeys to the Temple of Heaven for the imperial sacrifice took place in a closed palanquin, one of a number of identical conveyances. In which one he reclined, only his closest servants knew. At the age of twenty-two, more than a decade ago, he had built himself an elaborate tomb in the valley in the Western Hills where his ancestors lay, and when it was finished after six years of work and the expenditure of eight million ounces of silver, his morbid fancy caused him to give a large party in its vaults.[10] A sentence of Xavier's where he remarks that the empire of China is "governed by a single sovereign whose will is absolute" may have recurred in Ricci's mind as he prepared to leave the palace. By now he could begin to understand a little of what that meant in Peking. The auspices were hardly favorable for him or for his faith.

10. The recent opening of the Wan-li tomb, and the ingenuity with which this feat was accomplished, is a fascinating story. See the description in my and Brian Brake's *Peking*.

Such was the success of the clocks that Wan-li kept the smaller one in his private apartment and commanded two eunuchs to attend daily to wind it. He was curious about the donors of his new toy but refused to see them in person. Instead he ordered their portraits to be painted for him. These turned out to be full-length studies in the Chinese manner, and apart from the larger noses and wider eyes and the beards on their chins, Ricci says, they were little enough like Pantoja and himself. Viewing the scrolls, Wan-li decided they must depict Saracens (because of the beards), and he was only disabused of this idea when told that the Jesuits ate pork.

On another occasion Ricci sent the emperor an engraving of St. Mark's Square in Venice which doubtless showed the Doges' Palace. This elicited the imperial remark that it was terribly inconvenient for the poor European kings to have to climb stairs to the upper floors, where he understood they lived. Positively dangerous, the emperor thought. Nearly all buildings in China, except an ethereal pagoda rising delicately from the fields here and there, consisted of one story. Peking on its plain was a flat city, almost the only structures rising above its sea of gray-tiled houses being the yellow-tiled roofs of the imperial palaces visible from afar above the prison of their surrounding walls. Here and there a temple, a Tibetan-style dagoba (such as that on a hilltop in the imperial park), the pavilions crowning the artificial hill north of the palaces, broke the monotony of the city skyline. But apart from those and the gold bud on the Prussian-blue cone of the Temple of Heaven, only the great triple-roofed gates of Peking's outer walls rose above the common level. Hence Wan-li's surprise. His own rooms in the palace were small, stuffy, and shuttered.

The questions petered out after a time, and Ricci and Pantoja left the palace no further forward with their schemes. They took a house near the palace gates and were just installed there when a quartet of palace eunuchs visited them. These were royal musicians. They had been sent to learn how to play the clavichord, apparently convinced the instrument could be mastered in a few days. (Or was it Wan-li who thought so?) Ricci was willing to try, and Pantoja had learned to play with exactly this in mind. Ricci's opinion of Chinese music was not high. He had already attended an orchestral rehearsal in Nanking. "The priests who composed the orchestra were vested in sumptuous garments, as if they were to attend a sacrifice, and after paying their respects to the magistrates they set to playing their various instruments; bronze bells, basin-shaped vessels, some of stone with skins over them like drums, stringed instruments like lutes, bone flutes and organs played by blowing into them with the mouth. . . . These curious affairs were all sounded at once, with a result that can be readily imagined, as it was nothing other than a lack of concord, a discord of discords. The Chinese themselves are aware of this," he continues, misunderstanding what he is about to quote. "One of their sages said on a certain occasion that the art of music known to their ancestors had evaporated with the centuries, and left only the instruments." Neat though this interpretation is, what the sage actually meant was that the ancient musical modes and melodies were lost, but that the instruments were still in use.

Doubtless the imperial college of music in the palace was furnished with such ancient instruments. Certainly Ricci was glad to go there. He found the eunuchs chosen to learn the clavichord trembling with fear lest they prove unequal to the task and thus incur the imperial rage. They insisted on kowtowing to their teachers before each lesson, and also to "the clavichord, for assurance of progress, as if it had been a living thing."

But this pantomime was not without result. Soon Ricci and his companion "were being entertained at meals and visited by some of the eunuchs in high position. Gradually," Trigault adds, "they became known to the whole palace retinue with some of whom they formed permanent friendships." Ricci fostered the music lessons. The terrified pupils never attempted to learn more than one piece on the strange instrument; but they asked for words, so that they could sing if the emperor happened to command. Seizing the opportunity, Ricci sat down and wrote eight "Songs for the Clavichord," embodying "ethical subjects . . . aptly illustrated with quotations from Christian authors." They sound dull enough but in fact proved so popular with literati and others that they had to be printed "as a musical booklet, written in European lettering and also in Chinese characters."

Meanwhile Ricci was attempting to escape from a delicate situation—the sponsorship of the eunuch Ma T'ang, whose minions still shadowed him everywhere. Ma T'ang had overreached himself in detaining Ricci in Tientsin, for when it became obvious that the Jesuits were well received by such officials as the governing magistrate of the High Council (who came of his own accord to their house), Ma T'ang began bribing people to forget the matter. But the situation had further convolutions, for Ricci was sponsored (against correct precedent, though possibly he did not know this) by eunuchs. This displeased the Board of Rites—the proper authority. Those contra-rotating cogs began to grind him between them. He was sent with his companions to the Hui-t'ung-kuan, a sparsely furnished barracks, strongly guarded, where all "tribute-bearers" to Peking were kept under virtual house arrest until summoned, rewarded with imperial presents, and sent off to their own countries again.[11] In this cold caravanserai, they might have languished forever while the officials celebrated a victory over the eunuch factions who had sponsored Ricci. But once again it was the personality of

11. The name Hui-t'ung-kuan means "hostelry." It was sometimes called the "Palace of the Four Barbarians," probably signifying "foreigners from the four quarters." As Bernard says, the designation palace was an unwarranted euphemism. The interesting history of the place, which, in Ricci's time, seems to have been in what later became Peking's Legation Quarter, is followed at length by Paul Pelliot in *T'oung Pao,* No. 38, 1947.

the Jesuit that saved the day. A series of memorials was sent to the emperor and eventually the Jesuit position in Peking was tacitly recognized. The emperor never gave his explicit permission, but neither did he say no to their continued residence in the capital. On this shaky but workable basis they returned to their house and began their tasks.

But before the Jesuits emerged at last from the Hui-t'ung-kuan they had been summoned to audience at the palace. At least one writer dealing with this has succumbed to the temptation to dramatize it.[12] Ricci is portrayed as he went at dawn, dressed in the special clothes prescribed, holding before his mouth the ivory tablet which was supposed to restrain the breath in the emperor's presence. There, hoping against hope, Ricci is said to have gone through the ceremony, kowtowed to the empty throne and backed away after the sham was over, still hoping for a sight of the emperor. The fiction does him a disservice. Undoubtedly Ricci knew, well before he went to the audience, that Wan-li had not attended one of these ceremonies for many years, that the audiences were empty formality, conducted as tradition required but while the Son of Heaven probably still reclined in bed asleep with some concubine. Ricci went, however, and apparently performed the kowtow in good Chinese fashion without a qualm. Henri Bernard in his book on Ricci tells us that in the usual rehearsals for this elaborate ceremony of homage Ricci had been put through his paces in company with three Chinese Moslems because it was considered that these men were his compatriots!

Ricci never saw the emperor Wan-li, and very soon he renounced the impracticable dream of converting him. But from various stories and events it seems that Wan-li held Ricci in some regard. The emperor showed on more than one occasion that he did not want the Jesuits sent away from Peking. Doubtless at first this was on account of their skill with the clocks and the clavichord, but later there must have been other reasons. Possibly Wan-li was pleased to hold in his royal city as a kind of marvel the only foreigner from the West who spoke and wrote Chinese like a native. It was doubtless as such that the royal and birdlike ego of this latter-day Son of Heaven prized one of the best minds of the age.

Ricci now found his mission equipped with a stipend and food rations granted from imperial funds. Such was the power of powerful friends. And it was just those powerful men in Peking whom he now proposed to win over to Christ. During the remaining years he was astonishingly successful. At one time the reputation of the Jesuit mission was such that their house was the constant focus of visits from high officers of government and of scholars of wide renown. In fact, the impression created by this *succès d'intellect,* that theirs was a cult reserved for the mighty, had to be strenuously contradicted by Ricci before the poor could be induced to come.

12. Cronin, the style of whose book seems unsuitable for, and greatly at variance with, the feeling of Ricci and his time in China.

It may be the fault of surviving information, but it seems that the majority of converts in Peking eventually came either from the ranks of the great and powerful or from the mass of the poorest of the poor. For different reasons, neither had anything to lose in setting aside, as conversion required of them, some of the basic assumptions of Chinese life. Both apparently did so with sincerity, a remarkable feat in Ming China, where not to conform to the prevailing wind of Confucian social principles was as hard as practicing pacifism in World War II in the West. To scholars, Christian belief appears to have offered a mental catharsis in the constipation of changeless formulae which had been their intellectual state. To the poor, little solaced by the not notably attractive afterlife offered by other cults, with little assurance of descendants to attend their shrines when they died, the new religion revealed a way to ultimate release and joy for which they had never hoped. But among those other Chinese, the middle stratum of society, there seem to have been few conversions. These people were less secure, more conformist by circumstance, and might lose all, should the tide turn against the Jesuits. To dissect in this way is not to underestimate the effect of the new faith, or the evident release it brought to those who embraced it. That anyone at all gave it his whole heart in Peking is eloquent testimony to the genius of Ricci, to the success of his "cultural accommodation."

Ricci seems to have been all things to all men. He managed the all but impossible, maintaining a lively relationship with eunuchs in the palace (where, always excepting the innermost precincts, he was free to come and go as he pleased), and also with the bitterly opposed faction of civil servants. Only he, with his accommodating tolerance and humor, with his scholastic attainments, and with his transparent integrity, could have done this. In him an absolute spiritual integrity was unusually well combined with the most subtle powers of diplomacy. Nothing demonstrated Ricci's peculiar brilliance better than his views on the time-honored cult of ancestor worship, and also on the veneration paid to Confucius. His opinion that neither the one nor the other was idolatrous shows, in the words of a generous Jesuit scholar of our own times, Columba Cary-Elwes, "the gradual *rapprochement* growing between the great missionary and the tradition of China." According to Ricci himself, his view was founded on study of the classical texts, in which he found no taint of idolatry, in which, indeed, "one might be allowed to assert . . . there was no superstition either." Both practices were forms of reverence, nothing more. Therefore they were not abhorrent to Christianity.

But while taking his stand on the Classics, he disregarded the traditional commentaries on those same books, commentaries which in Chinese eyes had almost equal authority and stature. (Other Jesuits, in his lifetime and after, were not slow to point this out, and the subject became the basis of the long Rites Controversy that in the end was to destroy all Jesuit work in

China.) The commentaries appear to view the ceremonies of ancestor worship, at least, as real worship. The subject could not be dismissed, since it ran so deep in Chinese life, and Ricci's answer was, doggedly, that the ceremonies were simple respect, not superstition. He thought too that with deepening Christian conviction in the convert, the rites would become even further secularized. The force of his personality, and his recent appointment as head of the China Mission, now separated from the Eastern Mission, assured the acceptance of this dictum in his lifetime. He believed in "interpreting in our favor some texts which were ambiguous." And whatever the truth or error in the conviction, it was certainly the only way by which Christianity might have conquered the Chinese mind.

His nine years in Peking were great times for Ricci. With his "curly beard, blue eyes, and voice like a great bell," to quote a contemporary Chinese description,[13] he must have been a familiar figure in the capital. He himself was pleased that the dust of Peking streets in the dry season had sanctioned the general practice of wearing a veil over the face when in town, and that because of this one did not have to dismount every time one saw an acquaintance and greet him. But most likely, even behind the veil, he was recognized.

Never was a man more fitted for the work he had chosen, never did a man so labor to make himself even more fit to perform it. A mission house was bought and added to, a church built, princes of the blood royal (powerless but prestigious) and their families were converted, great scholars joined the flock, the poor crowded to the doors in hundreds, multitudes of abandoned and dying female infants were abruptly baptized, Western books (the *Elements* of Euclid, others on mathematics, astronomy, hydraulics, portions of the Bible) were translated, original compositions on religious and ethical subjects were written and printed. Meanwhile at the stages of that long journey Ricci had made over the years from Canton toward Peking, the missions he had founded flourished in spite of occasional persecution. At Shao-chou, Nanking, and other sites it was easier now that influential friends had influential friends who were magistrates there. Ricci was more than happy to live out his life in Peking. He survived a dangerous Buddhist attempt there to impeach him. He survived a scandal campaign in which he was innocently involved. At one point he even said he would leave the capital, an idea which was vetoed by Wan-li, as Ricci adroitly guessed it would be when he made the threat.

With his usual sagacity, Ricci had seen long before he arrived in Peking that one road to influence in China would be to correct the Chinese lunar calendar, which had been in error for some centuries. The rules for the necessary calculations had been lost for a long time, and only empirical

13. Quoted by Latourette in *A History of Christian Missions.*

calculations were made, producing serious mistakes. Since almost every event in China occurred in deference to calendrical time, and since in a sense the calendar was a political instrument, this was a serious and troublesome state of affairs. Writing to Rome from Peking in 1605, Ricci complained: "I do not have a single book on astrology [he means astronomy—the use of the terms was not yet so concrete as it is now], but with the help of certain ephemerides and Portuguese almanacs I sometimes predict eclipses more accurately than they do. For this reason, when I tell them that I have no books and . . . do not wish to start to correct their rules, few people here believe me. And accordingly I say that if the mathematician of whom I spoke came here, we could readily translate our tables into Chinese characters and rectify their year. This would give us great face, would open wider the gates of China, and would enable us to live more securely and freely." He goes on in the same letter: "I wish exceedingly to beg Your Reverence for something which for years I have been asking, but without response. . . . One of the most useful things that could come from [Rome] to this court [at Peking] would be a father . . . who is a good astrologer. . . ."

And in his second-last letter from Peking Ricci was still making the same request. It was not until years after his death that his requests were answered and suitably qualified priests were sent. When they were, the Jesuits came into possession of one of the most potent means of enhancing their influence in China—exactly as Ricci had predicted. In a very dramatic manner, later members of the Society of Jesus in Peking were to show the superiority of European astronomical science. But to Ricci must go the credit for seeing the opportunity and making the original requests for the Jesuit astronomers who would eventually come and regulate the calendar, an activity which, as it turned out, was to remain one of the very rare Western contributions to Chinese life.

Ricci's contribution to Western knowledge on the subject of China was much more substantial. In the second edition of Hakluyt's *Principal Navigations, Voyages, Traffics, and Discoveries of the English Nation,* published in 1599, the section dealing with China relies heavily on material already published by Alessandro Valagnani, which in turn relied heavily on the reports of Ricci. And although the conclusions of Ricci on the hoary old topic of whether Cathay and China were two different countries or one and the same did not appear until after his death, they finally put an end to that speculation. His narrative of the journey of Brother Bento Goes, who died on the far western borders of China at the end of a remarkable trek from Agra in India, proved once and for all what Ricci had already all but decided for himself, that the Cathay of Marco Polo and others was in fact the China he knew.

But perhaps Ricci's greatest impact on European knowledge of China came with the publication in 1615 of his journals translated into Latin and edited by Trigault. This book was soon appearing in French, German, Spanish, and Italian editions, and excerpts from it occur in *Purchas His Pilgrims,* published in 1625 in England. Sad to relate, the great mass of accurate and useful information in Ricci's journals seems to have been little read by those who attempted in the near future to trade with China. Almost the only practical use to which his writings were put in the century after his stay in Peking was by the Jesuits themselves. The Western world remained obstinately faithful to the China of Marco Polo and of that inspired literary robber and romancer Sir John Mandeville. It is fortunate that China was a civilization in which basic and radical change was, to say the least, rare; otherwise those accounts would have been still more out of date.

Ricci was infinitely more interested in people than in things. The personalities and possibilities of his converts and aspirants caught his imagination more than the grandeur of Ming Peking. The philosophy of China and his attempt to interweave in it the religion of Christianity occupied his thoughts more than did the beauties and peculiarities of the Chinese scene. He was a man single-minded, yet not bigoted. The governor of the province of Kweichow once said of him: "Ricci has been so long in China that he is no longer a foreigner, but a Chinese." It was true. He was the first and in many ways the best of all those later men dedicated to the Chinese (or sometimes to their own concepts of what the Chinese were or ought to be), men no longer quite European or Western who had found another home for themselves in the East. Anyone who has lived among the Chinese knows why this comes about. Most people who have not find it hard to understand. The stories of those many Westerners who were to come after Ricci may help to elucidate the mystery a little.

In his last letter to Rome, dated February, 1609, Ricci once more underlined the importance of sending to China men who were not only "good, but also men of talent, since we are dealing with a people both intelligent and learned." It is refreshing to read this sentence, and had later missionaries printed it in large letters and hung it with the pious texts in their mission houses, something more might have come of their efforts. Ricci's foremost converts were of this kind.

His closest friend among the Chinese was Hsu Kuang-ch'i, a brilliant scholar baptized Paul by Father Cattaneo in Nanking. In 1601, Paul Hsu came to Peking to sit for the imperial examination held every third year in the capital. From the successful candidates the future elite of the Chinese bureaucratic machine were chosen. In this test he came out seventh, but somehow the examiners made a mistake and announced one too many passes, one over the quota stipulated. Arbitrarily, they struck one name off the pass list. It happened to be that of the unfortunate Paul Hsu. He was thereby forced to wait another three years until the next examination session in 1604, when he came again to Peking, passed once more, and was

PAUL HSU, Ricci's powerful Chinese Christian friend, in a non-contemporary painting. (From d'Elia's *Fonti Ricciane)*

officially confirmed. Encouraged by Ricci, he went on to pass even more rigorous examinations which qualified him for the Imperial Academy. He was now in a position to be chosen for such key positions in the upper echelons of the state machine as an official historian, a writer of imperial edicts, or—which was what interested Ricci—tutor to the sons of the emperor. Hsu had a first-class brain and a lively intelligence. He became adviser and in a sense protector of the Jesuit missions. Through him Ricci met hundreds of officials from all over China when they came to Peking, contacts which enabled him to smooth the path of the missions in the provinces. Of Ricci, Paul Hsu said: "Today we have the True Man, learned and great, who brings our moral code to completion. . . ." Between them, Ricci and Hsu translated Euclid's *Elements*. Only Hsu and another convert, Li Chih-tsao, succeeded in mastering the subject.

But Ricci's happy association with Paul Hsu was interrupted in 1607 when Hsu's father died and the son was forced by tradition to retire from public life to the country and spend his three years of mourning there. Fortunately for Ricci, who was a man of significant friendships, in that same year the Peking mission was joined by Father Sabatino de Ursis in place of Pantoja, whom Ricci had always found difficult. The new father came from southern Italy and, for the first time in years, Ricci had with him a warm, very human European companion, one with whom he could speak Italian. Their friendship was soon very close and sustained Ricci in his last years.

There was also the other outstanding convert Li Chih-tsao, baptized Leo, whom we met briefly above mastering Euclid. He passed his examinations at the age of thirty-three in 1598, taking eighth place on the list, and was appointed to a post in the Ministry of Public Works in Peking. Ricci met him very soon after his arrival in the capital and their friendship never flagged. For nine years Ricci strove to convert Li, without success—not because the Chinese would not accept Christianity, but because, accepting it, he still could not bring himself to renounce his concubines. Li was a geographer and had published a map of the world that showed only China. Ricci's maps opened his eyes to geographical reality, and it was he who was responsible for urging Ricci to make a huge world projection five feet high and in six panels (see illustration pages 192-93). He also translated one of the books of Clavius and was responsible for introducing the first work by an English author to be published in China—the *Tractatus de Sphaera* of John Holyroods, to which Ricci added a poem. The poem was titled "Treatise on the Constellations" and cleverly delineated the characteristics of the twenty-eight Chinese constellations in easily memorable verse. Not long before Ricci's death, Li renounced his concubines and was baptized at last.

By the winter of 1609, Ricci, prematurely aged, and very gray, felt his strength flagging. But with his lifelong determination and zeal he continued to spend his force with the same generosity he had shown from the moment he set foot in China. He seems to have sensed that his life was nearing its end.

"One day," says Trigault—it was May 3, 1610—"when Father Ricci returned to the mission house thoroughly fatigued from consulting with visitors at the court, he took to his bed to rest. At first the fathers thought he was suffering from an attack of migraine, to which he was subject. . . . When questioned, he said it was something quite different, and told them that he was sick unto death from utter exhaustion, and he seemed to be so unconcerned about it that when one of them asked him how he felt, he said: 'Just at present I am wavering in doubt as to which of two things I would prefer; to accept my eternal reward, which is not far away, or to continue the routine of my daily labors on this Christian Mission.' "

Leo Li sent his doctor and the fathers sent for the six best medical men in Peking. Their prescriptions were of no avail in helping the dying Ricci. "The only one to be satisfied with it [the medicine] was the patient himself who felt that he was coming to the end of his labors. He seemed to be particularly happy about this and his almost joyful disposition served to lighten the grief of both fathers and converts.

"On the sixth day of his illness, he made a general confession of his whole life, and the father who attended him was so overcome by his lighthearted disposition, that he said he had never in his whole life experienced more spiritual joy than that which radiated from the gentility and the innocence of the soul of Father Matteo."

Soon after, when they approached him with the blessed sacrament, "he summoned up all his strength, and without assistance from anyone, got up from his bed and knelt on the floor." Two days later he asked for extreme

RICCI'S WORLD MAP OF 1602. The great Jesuit's contributions to Chinese map-making technique have often been exaggerated, but at least his Chinese names for many a place have remained as the official ones in China. Moreover, his

depiction of the American continents was the first to appear in China. The photograph reproduced shows one of the original copies of the map made in China and now in the Royal Geographical Society, London.

unction. Almost his last words to the assembled fathers were: "I am leaving you on the threshold of an open door that leads to great reward, but only after labors endured and dangers encountered." Toward the evening of the eleventh of May, sitting up in his bed, he closed his eyes "as if falling asleep. . . . Father Matteo Ricci was dead."

The coffin was provided by Leo Li, the ground (formerly the burial place of a eunuch) was donated by the emperor Wan-li himself as the result of a memorial supported by Leo. One of the pallbearers was Paul Hsu. The plaque over the tomb bore the words: "To one who attained renown for justice and wrote illustrious books. To Li Ma-tou, a man from the Great Occident. Erected by Huang Chi-shih, Governor of the Royal City of Peking."

So, fittingly honored by the Chinese themselves, Matteo Ricci was laid to rest in a Christian cemetery outside the walls. He was one of the noblest men of Peking, a citizen of that richest, perhaps most beautiful city of the seventeenth-century world. No son of a Ming heaven in his time, and none until that lingering dynasty had slunk out of its now ill-fitting robes of power, deserved even a place beside him. And, looking down the curious list of Western sojourners in that changeless and somehow radiant Peking in the centuries after Ricci, one is not likely to find his equal.

He was the first of the Jesuits and virtually the first of the many Europeans who were to beat their heads on the imperial and adamantine rock of China, convinced, each for his own particular reason, that the rock would split, would open and engender a familiar Western flower.

Chinese Diagram Explaining the Sphericity of the Earth, dating from 1648, from a book called *Scientific Sketches,* by Hsiung Ming-yu. The influence of Ricci's ideas is visible in the division of the sphere into 360 degrees instead of the Chinese 365.25. And although one land mass carries a Chinese pagoda and the ships sailing around the earth are of Chinese design, the antipodes are envisaged as a Western domain and show a building resembling a European church. (From Needham)

CHAPTER NINE

THE EUROPEAN RIVALS

IN 1610 when Ricci died, the Wan-li emperor[1] still occupied the Dragon Throne in name if not in actual fact. An obese and virtually powerless lump, his name was still blazoned on the edicts which issued in a futile stream from the Great Within—edicts intended to regulate the affairs of a now disintegrating state. In the warm paddies of the far south, under the hard sun, and in the keen cold of Tibet's borders, in the yellow hills of the Yellow River's mid-reaches and among the orchards of Confucius's Shantung, those edicts came to rest in the broken mechanism of a bureaucracy now falling to pieces. Peking, the emperor, had no power. Power had sifted like sand unheeded through the imperial fingers, and none of Wan-li's successors could gather it up again. Neither could the rival ranks of eunuchs and scholar civil servants trap enough of this vanishing power to rule. The Ming trembled to the close of its three hundred years, permissive instead of puissant, through lack of will, through contradictions. The dynastic story, with its cycle of vigor, decline, terminal ineffectuality, and final extinction by a new and vigorous rebel or invading regime, was not new to China. Such developments, though temporarily confusing, hardly surprised the Chinese.

If Prince Ivan Semenovich Kurakin, *voevodo* (provincial governor) of Siberia in the year 1618, had known how weak the Ming dynasty actually was, it is doubtful that he would have sent his two illiterate *sluzhilie liudi* (military officials) to Peking in an effort to exchange representation between Russia and her ancient neighbor. It was a pity, too, that in the manner of others before and after him he did not take the trouble to discover the usages of the Chinese court, and failed to equip his envoys with presents for the

1. It is always a problem to decide how to render the names of Chinese emperors in a Western language. Wan-li, for example, is the name of the ruler's reign period, I-chun his personal name, and Shen Tsung his posthumous name. Some emperors are more commonly known in literature by one or the other of these three names. Inaccurate though it may be, I have generally used the one best known in the West and, in the case of a reign name, either the more accurate style "the Wan-li emperor" or, by analogy with an actual name, "the emperor Wan-li."

emperor. Nor did he understand that to pay a visit to the emperor of China meant in fact that your country was from that day onward regarded as one more on the list of places tributary to the Great Ming.

So Ivan Petlin and his companion Ondrushka Mundoff arrived in Peking not only empty-handed, but with their heads stuffed with instructions on how not to compromise in the very slightest degree the name, honor, and integrity of their sovereign lord the tsar of Russia. Even if their scope was thereby fatally limited, they were at least the first recorded emissaries from a Western power to reach Peking and return home safe. Which was more than poor Pirès had done.

Petlin's account of the trip is contained in the *stateini spisok* (report) he dictated in Tobolsk the following year.[2] He describes Peking, calling the city "Great China" for some muddle-headed reason of his own.[3]

"There dwells the Tsar Taibun [he means Wan-li] himself; a very great city white as snow, and standing foursquare, four days' journey around, and at the corners and in the midst of the walls are great towers, high and white as snow, picked out with different-colored paint; and in the towers great cannon at the embrasures and twenty men on guard at each gate. And . . . inside the White City is the Magnet City in which the Chinese Tsar Taibun himself lives." All of which is an example of the futility of sending the uninformed to find out. Why Petlin thought Peking a "white city" must remain his secret, unless we may guess he was referring to the yellowish gray dust that caused the wearing of veils as Ricci reports. The "different-colored paint" may be the tiles of yellow and the bracketing of eaves painted in many colors. The twenty men on guard at each gate is something of an understatement. The "Magnet City" is a mistranslation of the Chinese character for the polestar—the emperor's paradigm in the sky.

Hardly surprising, then, that "we were not presented to the Tsar Taibun, nor did we see him, for we had no gifts to offer. And the Secretary of the *Posolski Prikaz* [presumably the Board of Rites] said to me: 'Our land of China has the following custom—without gifts no one can make his appearance before our Tsar Taibun. If only your white Tsar had sent . . . some small thing. It is not important that our Tsar admits to his presence [or not] but it is important that the White Tsar should send gifts. . . . Then our Tsar . . . would have received you and sent back the equivalent, and moreover, would have made presents to all you envoys, and dismissed you. However, the Lord Tsar Taibun will now give you a letter to your

2. The curious story of Russian overtures to China is told by Baddeley in *Russia, Mongolia, China.* Long out of print, this two-volume work contains the sole English translation of rare reports which Baddeley found moldering in the Moscow state archives before the Revolution.

3. Russian knowledge of China at the time was extremely limited. A contemporary estimate quoted by Baddeley reads: "As to China, the boyars state that it is completely surrounded by a brick wall, from which it is evident that it is no large place."

Tsar, but envoys [he] will not send to your Muscovite lord. . . .'" The letter rejected all Russian overtures.

So Petlin and Mundoff returned over the Siberian wastes to Tobolsk with something of a flea in their ears. And that, for the time being at least, was that.

Meanwhile the Jesuits, with the legacy of Ricci in their hands, were still at work in Peking. Longobardi, Ricci's designated successor, was in charge. He was a man of courage and intelligence, but of a certain narrowness of outlook. Moreover, he was an incurable optimist. His attitude to the Chinese rites problem conformed to that of Ricci in the latter's lifetime, but later he came to agree more with the opposing view that Christians could make no compromises with Confucianism, not even to the extent of performing the prescribed rites of ancestor worship. With this shift in position, it became more difficult to convert Chinese—of which the later missionary Huc was to say with feeling: "A melancholy trait it is in the character of this people, that Christian truth does but glide over its surface!" Longobardi was a hardy perennial, not relinquishing his hold on life until 1654, when he died in Peking, aged ninety-eight, after fifty-seven years in China. Ricci's real successor was, appropriately enough, one of the astronomers for whom he had entreated Rome, Adam Schall von Bell (1591–1666), who reached Peking in 1622. We shall hear more of him and his calendar reform soon.

For our story, the chief interest in those distracted years between the death of Ricci and the fall of the Ming thirty-four years later lies in two aspects of Jesuit life during that time. The first is the persecution which came upon them and some of their converts, and lasted for about twelve years.

By 1610 the strength of Jesuit missions in various parts of China had greatly increased. Ricci's major disciples, Paul Hsu and Leo Li, had left Peking, each to found a mission in his home town and to exert powerful local and, later, central influence in the Jesuits' favor. In his *History of Christian Missions in China* Latourette estimates that shortly after the end of the Ming there were 150,000 Chinese Christians; but there is no sure means of checking the figure and one is at liberty to suppose (with another authority, Bishop Stephen Neill) that it includes every soul saved since the beginning of the Jesuit mission. It was in part this numerical strength, in part some indiscretions of Jesuit fathers, and in part also the perpetual Chinese mistrust of foreigners that brought about the persecution.

One example will serve for all. In 1615, fathers Semedo and Vagnoni were working in Nanking. In that year a renowned Confucian scholar called Shen Ts'ui was appointed to the post of assessor of the Board of Rites in the same city. This man disliked Christians because he felt he had been humiliated by them in arguments over the question of religion. He submitted a memorial to the court alleging that Semedo's teachings in Nanking were contrary to Chinese custom, even to Chinese concepts of the heavens—

Some Events in Europe and China, 1610–44

A glance now and then at the following chronology will help the reader follow the press of events and people in this chapter.

	Europe	China
1610	Charles I king of England. Galileo publishes *Star Messenger* in Italy.	Wan-li emperor of China. Death of Ricci in Peking, his requests for astronomers still unanswered.
1611	Galileo feted in Jesuit College at Rome.	
1613		Trigault leaves for Rome to bring astronomers.
1614		Galileo's book discussed in Peking.
1615		Jesuits seriously persecuted at Nanking. Shen Ts'ui obtains expulsion order against Jesuits.
1616	Roman church condemns Galileo's heliocentric theory of the universe.	Nurachi made khan of the Manchu with capital at Mukden.
1618		Russian embassy at Peking.
1619		Trigault returns to China with two astronomers.
1620		Death of Wan-li, and then of his successor. T'ien-ch'i becomes emperor. Paul Hsu imports Portuguese cannon-founders, but they are expelled.
1622	Dutch sea power in East growing.	Dutch attack Macao. Persecution of Jesuits gradually abates. Manchu aggression in north; rebellion in south; civil war.
1623		Jesuits Longobardi and Diaz make a terrestrial globe.

a statement which, in essence, was quite true. Semedo and Vagnoni were forthwith thrown into prison in August, 1616, and Vagnoni was beaten. They remained incarcerated for a year or so until Shen Ts'ui obtained a decree expelling all Jesuits from China. "They put us in narrow wooden cages," Semedo says, "chained at the neck and manacled at the wrist, our hair and dress in disarray, to show that we were foreigners and barbarians. Shut up like beasts we were taken to a tribunal on 30th April, 1617 where the cages were sealed with the imperial seal. In front of us they bore three big placards with the decree on them, warning all Chinese to have nothing to

1625	Pope Paul V permits celebration of Mass in Chinese language.	Semedo hears of the Nestorian Stone recently discovered near Sian.
1627		Last Ming emperor enthroned.
1628		Paul Hsu becomes vice-president of Board of Rites.
1629		Manchu break through the Great Wall. Jesuits correctly predict solar eclipse.
1630		Portuguese military forces under Texeira invited into China, but are sent out to Macao before reaching the north.
1631		Paul Hsu becomes president of Board of Rites.
1632	Galileo on trial at Rome.	
1633	Roman church condemns Galilean-Copernican heliocentric theory and opens China to missions of all religious orders.	Various condemnations of Ricci's maps and geographical ideas.
1634		Galilean telescope presented to emperor.
1635	Courteen Association established and licensed by King Charles I. Captain Weddell appointed to command expedition to China.	
1637		Weddell at Canton with Courteen ships.
1640		Rebels under Li Tzu-ch'eng control most of south.
1642	Death of Galileo—"and yet it moves."	
1644		Capture of Peking by rebels. Fall of Ming dynasty to the Manchu.

do with us."[4] For thirty days on the journey to Macao they were caged in this way, carried along amid the jeers and vilification of the populace. Meanwhile the Nanking mission was destroyed. And, in Peking, Longobardi and others went into hiding.

But Shen Ts'ui's triumph was short lived. In the same year, 1617, he fell from grace for other reasons. The Jesuits crept back, only to find that a mere four years later, in 1621, the redoubtable Confucian had risen again to

4. Quoted by Pfister. Chinese methods altered little over the centuries, as will be seen from the illustration of a caged Englishman (page 201).

THREE VIEWS OF MACAO, the center of Portuguese activities in China. *Above,* an engraving showing the city in 1646. Perhaps the scene is a bit fanciful, but the new cathedral at the top of a broad flight of steps is recognizable. (A Dutch engraving from *Begin ende Voortgangh,* as reproduced in Boxer's *Macau*). *Lower left,* a nineteenth-century view of the cathedral after it had been burned and only the facade remained. (Museum and Art Gallery, City Hall, Hong Kong). *Lower right,* the cathedral facade in 1965. (Photo by the author)

A Caged Englishman. Although the incident depicted here occurred almost two hundred years later, the scene was doubtless much the same in the case of Father Semedo's encagement, a not unusual form of punishment in China for many centuries. (From Scott, showing a junior officer of the English armed brig *Kite,* wrecked off the coast of South China during the Opium Wars, imprisoned in a cage)

favor. Whereupon they returned to hiding, and a large number of Nanking Christians were tortured.

The story of the intermittent Jesuit persecution in these years is a morass of conflicting evidence. It was not until about 1622 that the violence of anti-Christian feeling began to die down. This persecution of the Jesuit Christians in the early seventeenth century seems to have been caused by a combination of factors, bulking large among which was Chinese apprehension that the fundaments of their traditional way of life were threatened by the foreigners. What they taught—a religion (in itself dubious to Chinese cultured opinion)—seemed to challenge the Confucian orthodoxy that had ruled for hundreds of years.

During this decade there were other alarms and excursions which added their fuel to the smoldering fire of orthodoxy and xenophobia. Soon after Ricci's death the Portuguese at Macao, assisted by Jesuits, began to erect fortifications to protect themselves against the Dutch. For by this time the power of Portugal was on the wane in the East and that of the Dutch on the ascendant. The new fort both displeased and scared the Chinese. They felt it presaged some Portuguese assault on China. The situation grew electric as the fortifications rose.

The Dutch had already come to Chinese attention when, in 1602 and 1604, their attempts to trade at Canton had been thwarted by the jealous Portuguese. But in 1607, as a Chinese history records, the Dutch showed their colors clearly: "In the eleventh month Hsu Hsiu-tsu, viceroy of Fukien [a southern coastal province] reported to the Court that the Hung Mao had slain some Chinese merchants and pillaged their vessels, and that thereafter they had landed as if they proposed to establish themselves on the mainland."[5] Hung Mao, literally "redheads," was an opprobrious name applied to the Dutch, and later to the English. Since red hair is the distinguishing mark of devils in the Buddhist hell, the term "foreign devils," also came into use as an alternate to the already existing "ocean devils." Presumably the Dutch were driven off, but their aggressiveness left an indelible memory.

The situation was far from clear in Chinese eyes since they were unable to sort out the foreigners one from the other and did not know at first that the Portuguese were quite as anxious as they themselves to see the backs of the Dutch. For the Hollanders threatened the Portuguese monopoly in the lucrative China trade and were possible Protestant rivals of the Jesuits—which accounts for Jesuit help in the fortifications. At last a high Chinese official visited Macao and saw for himself that the battlements were designed to ward off seaborne attack. Thus were the Chinese mollified for the moment.

During these years of Jesuit-baiting the situation within the disintegrating realm of China went from bad to worse. Some men of high rank knew well enough that the union of the Manchu tribes in the northwest beyond the Great Wall was significant, that, under their khan Nurachi, they represented a serious threat. The death of Wan-li, the month-long reign of his successor, the enthronement of yet another emperor, T'ien-ch'i, did nothing to instill order into the disarray of the state administration.

Paul Hsu managed in 1620 to import some Portuguese cannon-founders in an effort to bolster the Chinese armies with modern ordnance in their battles with Manchu armies; but feeling in China ran high against such foreign ideas, and the founders were expelled. In 1622 the Dutch attacked Macao, were beaten off, and settled in Formosa, not then a part of the Chinese empire.[6] And in the same year serious Manchu attacks in the north were accompanied by rebellion in the south of the country. From this time until the end, civil war raged in China.[7]

In Europe, meanwhile, as the last Ming rulers succeeded one another in the Forbidden City, things were in a ferment. Charles I of England sat

5. Quoted by Fitzgerald in *China*.

6. Semedo gives a spirited description of this event, obviously deriving some pleasure from the Dutch stupidity and defeat. He gloats over the spectacle of the Dutch falling under withering fire from Portuguese muskets, and the retreating soldiery capsizing a boat in their panic to escape; and records with satisfaction the four hundred dead.

7. There is a vivid picture of social life in late-Ming China in the novel *Jo Pu Tuan* (The Prayer Mat of the Flesh) by Li Yu. Published in 1634, it is erotic in the style of its times but succeeds in much more than simply following the adventures of its hero in the beds of other people's wives. The recent English translation by Richard Martin is called *The Before Midnight Scholar*.

uneasily on his throne; the only thing he had in common with his Ming contemporaries was an acute shortage of money. In Italy, the year 1610 saw the publication of Galileo's *Star Messenger,* in which he developed the heliocentric theory that the earth revolved around the sun and not (as popularly supposed in both China and Europe) the reverse. The distinguished minds of many Jesuits took intellectual pleasure in the theory and in Galileo's demonstration of its correctness. But in 1616 the Roman church condemned the theory as being contrary to its teachings and therefore heretical. Already, however, the book had been the subject of lively discussion among the Jesuits of Peking.

But then came the second aspect of Jesuit life in Peking—the remarkable era of their scientific propaganda in the heart of the Chinese scene. Perhaps the very uncertainty of the times, the crumbling of the old-established forms of Chinese society and order and government, may have been in a sense propitious for the Jesuit effort. Those breaches in the wall of the Chinese ethos were the very places into which they inserted the new and surprising plugs of Western science.

Even before he went to Peking, Matteo Ricci had repeatedly pleaded with Rome for astronomers. And now, after the return of Trigault from Italy in 1619, two highly skilled priests arrived—Adam Schall von Bell and John Schreck, or Terrentius, as he was more often called. They reached Peking in 1622. A mere four years later Schall published there the first description in Chinese of the Galilean telescope, an instrument quite as shattering in its effect on Chinese traditional beliefs as it was on those of the Europeans at much the same time. As usual, the learned and subtle Jesuits missed no opportunity of recommending the new in terms relevant to the existing facts of life in China. Among the advantages, Schall wrote in his book, was the fact that "now with the telescope there is no longer either small object or distant object. . . . Both heaven and earth become part of our visual field. In the mountains or on the sea . . . one can see ahead of time incursions of brigands or pirates. . . . It is truly an instrument which unexpectedly renders sight acute, and it is a joy to the scientist."[8]

It was much more. It was the Jesuit scientists' most potent weapon in Peking. The illustrations in Schall's text showed the telescope and also some of the "celestial revelations" it was capable of making—the constellations familiar from antiquity to the Chinese but now for the first time in stark and dramatic close-up. The final section of Schall's book dealt with the optical components of the telescope and must have proved as baffling to the Chinese as would the programing of a computer to most of us today.

But a year later a Chinese scholar was amplifying the book in one of his own, called *Illustrated Explanation of the Instruments of Mechanics of the Far*

8. Quoted in d'Elia's *Galileo,* a short book dealing brilliantly with the Jesuit introduction of Western astronomy into China.

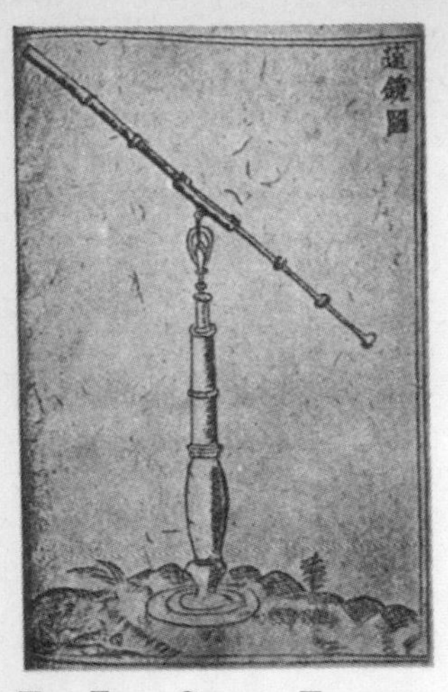

THE TELESCOPE IN CHINA. This was the first depiction of the telescope to appear in China, in a book published in Chinese in 1626. (From Schall's *Historica narratio*)

THE FIRST CHINESE TERRESTRIAL GLOBE, made in Peking in 1623 by the Jesuits ▶ Nicolo Longobardi and Manuel Diaz the Younger. The entire surface, with its seas and continents, its European-type sailing ships, its whalelike fish, its place names and cartouche of description in Chinese characters, is of Chinese lacquers in various colors. The Asian continent, including Japan and Southeast Asia, is colored imperial yellow (now much faded), and the coasts are outlined in red. The importance of China to the Chinese at the time when the map was made is stressed by putting the principal explanatory cartouche directly under China. This placement has been explained as the Jesuits' way of concealing their ignorance of the outline of the supposed "Southern Continent," but it would seem more likely that they intended to forestall Chinese criticism at finding China merely one of many great land masses. On his map of 1602 Ricci had written: "The earth and sea in fact are round and together form a sphere set in the middle of the heavens like the yolk of an egg in the white." Here the analogy is further elaborated, in the cartouche: "Earth and sea, linked together, are suspended in the heavens. Now the heavens are uppermost, and completely surround the earth, so that they are always above it. . . . The earth is like the heavy turbid yolk of an egg concentrating in one place . . . namely in a point which is the centre of the earth. . . . All objects having mass by their nature tend toward [this point]. A needle being attracted downwards to a lodestone is a rather inadequate explanation of this. Chinese scholars, seeing only a flat surface, said the earth was flat. Westerners, using the principle of parallels, travelled far and wide over the oceans. Some went from west to east without interruption until they finally returned to their starting-point. . . . They found the earth was spherical, corresponding to the heavens above it. Thereupon they took the poles as a base for the meridians and the equator as a base for the latitudes, defining north and south and calculating east and west. . . . We have made a model in the shape of a spherical ball." (Translation by H. M. Wallis and E. D. Grinstead, in *British Museum Quarterly*, Vol. XXV, No. 3–4, 1962. The globe measures 59 cm. in diameter and is in the British Museum, London.)

West. He too took the opportunity of underlining the telescope's practical uses. "If there should break out unexpectedly a military revolution," he says with prophetic pertinence, "whether by day or by night . . . one can look at . . . the place of the enemy, the encampments, the men, the horses, whether the army is armed or not . . . ready or not, and also whether it is fitting to discharge the cannon."[9] The Manchu hordes massing beyond the Great Wall are implicit in the statement. Soon they were to break through and descend on Peking.

Terrentius was not far behind Schall. In 1628 he printed in Chinese a treatise on the movements of the stars and planets, in which he frankly labeled the old astronomy "fantastic." He went on to show how, with the telescope, one could see the error of such ancient fancies; how, indeed, Venus and Mercury were satellites revolving round the sun, not embedded in the layers of a solid crystalline heaven (as had been thought in Europe but never in China). Probably, as they peered through the lenses of the telescope and saw the rotundity of those planets, some Chinese were better able to comprehend the truth of Ricci's maps, with their contention that the earth also was a sphere; and maybe they were better able to accept that noble sphere, its seas and continents displayed in Chinese lacquers, which Longobardi and Diaz had constructed for them in Hangchow in 1623.

The moment of truth came with an approaching eclipse of the sun in Peking, due on June 21, 1629. As usual the astronomers of the Bureau of Astronomy attached to the court were officially requested to give their predictions. (The true explanation of eclipses, the fact that one heavenly body is covered by another coming between it and the earth, had been discovered by the Chinese in the first century A.D., but the phenomenon was nevertheless still regarded as an evil omen.) The Chinese and Moslem-Chinese experts of the bureau predicted that the eclipse would occur at 10:30 and last two hours. The Jesuits calculated that it would begin at 11:30 and last a mere two minutes. Precisely at 11:30 it happened, as the Jesuits foretold.

Two days later the highest officials of state presented a memorial to the emperor castigating the indigenous astronomers and eulogizing the Jesuit work. Two months later the Board of Rites, with Paul Hsu now its vice-president, memorialized the throne on how the lunar calendar could be corrected (it had been in error for about 350 years). And finally, on September 1, an imperial edict replied by entrusting the reform to the Jesuits. A fortnight later still, Paul Hsu in collaboration with Terrentius put forward a far-reaching program for translations from Western scientific literature on mathematics, optics, hydraulics, even music, and containing provision for the construction of dozens of up-to-date instruments of astronomy and time-

9. Philip Wang Chen, a Christian convert, quoted by d'Elia.

THE JESUIT PRIEST SEMEDO, who translated the Nestorian stone in Peking. The engraving comes from the English translation of his book, made in 1655. The book's subtitle is of interest: "Lately written in Italian by F. Alvarez Semedo, a Portuguese, after he had resided 22 years at the Court, and other Famouse Cities of the Kingdom. Now put into English by a person of Quality, and illustrated with several maps and figures, to satisfie the curious, and advance the trade of Great Britain." (British Museum, London)

measuring. And, of course, they advocated the making of telescopes too. Intellectual Peking was stirred with a deep enthusiasm. The implications of Western astronomy had begun to open closed mental doors. Eulogies poured forth on the subject of Western science and Jesuit skill. Even some totally un-Confucian speculations went on Chinese record!

But the strands of history, East and West, were now tangling, with consequences unforeseen by any at the time. Pope Paul V had assented to the China mission's celebrating Mass in Chinese instead of in Latin. By 1632, Galileo, the inventor of the much admired telescope, which was furthering the Christian cause in Peking, was a wretched man on trial in Rome, about to be broken over the knee of Catholic orthodoxy. Far away from that ignoble scene, in Peking, Father Semedo, who was to become the historian of the Jesuits in China, had added the curious note of his translation of the Nestorian Stone to the ferment of ideas in the capital. Suddenly, as if divinely revealed to Jesuits and Chinese alike, the existence long ago of legions of Christians in the Celestial Empire was discovered. The inscription made it evident that those Christians had been held in great respect. Now the Christian cause seemed once again to prosper a little. Paul Hsu, staunch Christian official, was elevated to the presidency of the Board of Rites.

But other events, in China as in Rome, were pulling in a contrary direction. Ironically, a few months before the emperor received his personal telescope (1634), which deeply pleased him, there was a strong reaction in certain circles against the geography taught by Matteo Ricci with the aid of his maps. At the very moment when his new geography might have been thought ready to take its rightful place, buttressed by the revelations of the new telescope, Ricci's contribution to Chinese knowledge of Western science came under fire from the Chinese: "Lately Matteo Ricci utilized some false teachings to fool people, and scholars unanimously believed him. . . . The map of the world which he made contains elements of the fabulous and mysterious, and is a downright attempt to deceive people on things which they personally . . . cannot verify for themselves. It is really like the trick of the painter who draws ghosts in his pictures. . . . Just take, for example, the position of China on the map. He puts it not in the center but slightly to the west. . . . This is altogether far from the truth, for China should be in the center of the world, which we can prove by the single fact that we can see the North Star resting at the zenith of the Heaven at midnight. . . . Those who trust him say that the people of his country are fond of travelling far, but such an error as this would certainly not be made by a widely travelled man."[10] The roots of this and other such diatribes were twofold: first, indignation at the off-center placing of the Middle Kingdom among the countries of the world: second, the slavish, illogical pursuit of popular concepts or superstitions which Chinese astronomers themselves would have rejected as untrue. But scholars in the literary arts were the only acknowledged scholars of China; and however much a number of them might from time to time be intrigued by science, fundamentally they mistrusted it and its exponents. The old steadfast traditionalism was beginning to creep back, outraged at the temerity of mere foreigners and their trumpery ideas in setting it aside. If Abbé Huc, who said that Christianity seemed only to "glide over the surface" of Chinese life, had thought of it, he might have included Western science as well.

As if to support the Chinese traditionalists, the Roman church in Italy

10. Quoted from Wei Chun by Kenneth Ch'en in the *Journal of the American Oriental Society,* Vol. LIX, 1939.

now condemned the Galilean extension of Copernican theory. Irony within irony—while Italian churchmen were looking through that revealing telescope of Galileo's and saying that although they saw what Galileo said they would see, the fact that they saw it had no bearing on the truth, the hidebound Chinese, on the other hand, were placing their acute little eyes to similar lenses and were much more open to accept the evidence of their optic nerves. For the moment, at least.

The second act of the Roman church about this time (1633) was to decree China open to the ministrations of *all* missionary orders. The Jesuits' preserve, by a stroke of the papal pen, was theirs no longer. Having done the spadework, they were quite naturally annoyed. When the Dominicans arrived in China, the Jesuits therefore received them pleasantly enough, but behind their backs made numerous efforts to cut the inexperienced order's feet from under it. There are unpleasant little stories scattered here and there, glossed over for the most part by Jesuit historians. For the first time—shades of things to come!—Westerners in China began to squabble and bicker among themselves. To Chinese it must have been a shock to learn that there were cracks in the faith which they had been told was one and indissoluble, that this allegedly shining concept was not faithfully reflected in a perfect temporal order in the West.

Beyond all this flurry of new science in Peking, and beyond the undeniable fact that it was Jesuit science which finally corrected, and showed how to keep correct, the Chinese lunar calendar—beyond and underneath there lies one unresolved little doubt. Did the Jesuits purposely hide the full facts, which they intellectually accepted, of the heliocentric theory? Did they, not to put too fine a point on it, deliberately purvey to the Chinese what was in truth out-of-date astronomy, simply because their church had condemned what they knew in their minds was the correct theory? *In effect* it appears that, at the moment when Western science was making one of its most audacious leaps into the future, what the Jesuits of the Ming gave to the Chinese was more or less the Western scientific past. Their attitude was—to put it mildly—ambivalent. It was almost bound to be. They contented themselves with teaching in China that the planets moved round the sun and were its satellites. But they evaded the logical, forbidden corollary that the earth did too. A pity. Because Chinese ideas were, broadly, much more in line with a Galilean universe than were the received ideas of European tradition and Catholic orthodoxy. To be fair to the Jesuits, it ought to be said that sometimes it seems their brilliant intelligence was very tenuously contained by the rigid box of Christian dogma.

The situation in China had by this time worsened. In 1629 the Manchu had broken through the Great Wall (it was of course the year of the eclipse, so traditional Chinese were doubtless prepared for the worst). Paul Hsu, becoming president of the Board of Rites in the following year, was instrumental in inviting a force of Portuguese soldiers to come north to Peking. There were high hopes of this force. It was tiny—a mere four hundred men—but equipped with small arms and with cannon of a type much superior to that used by the Chinese. But the force never reached the scene of the trouble. The merchants of Canton, fearing that their lucrative trade with the Portuguese, now firmly established in Macao, might suffer should success attend the expedition and the foreigners be granted privileges of trading in other places along the coast, intrigued with the local officials, bribed everyone who could use his influence at court, and had the Portuguese stopped south of the Yangtze. The force was turned back to Canton, where the merchants themselves paid the wages of the soldiers, and returned them safely to Macao. The story is interesting as the first instance of foreign mercenaries employed by a Chinese government, at any rate since the T'ang dynasty. Those Portuguese troops were in fact only the vanguard of the many mercenaries, including Chinese Gordon and a number of American adventurers, who would swagger through China in the troubled years of the next dynasty.

At this moment we have to turn the lacquered globe of Longobardi and look at the small island of England on its other side. For there Charles I in 1635 was granting a charter to the Courteen Association to sail to China in the hope of trading.[11] The East India Company, which held the monopoly of Eastern trade, had failed to produce money to fill the depleted royal treasuries, and Charles simply put it on one side and licensed the Courteen Association. He "put into the Joynte stock" the sum of ten thousand pounds as his share of the capital, and issued a royal commission to a certain Captain John Weddell as commander of the squadron of four ships and two pinnaces. He also gave letters of instruction to Weddell and to the "Chief Merchant" Nathaniell Mountney. With the expedition went Thomas Robinson, who spoke Portuguese, and Peter Mundy, who was to be the diarist and historian of this venture.[12]

John Weddell, previously in the service of the East India Company, was a useful renegade. His skill as a commander was nicely balanced by his inveterate breaking of the company's rules forbidding private trading while in their employ. King Charles's opinion of him was apparently glowing. In a letter to the president and council of the East India Company in India he says: "Trulie and well beloved, wee greet you well Whereas the good Shipps the 'Dragon' the 'Sunne' the 'Katherine' the 'Planter' and the 'Anne' and the 'Discovery' are set forth by our spetiall command for a voyage and

11. Almost forty years earlier, in 1596, Queen Elizabeth had sent a letter from England to Wan-li which seems not to have reached him. Charles was in too much haste to make money for any such diplomatic approach.

12. A good brief outline of this first episode in Sino-British relations is in Morse's *Chronicles*. Morse, an American, headed the Chinese Maritime Customs, 1886–1907.

"ENGLISH WAR-SHIP SHOWING A ROW OF CANNON," engraved by Wenceslaus Hollar in 1653. It was probably in this type of ship that Weddell sailed to Canton. (British Museum, London)

discovery to the South Sea under the principall charge of our trustie and well beloved subjects Captain John Weddell and Nathaniell Mountney, in which adventure we have a particular interest. . . ."

In April, 1636, the ships left England for the East. In India they fell foul of the East India Company, which little cared to see its preserves poached upon, even with the king's consent. At Goa, Weddell received irate letters from the company, and replied: "We are not taken with a few flashes and peremptory jeering menaces, nor ledde away with unnecessary verball complementall congratulations. But bee assured that if you attempt the least underhand injury by your suppostitious tricks, though it never come to perfection . . . eyther you or yours shall answer it to no mercenary man. You doe me wrong to taxe mee with cutting the Company's stringe of trade . . . but every cocke will crowe etc." And much more to the same effect.

The ships sailed on to China. By the time they arrived off Macao, Weddell was fifty-four, well versed (he appears to have thought) in every trick of the Eastern trade. He seems to have had no hesitation in heaving to within sight of the jealously guarded Portuguese trading station. He was that kind of man. "Every cocke will crowe," his own phrase, describes him well enough. And doubtless he would have been able to deal with the Portuguese at Macao had it not been for the economic facts of South Chinese life. The

PETER MUNDY'S CHINESE SKETCHES. Mundy, who accompanied Weddell to Canton in 1637, kept a lively diary that included several sketches. His various types of Chinese boats are little different from many to be seen in China even today, and how frail they look in comparison with the preceding illustration. The sketch of a Chinese magistrate at work is reminiscent of Ricci's interviews with the governor of Chao-ch'ing. The dark shading of a figure who is apparently out in the rain is doubtless intended to show the straw raincoats that peasants still wear in South and Central China. (From Rawlinson MS, Bodleian Library, Oxford)

Portuguese were there on sufferance, a monetary sufferance in many ways welcome to the merchants of Canton, and therefore also to the governing officials of the province, who were bribed to turn a blind or at least indulgent eye on this irregular trade. The combination of the two nations, Chinese and Portuguese, proved to be Weddell's undoing.

Weddell's arrival put the Portuguese governor of Macao in a difficult position. He had insufficient force to drive Weddell off, and every reason to fear for his trade with Canton should the English get a footing there. He therefore adopted delaying tactics while, we may presume, seeking means to circumvent Weddell. The impatient Englishman, however, soon sent one of his boats, the *Anne,* to find the entry to the river and so to Canton. The *Anne* sailed up to within fifteen miles of the city. Weddell then took his fleet as far as the river mouth. Poor winds and unknown currents hampered him and made the journey very slow. But the Chinese were even more discouraging than the weather.

According to Peter Mundy, whose freely spelled record we will follow in part, "a greatt Mandareen att Cantan" came aboard the *Dragon* to "Know our intentts and Demaunds." Two weeks later, when the ships were still at the river mouth, came four more mandarins (one of whom had traveled from Canton) to take notes on the number of crews, armament, cargoes, and quantity of money the English had to invest in trade. It was indicated that the English could trade—but only at Macao. Weddell was asked to stay where he was for the time being; but "thatt nightt wee wayed" and went on up river. Then "there came to us . . . another fleete of greatt China Juncks, The Kings Men of Warre, about 40 saile," who told Weddell to anchor. He did, but asked permission to go farther upriver for shelter. And when the answer to this did not come, he sailed onward.

The Chinese were now evidently preparing to attack the squadron of foreign ships. Weddell hoisted the "Kings coullours on our Mayne toppes" and the Chinese, apparently understanding the significance of the flag, sent to ask Weddell to wait a few days for Canton's authorization before proceeding. Weddell agreed. But soon he sent off his barges to take soundings. The boats were shot at for their pains. This was more than Weddell's temper could withstand. "Then outt again went our Kings Coullours . . . and wee came uppe, Anchored Near unto [the fort] and beesett it with our four shippes." The English fire was accurate and they landed and took the fort, "tooke Downe the China Flagge," and hoisted the king's colors. They removed quantities of Chinese cannon as prizes of war. Capturing a few passing Chinese vessels, they sent a note to Canton with a fisherman "shewing therein a reason of our thus proceeding with the Chinese and thatt . . . our Desire was to have their Freindshipp and Free Commerce in their country."

By this time the Portuguese had had time to get in touch with Canton.

THE APPROACHES TO CANTON at the time when Portuguese and later English ships were making their early trading ventures to Canton. Actually, the cities of Kowloon and Victoria did not exist until the nineteenth century.

We shall never know *if* they did, but it seems likely that the unsavory Pablo Noretti, who arrived from there at this time, was a not unwilling tool of the Portuguese, possibly doing a double deal with the Chinese merchants also. He brought to Weddell what he said was a "warrant from the Great Mandareens at Cantan" informing the English that if they would surrender the captured guns and pay the Chinese duties as did the Portuguese, doubtless they would be allowed to trade. Noretti took Mountney and Robinson back to Canton. Later he returned with yet another document in Chinese which he said appointed him as broker in any trade. Weddell seems to have been satisfied with this and sent three merchants and a quantity of silver to Canton.

Noretti, however, had lied. The correct translation of the document exists in Portuguese records in Lisbon (a fact which tends to confirm Portuguese complicity with the Chinese in getting rid of Weddell). The documents actually complained of the high-handed acts of the English and directed that the ships of the red-haired barbarians put out to sea forthwith. "Should you have the great boldness to harm so much as a blade of grass . . . I promise you that my soldiers will make an end of you, and not a shred of your sails shall remain. . . ."

Seeing signs of a "Tuffaon,"[13] Weddell moved, contrary to instructions, further upriver. He then sent more money to Canton and replied "in a slighting Manner" to a Portuguese note of protest. But his hopes were soon dashed. At two in the morning the crews sighted three fireships bearing down on them. Weddell only avoided destruction by slipping his cables immediately. Incensed, he debated what to do, for his merchants were still in Canton. In the end he decided to "doe all the spoile wee could unto the Chinois." He began to burn junks, he set a small town on fire and carried off its pig population.

Mountney sent a letter from Canton saying he was a prisoner and had lost sight of the money. Weddell then went ashore and blew up a fort. Then he dropped downriver to within four leagues of Macao, sending a letter to the authorities there accusing them of providing and fitting out the fireships. This was repudiated by the governor, and Weddell (who appears to have been scared of going ashore at Macao) eventually had to climb down and humbly request the Portuguese to rescue the merchants from Canton. The governor acceded in return for Weddell's promise that when the merchants returned he would "depart peacefully from Chinese waters, without injuring anyone" and "would never return to these shores." Two days later he was given permission to conduct a "limmitted trade" in Macao. The merchants returned ignominiously. Weddell and Mountney signed a document of apology to the Chinese in the most abject terms. Perhaps to restore somewhat his deflated ego, Captain Weddell wrote to the Courteen Association recommending a wild scheme to capture the large island of Hainan as a base for future operations. And so the English sailed away.

The king in England got nothing from the venture. The English had only succeeded in establishing themselves in Chinese eyes as barbarous people,

13. I.e., a typhoon, a word possibly derived from the South Chinese pronunciation of two Chinese words meaning, literally, "great wind."

like the Dutch, and the Portuguese position was strengthened since the Chinese now knew that at least they could deal with them. Thus the three major Western seafaring powers had each begun its relationship with China under clouds of arrogance, trickery, and willfulness. The Chinese contained the Portuguese and managed to dismiss the others. Virtually all that had been accomplished was to confirm in Chinese opinion the savage, uncultured, and unpredictable characteristics of all Westerners in ships—and therefore of Westerners in general. They were right—and also wrong. But that was not entirely the Chinese fault. The West sent adventurers, and as such the West was categorized by the Chinese. It was not unfair.

Three years later the end of the Ming was in sight. Rebels under their leader Li Tzu-ch'eng were in control of most of southern China, moving from strength to strength. A great dynasty was about to fall. Not the greatest—for that we must look, as the Ming themselves did, to the T'ang—but a very Chinese three hundred years.

Even in their moments of closeness to a few of Europe's best minds, and in their contact with the marvels and advantages of the science these Europeans brought and wanted to give away, the Chinese of the Ming could not quite bring themselves to accept. They admired, sometimes they were excited; then they copied, as they had for centuries copied their own discoveries and texts; and the heart went out of the matter, and they eventually turned on most of those Western things. A sentence or two of Galileo's own thought makes clear the divergence of the Chinese and the European outlook at this time: "Philosophy is written in that great book which ever lies before our eyes—I mean the Universe—but we cannot understand it if we do not first learn the language and grasp the symbols, in which it is written. This book is written in the mathematical language, and the symbols are triangles, circles and other geometrical figures, without whose help it is impossible to comprehend a single word of it; without which one wanders in vain through a dark labyrinth."[14]

The gulf was opening between science and the ancient way of religious thought in Europe. With Galileo and others the West began to consider that the book of nature was an authoritative source, that man could stand outside himself and look both at the world and himself from that external vantage point. The Chinese opened no such gulf between themselves and their ancient way of thought. They seem to have feared to be other than they already were. Within that wonderful system of morality and polity stemming from ancient times (a system just as splendid as any Christianity ever raised) they felt safe, if a little dull. They thought—and here they were undeniably right—that foreign philosophies were not only basically antagonistic to theirs, but that those philosophies would eventually prevail over them if

14. Quoted by Bronowski.

they were allowed free rein in China. When we recall the violence of the Christian rearguard action in the face of scientific "erosion" of its antique outlook on the world, the universe, and human relations in Europe alone, the Chinese reaction appears quite natural. Instinctively the Chinese allowed no such invasion of their philosophical domain. What they could not eject, they contained and neutralized. What they could not contain (and it was little) they ridiculed in a gentle, Confucian, whimsical way. For about two hundred years after the end of the Ming their smiling passive resistance succeeded in excluding the central message of science. But this was at their extreme peril.

PART FOUR

SUITORS OF THE MANCHU

Ambassadors, Intriguers, Jesuits, a Scotsman, and an American

CHAPTER TEN

A GERMAN, A RUSSIAN, AND TWO DUTCH MERCHANTS

MANCHU horsemen took Peking. A virile horde from the steppes and forests, from the hard mountains and sharp skies of the northwest beyond the cultural barrier of the Great Wall, they swept down on the North China Plain. They took Peking almost by accident—through the perfidy of some of the Ming defenders and the miscalculation of a Ming general. Doubtless they would have captured the city in any case, even if General Wu of the Ming had not sided with them in the mistaken opinion that he was thereby driving out the rebels from the south who had recently occupied Peking. But once in Peking the Manchu armies stayed on and set up their own emperor as the first of the Ch'ing—the Pure—dynasty, and began the task of subjugating the rest of the country. If we accept this interpretation of General Wu's actions (and some authorities disagree), the stalwart soldier must perforce join the ranks of history's dupes along with his twentieth-century companion Neville Chamberlain. But his Chinese sword seems somehow less pathetic than the latter's Establishment umbrella.

The days of the Great Ming were over. But the days of the diligent Jesuits at the Peking court and elsewhere were not. They were nothing if not painstaking and courageous. Many in the provinces lost their lives in disturbances as the Manchu conquest swept south—one, for example, was drowned with 200,000 Chinese when a patriotic general breached the Yellow River's banks (an act repeated with even greater loss of life by Chiang Kai-shek three centuries later) in a misguided effort to oust the besiegers of Kaifeng. In a sense the Jesuits in Peking, led by Adam Schall, were in a position similar to that of ambassadors in a modern state when that state suffers a *coup d'état*. But in the seventeenth century they had no way of consulting their home government—the pope—since letters took a minimum of a year to reach Europe. With the hazards of piracy, tempest, and other catastrophes generally attributed to a benevolent God, letters often did not arrive at all. So the

EMPRESS DOWAGER HELEN in court dress. (From Favier)

Jesuits had to make up their own minds on the question of recognizing the *de facto* Manchu regime. It must have been a hard choice. Their position was insecure, their only defenders their own clever tongues. Some bent with the Manchu wind, staying in Peking to serve new masters; others fled with the Ming to a tattered exile court, a pasteboard replica of the luxury of the former Ming court at Peking.

The handful of decamping Ming royalty consisted of the late emperor's brother, Prince Kwei, who now set himself up as emperor at Kweilin amid that dreaming landscape of sudden hills wreathed in mists, that landscape so greatly beloved of Chinese painters through the centuries. And with him there in Kwangsi province far to the south were two genuine empresses dowager, his wife (now styled empress), and his son and heir. The sham little court spent a harried existence far from the seat of real power, the "emperor" issuing decrees for the government of a country from which Heaven had manifestly withdrawn his Mandate to Rule.

The ladies of the court appear to have taken comfort from the services of a German Jesuit named Andrew Xavier Koffler. His single claim to fame is that, of all those courtier Jesuits over the years, he was the only one who managed to convert to Christianity an empress, two dowager empresses, and an heir apparent. By the time he achieved this feat the conversion was a hollow one. For the royal family ruled nothing, not even, one may suspect, their own aristocratic heads. Koffler it must have been who persuaded the

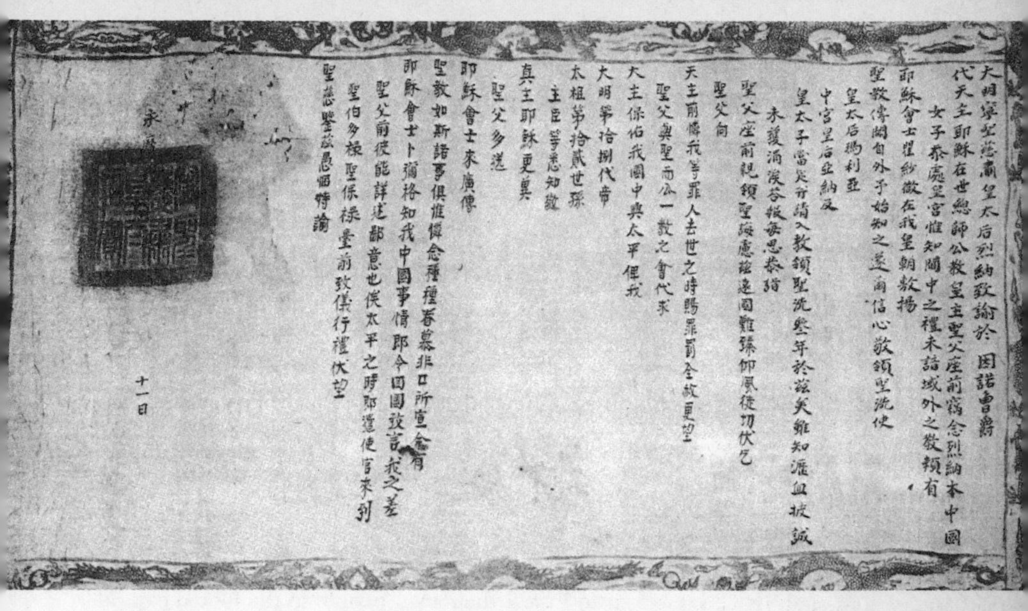
大明寧聖慈肅皇太后烈納致諭於 因諾曾爵
代天主耶穌在世總師公教皇主聖父座前竊念烈納本中國
女子忝處皇宮惟知閫中之禮未諳域外之教賴有
耶穌會士瞿紗微在我皇朝敷揚
聖教傳聞自外予始知之遂爾信心敬領聖洗使
皇太后瑪利亞
中宮皇后亞納及
皇太子當定並請入教領聖洗三年於茲矣雖知瀝血披誠
未獲涓涘答報每思恭詣
聖父座前親領聖誨慮茲遠國難臻仰風徒切伏乞
聖父向
天主前憐我等罪人去世之時賜罪罰全赦更望
聖父與聖而公一教之會代求
天主保佑我國中興太平俾我
大明第拾捌代帝
太祖第拾貳世孫
主臣等悉知敬
真主耶穌更冀
聖父多遣
耶穌會士來廣傳
聖教如斯諸事俱惟憐念種種眷慕非口所宣今有
耶穌會士卜彌格知我中國事情即令回國致言我之差
聖父前彼能詳述鄙意也俟太平之時即遣使官來到
聖伯多祿聖保祿臺前致儀行禮伏望
聖慈鑒茲愚悃特諭
永
十一日

Helen's Letter to Innocent X, dated "the fourth year of Yung-li, tenth month, eleventh day," or November 5, 1650, reads in part: "The Empress Dowager Helen of the Great Ming Dynasty conveys her commands to the Throne-front of the Holy Father. . . . She humbly reflects that she, Helen, beginning life as an ordinary Chinese girl, is now an unworthy occupant of the Imperial Palace. Ignorant of the Faith of Foreign lands, she is only cognizant of the established rites governing the female apartments. But, thanks to the Jesuit Xavier Koffler's preaching the Holy Faith from abroad at the Imperial Court, she has now first heard of it, and, accordingly, with believing heart she has received baptism. . . . Often she has thought of proceeding in person to the Front of the Holy Father's Throne [but] she fears difficulty of access from this remote country, and she can therefore only indulge in a vain longing. . . . She . . . trusts that the Holy Father . . . will vicariously entreat the Lord of Heaven to protect her country, to its restoration and peace, so that the eighteenth Emperor of the Great Ming . . . may, together with his subjects, know how to revere the true Lord Jesus. . . ." (Translation by E. H. Parker in the *Contemporary Review,* January, 1912. Photo by courtesy of the Vatican Archives, Rome.)

dowager empress who took the baptismal name of Helen to try to enlist papal support for the already lost Ming cause. In the winter of 1650 she sent a letter to the pope in her own indifferent handwriting, the original of which is preserved in the archives of the Vatican.

Attended by Koffler, the exiles were driven from city to city before the

Manchu armies. Expelled from Canton on its capture, they briefly returned when the city was recaptured for the Ming by a Christian Chinese general with Portuguese guns in 1648. But in 1651, Canton was finally retaken and the last demoralized emperor of the Ming fled to Yunnan, and later, it seems probable, to Burma. But even in his extremity he did not succumb to Koffler's blandishments, remaining faithful (as all Chinese emperors of necessity did) to the beliefs of his ancestors. The conversion of his women probably meant very little, for Christianity had been something of a fashion in high society for many years in Peking. It was more respectable than Buddhism, if only because its priests were better educated and more cultured.

Jesuits in Peking, meanwhile, were engaged in the much more weighty task of entrenching themselves in the favor of the new rulers, continuing what we must see as their real destiny in China—the exposition of Western science and culture (as distinct from religion) to the Chinese. The parlor games they had to play at court may seem comical and even degrading for men of their intelligence, but (as intellectual courtiers have found elsewhere both before and since) only in such ways could their position be maintained. Their reward was not far distant, though it was hardly the one they primarily sought.

One of the first acts of the Manchu rulers was to evacuate the city of Peking of all its Chinese and to install their own people in the looted capital. Adam Schall, who had been persuaded by the last Ming ruler to set up a foundry for casting Portuguese-style cannon, was in a dubious position with the Manchu. But by some means he managed to preserve his house, books, and astronomical instruments intact in the general melee. Courageously, Schall composed a memorial to the boy-emperor Shun-chih pointing out that his life work as a Jesuit astronomer was the reform of the confused Chinese calendar, without which even the Manchu knew no dynasty could hope to regulate the affairs of agricultural China; and showing, moreover, that the propagation of the True Religion would be retarded if he were forced to leave his house and books. Oddly enough, the Manchu officials overlooked the second point (probably deciding it was mere irrelevance) and took up the first. Schall found himself protected, even provided with a decree to adorn his front door. He proceeded with reform of the calendar.

In Shun-chih, the first Manchu emperor, the Manchu found themselves saddled with a child ruler, five when they enthroned him, six when they took Peking. His uncle acted as regent until 1650 when Shun-chih began to reign. He was then twelve. Already his real nature, which was contemplative, neurotic, and introverted, was apparent. One of his first edicts concerned the rules for admission to the Buddhist priesthood. His court was marked by a rigid respectability which, in its complete reversal of the profligacy of the last Ming courts, is reminiscent of what happened in the court of

England at Queen Victoria's accession. The hordes of palace eunuchs, whose presence had done as much as any other factor to vitiate dynastic rule in China, were banished for the time, and a certain Manchu order, crude by Chinese standards, prevailed. Shun-chih married at fifteen. His third son was to succeed him as the illustrious K'ang-hsi, of whom we shall hear more later. Shun-chih reigned eighteen years, until 1661, the officially proclaimed year of his death. Whether—as some Chinese historians aver—he secretly handed over power at that time to a council of regents and withdrew to monastic life, or whether he actually did die, is one of the minor riddles of Chinese history.

At this time, far away in Moscow, the tsar Alexis, second of the Romanov dynasty, had begun to show that duality of interest—deep attachment to traditional Russian roots combined with a strong attraction to Western European culture and progress—which was to modify the medieval state of Russia as the seventeenth century went forward. By 1651 news of the Manchu effort to conquer Siberia, together with reports of that area's wealth, reached Moscow. Alexis sent a force of three thousand men toward the Amur River to occupy the territory. The force reached the Amur in August, 1653, but encountered severe Manchu opposition. It was unable to build forts to defend the area and retired the following year. Alexis was urged by his advisers to abandon the idea, but instead ordered his forces eastward again. There on the Amur in 1657 they encountered a fleet of about fifty Manchu barges armed with cannon. In the ensuing battle the Russian troops deserted in large numbers and their commander was killed.

Communications across the wastes of Central Asia being what they were—similar to those encountered by William of Rubruck long ago—the tsar was doubtless ill-informed of the exact military situation. His news was probably six months old. Hence, his indecision. While with his right hand he sent troops to the Amur, with his left he appears almost simultaneously to have launched one Theodor Isakovich Baikoff to Tobolsk (the city from which Petlin and Mundoff had previously departed) to discover what sort of goods might find a market in China, to ascertain how to take them there, and to seek details on the armed strength of China and its internal commerce. In 1654, Baikoff was dispatched toward Peking with a letter from Tsar Alexis to the emperor. Baikoff must have quailed at this command, for none knew better than he did, living in Tobolsk, the facts of the undeclared Russian war on the fringes of the Chinese empire and the probable reception he could expect in Peking—if he ever reached it.

The contents of the tsar's letter could have done nothing to ease Baikoff's fear. Alexis begins by recounting a rigmarole of his ancestors, their fame, and their dominant relationship as tsars of Russia with their neighbors (a gentle hint to China to follow suit). He comments that no previous tsar has had any written communication with a Chinese emperor, hence the proper titles of the Manchu sovereign are unknown to him. A reply, he intimates, will settle that, as it will also cement the friendship between the two rulers and promote their "loving intercourse."[1]

Baikoff's instructions were explicit. He was utterly encumbered by them. But they probably *had* to be explicit since, like Petlin and Mundoff, by his own admission he was illiterate. In the sending of such a man we may perhaps sense the veiled contempt in which the emperor and people of China were held by a still semibarbarous Russian tsar who considered Baikoff an appropriate envoy. He was told to go by the shortest route, forbidden to bow to the palace or to any threshold—the Russians confusing the Mongols, to whom the threshold (shades of William Rubruck!) was sacred, with the Manchu, to whom it was not. He was not to kiss the emperor's feet (no one ever did), not to give the tsar's letter and gifts to anyone except to the emperor himself, and not to explain except in most general terms the object of his mission to any but the emperor himself. And with the emperor he was to elaborate on trade and friendship, to invite Chinese envoys to Russia, assuring the Chinese of customs and tax exemption if they came. Secretly he was to ferret out the emperor's private opinion of Russia, China's strength in armies and money; to discover if a war was in progress and with whom; to enumerate precisely the articles China produced, its harvests, transport systems. And finally to describe minutely the best route to China.

Baikoff was halted at Kalgan (Chang-chia-k'ou, at the Great Wall, a hundred miles northeast of Peking), while the officials there reported to Peking. When Peking's instructions came, Baikoff was forwarded in the prescribed manner as a tributary to the emperor. A sentence or two of his naïve description of what he saw en route is worth quoting, for it gives a little of the flavor of the man: "And in the towns they carry before the *voevodos* [i.e., high officials] great yellow cotton umbrellas, borne on poles each by one man; and on either side of him go one or two men; and in their hands they carry sticks, gilt at the ends; and as they go along they cry out in their language, but what, one knows not; and the *voevodos* are carried by four or five men, as it were in huts on poles."

On March 3, 1656, a Monday, Baikoff arrived at Peking, met "half a verst beyond Khanbalik [a long-disused Mongol name for the city]" by ten men, who ordered him "to dismount and bow down to the palaces." He patiently explained he could not. Finally they took him on and lodged him poorly in Peking. Officials came to him the following day to take the gifts he had brought for the emperor, in order to make the usual inventory. Again Baikoff demurred and explained that he could give them only to the emperor himself. The officials took the gifts by force.

1. Details and quotations concerning the Baikoff embassy are taken principally from Baddeley.

Two days later he was commanded to bring the tsar's letter, presumably to the Board of Rites. Six months later, the Chinese being nothing if not patient, it seems the board was still asking for the letter and Baikoff was still refusing, doggedly, to surrender it to anyone but the emperor. Finally, on August 31, the Chinese brought the matter to an end by returning the tsar's gifts, thus curtly dismissing Baikoff for his many refusals. Harking back to the first of them, the board pointed out that since the messenger whom Baikoff himself had sent ahead to warn them of his coming had bowed down to the palaces, why then had Baikoff refused?

"And Theodor Isakovich Baikoff [as he later dictated to a Russian clerk] was dismissed from Khanbalik with the Tzar's letter . . . on the 4th September, 1656, and he was sent away by no means politely, nor was any transport furnished to carry the Treasury Goods [the presents]." In view of the known fact that his countrymen were giving serious trouble to the Manchu government along the Amur River, the treatment Baikoff received actually seems quite lenient.

There, as he quits the capital, we must leave Baikoff for the moment to describe another event that had occurred while he was at Peking, one that reveals more about his embassy, and about the Chinese reaction to incoming Westerners, and finds the Jesuits in a role of doubtful propriety.

On July 17, 1656, a few months after Baikoff arrived in Peking, a party of solid Dutch merchants was escorted into the city and put up in that Palace of the Four Barbarians from which Ricci had cleverly extricated himself years before. Sent by the Dutch governor of Batavia, the merchants, like Pirès, had experienced the usual delay of many months at Canton while the administration made up its mind what to do in the case of the first "embassy" from a people who had already fully earned the unpleasant vernacular names applied to them by the Chinese. Led by two merchants, Peter de Goyer and Jacob de Keyzer, the party consisted of "six waiters, a steward, a surgeon, two interpreters, a trumpeter, and a drummer." The last two must have seemed to the Dutch a trifle redundant when they reached Peking and encountered the pomp of the Manchu court.

The "steward" was John Nieuhof, an inaccurate chronicler of what was to pass in China, but a fair draughtsman, from whose sketches originated the first engravings of Peking to be published in England. The single paragraph on his life in the Dutch National Biography lends a touch of the bizarre to a figure who, from his prose, we might otherwise dismiss as a sluggish enough being. A traveler and author, he "lived in Batavia from 1667–1670. His heavily illustrated works are not of much value and not always reliable. . . . His verses about Batavia are not amusing. He tells a curious tale of climbing on a stranded whale in Table Bay. When the author [and two others] were on top with a trumpeter [perhaps it was at Nieuhof's behest that the embassy trumpeter came along to Peking?] they got the latter to blow a lusty rendering of 'Wilhelmus van Nassouen' [perhaps a patriotic song?]. . . . In 1693 there was published 'An Embassy from the Netherlands East India Company to the Grand Tartar Cham.'" It is from this fancifully decorated volume by Nieuhof that our information on the Dutch embassy is derived.

As with Baikoff, the day after the Dutch arrived, officials of the Board of Rites came to list the presents. The merchants were bombarded with a series of questions about the Dutch and their motives. The merchants avowed that the reason for their trip was "to establish a firm league with the emperor and to obtain free trade for the Dutch throughout his dominions." They had said this before, in Canton, and had been sent on to Peking, so now they apparently thought the request was about to be granted more or less automatically. Consumed with a sense of their own importance, the Dutch and others before and after them refused to understand the Chinese tributary system. Bringing their presents, the Dutch left behind their sense of reality. They supposed that China *must* want to trade. All available information should have demonstrated that Chinese did *not* desire to trade.

The Chinese, for their part, wanted all the information they could get. It took some time, however, before they were convinced that the Dutch had a permanent homeland on a continent, since they had only heard of the Dutch as sea nomads. The Chinese elicited the information that the Dutch form of government was a federation called the United Provinces, a political form utterly incomprehensible to dynastic Chinese. But they pursued the interviews with queries on Dutch armaments and how they were made, and on Portuguese-Dutch relations. Both sides were reasonable in their attitudes to one another. The Dutch in particular seem to have made a good attempt to allay suspicion of their compatriots. Soon they were summoned to bring their presents to the "emperor's council."

"The chief commander sat at the upper end of the hall on a broad bench, with his legs crossed like a taylor; on his right . . . sat two Tartar lords, and on his left Adam Scaliger [Schall], a Jesuit, and a native of Cologne in Germany, who had lived in great honour at the court of Peking near thirty years. He was a very comely old man, with a long beard, and went shaved and clothed after the Tartar fashion."

Taking out the presents, the merchants described each, Schall interpreting for them. "And when any that was curious appeared, he fetched a deep sigh."[2] The moment is vividly reminiscent of a Chinese exile who, like Schall, had long been away from his country, bereft of his friends who had died: "Beholding only his own shadow, he was constantly sad at heart; and when suddenly [in India] he saw a merchant make offering of a white silk

2. The foregoing quotations come from Pinkerton.

fan from China, his feelings overcame him and his eyes filled with tears."[3] The Jesuit had been in China for thirty-four years at the time.

The Chinese had given Schall the task of interpreting for the Dutch embassy and of making a report on the conduct of the whole affair. This he did, but, says Nieuhof, he "deceitfully added of his own accord that the country which the Dutch then possessed was formerly under the dominion of the Spaniards, and did of right still belong to them." This was true in so far as Holland had been conquered by Spain at one time. But the Board of Rites ordered the passage to be deleted from the report, since it might offend the emperor by causing him to doubt if the embassy actually represented a real country.

Nieuhof, with a rare spark of humor, then entertains the reader with an amusing scene. "Whilst the clerks were taking several copies [of Schall's report] . . . his Highness [the president of the board] found himself hungry, and sent for a piece of pork . . . which was half raw, whereof he did eat most heartily in so slovenly a manner that he looked more like a butcher than a prince." Soon after, dinner was served. "His Highness and the rest of the Tartar lords fell on again as greedily as if they had nothing to eat all day; but neither the ambassadors nor Father Adam could eat of their cookery, most of the meat being raw." Manchu manners had not yet been refined by contact with the Chinese.

At this juncture Schall informed the Dutch that a Russian embassy under Baikoff had forestalled them in Peking by four months. And then he left, impressing the merchants by the grandeur of his palanquin and attendants.

Soon Shun-chih told the board to inform the Dutch that when his palace was repaired (it had suffered from the conflagration at the end of the Ming) he would send for them. But at the same time the officials began to debate about how often it might be prudent to allow Dutch trading visits to China. They mentioned an episode many years ago when ships, perhaps those of Weddell, had shot up the forts at Canton. Anticipating a favorable answer to their requests, the ambassadors were surprised; the more so because they had laid out 3,500 taels of silver to the Canton officials to ensure that their wishes were pressed in Peking. This money, they now feared, had in the quaint phrase of the English translation been "trepann'd by the viceroys." Moreover, they heard tales that Schall and other Jesuits in Peking had been bribed by the Portuguese to oppose the Dutch suit. This is possible though not probable. The Jesuits were basically opposed to the Dutch on religious grounds, and perhaps on racial grounds too. They did not need to be bribed.

Still later, the merchants were taken "in great state into a room of the old

3. From *The Travels of Fa-Hsien,* translated by H. A. Giles. Fa-Hsien was a Buddhist priest who walked all the way from China to India, stayed there many years in search of the true Buddhist scriptures, and eventually managed to return to China in A.D. 414.

palace, much like a library; for we saw none but scholars and gown-men with their books in their hands." In an open high-walled court "we were commanded at the voice of the herald to kneel three times and to bow our heads to the ground: after a short pause the herald proclaimed aloud in the Chinese language, *Cafchan,* which in English is 'God hath sent the emperor.' Afterwards he cried aloud, *Quee,*—'Fall upon your knees': then . . . the word *Canto*—'Bow your head.' After that *Coe,* bidding them stand up. And this he did three times . . . one after the other, wherein we also conformed." The Dutch made absolutely no bones about the kowtow. They were not emissaries of a proud king, not charged with an attempt to set up diplomatic relations between states. They simply wanted to trade.

Having achieved at least the promise of an audience, they must have been delighted to learn that the Russians were leaving without one. Nieuhof writes: "One of his [Baikoff's] gentlemen came . . . whilst the ambassadors were at dinner, and took leave in name of all the rest; and he desired . . . the favour of a letter to show in Russia that he had found us here; which was presently granted." Baikoff, doubtless fearful of the tsar's rage when he returned empty-handed, was taking no chances of being accused of not at least reaching Peking.

"Upon the day appointed for the long expected audience" officials came "in very rich habits to our lodgings about two o'clock in the afternoon" to conduct the embassy to court. The ambassador and his suite "came to court. [They were] placed upon the second plain of the court where [they] sat all night in the open air, upon the bare stones, till morning, when His Majesty was to appear upon his throne." They were joined by the ambassadors of the Great Moghul of India, a Tibetan embassy, and others. Then: "At the Court gate in which we sat expecting the dawn, we saw first three black elephants gallantly adorned . . . standing there for the greater state as sentinels. They had upon their backs gilded towers . . . beautiful with carved works and figures. The concourse of people was here so great; as if the whole city had been thronged together. . . .

"By daybreak all the Grandees . . . came gazing and looking upon us with great admiration as if we had been some strange Africk Monsters; but without giving us the least affront. About an hour later a sign was given, at which all started up on a sudden. . . ." Led to another courtyard, "guarded round with Tartar soldiers and courtiers," and then to a third court, the innermost, they saw "the House of the emperor's Throne. . . . This court . . . four hundred paces [square] was lined on all sides with a strong guard . . . in rich coats of crimson coloured sattin. On either side of the throne stood a hundred and twelve soldiers" with flags and different colo red uniforms. "Next to the emperor's throne stood twenty-two gentlemen, each with a rich yellow screen [perhaps a circular, long-handled fan] . . . in his hand, resembling the sun; next to these stood ten other persons each holding a gilt radiant circle in his hand, resembling the sun; next to these stood six others with circles imitating the moon at the full." And there were pikemen with tasseled weapons, thirty-six standard-bearers with dragon flags, "besides an infinite number of courtiers." (Nieuhof does not tell us, but one feels that by this time the Dutch had dispensed with the services of their paltry, single trumpeter.)

"Before the steps [of sparkling white marble, still there today] leading to the emperor's throne, stood on each side six snow-white horses, most curiously adorned with rich-embroidered trappings, and bridles beset with pearls, rubies. . . . Whilst we were beholding with admiration all the pomp and splendour of this court, we heard the noise and jingling of a little bell, sounding sweet and delightful to the ear." Then "thirty of the most eminent persons and chief councillors of the empire" appeared and went to make their kowtows in great state and humility to the throne, bowing nine times at the words of the herald, while delightful vocal and instrumental music was heard. Then came the tribal ambassadors, and then the envoys from Tibet. Before the great doors opening onto the throne room were brass markers set in the flagstones of the courtyard, each inscribed with an ideograph. The Dutch were now directed to advance to the tenth marker—one below the lowest grade of official.

"The herald called to us aloud: 'Go stand before the throne!' " They went. "The herald called again. . . . 'Bow your heads three times to the ground!' Which we did. And at last he called to us. . . . 'Rise up!' And we rose; and this happened three times. . . .

"Soon after, the bells tinkled again, which called all the people upon their knees. We endeavoured what we could to get a sight of the emperor in his throne as he sat in state, but the crowd of his courtiers about him was such that it eclipsed him from us in all his glory. He sat about thirty paces from the ambassadors; his throne so glittered with gold and precious stones it dazzled the eyes of all beholders. The ambassadors could discern very little of him." They drank tea with milk in wooden (probably lacquer) dishes. The "grandees" were sumptuously clad in gold, and the "lifeguards" (forty of them on either side of the throne) armed with bows and arrows. All these hindered the Dutch view of Shun-chih.

"This mighty prince having sat thus in magnificent state for about a quarter of an hour, rose up with all his attendants; and as the ambassadors were withdrawing, Jacob de Keyzer observed the emperor to look back after him, and for as much as he could discern of him, he was young, of fair complexion, middle stature, and well proportioned, being clothed and shining all in clinquant gold." It is one of the few descriptions we have of the young Shun-chih, and only the "clinquant" saves it from mediocrity.

Jacob de Keyzer was probably not mistaken in thinking the emperor looked back; for no sooner were the Dutch returned to their lodgings than a message arrived from Shun-chih desiring to see some Dutch clothes. So they sent him a "black velvet suit and cloak, boots and spurs, a pair of silk stockings . . . drawers, a band [perhaps a cumberbund], a shirt, a sword, a belt, and *bever* [a fur hat]." Now that the great moments were all over, the ambassadors were elated, astonished, impressed—but also naïvely crestfallen that the emperor had not spoken to them.

Meanwhile a first-class Jesuit intrigue had been afoot. While Baikoff was refusing to treat with anyone except the emperor himself and further scotching his chances by refusing to kowtow—and the hopeful Dutch were complying with all Chinese formalities—four Jesuits in Peking were making quite sure that the Hollanders' commercial enterprise would fail. Despite reports that the Russians went whoring in Peking brothels (for which their liberties were curtailed by the Chinese), and despite their heretical Russian Orthodox religion, the Jesuits had no scruples in taking their side against the Dutch. The reason is not hard to find. The Russians constituted no threat to the Jesuits' religion, nor to Portuguese trading interests, since they came overland and not by sea. But the Dutch were another matter. The expanding sea power of Holland in the South Seas, Dutch Protestantism, Dutch history of armed intervention at Portuguese Macao added up in Jesuit eyes to a threat to themselves and to the Catholic faith. Schall himself seems at first to have been reluctant to use his influence against the Dutch, but the Portuguese fathers Buglio and Magalhaens showed almost desperate zeal in what may have been a nationalistic as well as a religious cause.

In a letter, Magalhaens says: "Four of us brethern of the Society of Jesus . . . resolved to leave no medium unessayed to overthrow these Hollanders' designs, and with all diligence . . . to vacuate their undertakings. . . ."[4] They decided it was impossible to prevent the Dutch appearance at court "at least without vast bribery, because those viceroys of Canton had corrupted the great mandarins and opened the doors at the Court to them." But the Jesuits managed to have the liberties of the merchants restricted. "Whatever Father Buglio and myself have done to defeat the Hollanders, if it were not according to our wishes it was according to our power. . . . We went often . . . to confer with Father John Adam [Schall]. . . . Your Lordship [the recipient of the letter] owes much to Father John Valleat [a French Jesuit in Peking]," who exhorted Schall "to improve his interest in the emperor about our business." Here Valleat succeeded. There is a letter from Schall in which he describes how he intervened from an early date, warning Shun-chih that "if these people ever get a footing upon pretense of commerce in any place, immediately they raise a fortress and plant guns (wherein they are most expert) and so appropriate a title to their possessions." Listening to this recital of Dutch character, Shun-chih "stood musing

4. The quotations from Magalhaens and Schall in this paragraph are taken from Baddeley.

ADAM SCHALL: an engraving from his book on the history of the Chinese missions, 1665. He is shown wearing his mandarin robes. The drawing of the bird on his breast is not quite clear enough to identify it as an egret (which would mean he was an official of the sixth of the nine grades). He was a worried man during most of his time in China, and perhaps this portrait with its slightly apprehensive regard in the eyes is a good likeness. (British Museum, London)

for a small space . . . signified his clear apprehension of what I delivered, and presently asked me if the Muscovites were of the same temper. . . . I answered quite the contrary, very faithful and just people"—apart from imperfect religious views—"governed by a potent prince who could have no other design in his embassy hither but a mere congratulation of the emperor's fortunate conquest of his empire and happy inauguration on the throne. . . ."

How Schall squared these tendentious words with his conscience we shall never know. For Jesuits, preservation of the Faith was sometimes more important than strict adherence to truth. The wily priests Buglio and Magalhaens had meanwhile failed to bribe the "Master of Requests" (perhaps the president of the Board of Rites) and turned to his junior, who also refused. But, resourceful men, they "produced two rich vests which the emperor had bestowed on us. . . . These he accepted . . . with all the symptoms of great satisfaction." But alas, when the official realized that the "rich vests" were gifts of the emperor, he was too terrified to accept and repudiated the bargain, castigating the fathers for *lèse majesté* in giving away what the emperor had bestowed.

Not all Schall's eloquence could save the uncompromising Baikoff (and it probably was not really intended to do so). The Russian was sent away with nothing but the presents he brought. He stopped at the Great Wall—his fear of the tsar making him think again—and sent back to Peking an emissary (his cook!) to say that "if . . . your Tzar [Shun-chih] will be so gracious as to order me back to Khanbalik . . . I will obey your Tzar's commands in all respects." Word was sent from Peking finally dismissing him because he "has not the slightest inkling how to show respect to a sovereign." Baikoff could do nothing but take his long way back to Tobolsk and an impatient tsar with empty hands.

The Hollanders, having kowtowed, were more politely treated, but fared little better. The inkling of propriety they had shown produced an imperial letter to the governor of Batavia, who had sent the embassy: ". . . Our Territories are so far asunder as the East and the West, so that we can hardly come near to one another; and . . . the Hollanders were never before seen by us. But those that sent Peter de Goyer and Jacob de Keyzer to me" show their wise and noble minds in so doing. "For this reason my heart does very much incline to you, therefore I send you two rolls of satin with dragons. . . ." And Shun-chih enumerates fifty other items of silk and "three hundred taels of silver. You have asked leave to come and trade in my country, to import commodities into it, and to export others from it, which will make very much to the profit and advantage of my subjects; but [since] your country is so far off, and the winds very high here, which will very much endanger your ships, it would very much trouble me if any of them should miscarry on the way. Therefore if you please to send thither, I desire it to be but once in eight years, and no more than a hundred men in a company, whereof twenty may come up to the place where I keep my court. . . ."[5]

The thwarted Hollanders left almost at once. The Jesuits doubtless congratulated themselves; but it is hardly surprising to learn that in 1668, within a decade, the Dutch in Moscow successfully interfered to prevent Jesuit attempts to gain permission to travel across Russia to China.

The Manchu court of China, if it thought any more of these incidents, probably decided it could afford to relax and forget them. Troublesome little embassies that they were, they had nonetheless responded more or less as expected when the usual measures designed for other tributary peoples were applied. But in fact the West was knocking much harder at the Chinese door than the tradition-stopped ears of the Chinese could hear. The Chinese learned almost nothing of the West from Baikoff or from the Dutchmen, partly because neither had much to teach them, and partly because the Chinese did not consider they had anything to learn from the outer world. The ideas propounded by Jesuits in Peking—that the West

5. The reader may like to compare the letter of the Shun-chih emperor with others from later Ch'ing emperors to Western ambassadors or kings. See pages 312–13.

was the possessor of an advanced civilization—found no confirming echo in the quality of the embassies received.

But most serious of all, from the point of view of later Chinese history, was the Chinese blindness to an obvious fact. They had just seen the first important example of European powers, in the shape of their representatives, jockeying for position and influence in China. Of the three parties—the Jesuits representing papal power, the Dutch representing the burgeoning sea and commercial power of Holland, and the Russians representing an expansionist country beginning to emerge from medievalism—it was the Jesuits who emerged on top with a victory for the Catholic faith and for Portugal.

Nor did Europe learn anything significant from these events. Nieuhof's book, in Dutch, Latin, and English versions, appeared swiftly (in English in 1673). It was embellished with engravings by Wenceslas Hollar. These were probably based on Nieuhof's sketches, to which they are inferior in terms of actuality, but they were the first in England to show the imperial palaces at Peking. Their Chinese scenes belong to that never-never land of chinoiserie—a fabled place imbued with a kind of mad elegance that owes as much to Europe as to China. The West was quite as isolationist as China. Chinoiserie was just as far from Chinese reality as was the Chinese concept of the West as a rabble of barbarians.

In these clear-cut postures the Chinese and the Europeans firmly stood. Apart from a period of exercises in apotheosizing Confucianism, which occurred in the French Enlightenment and died a natural death, neither European nor Chinese attitudes to each other were to alter significantly until almost another three hundred years had passed.

SUITORS OF THE MANCHU

Principal Western Embassies to Peking

Unless otherwise indicated, the dates shown are those of arrival and departure.

PORTUGUESE	Pirès. Late 1520 to early 1521
RUSSIAN	Petlin and Mundoff. 1618
RUSSIAN	Baikoff. March 3 to September, 1656
DUTCH	De Goyer and De Keyzer (the Nieuhof embassy). July 17 to October 16, 1656
DUTCH	Pieter van Hoorn. June 20 to August 5, 1667
PORTUGUESE	Late 1667
RUSSIAN	Milovanoff. 1670
RUSSIAN	Spathary. May 15 to September 1, 1675
	(Treaty of Nerchinsk between China and Russia, 1689, negotiated with the help of the Jesuit Gerbillon)
RUSSIAN	Isbrant Ides. November 5, 1692, to February 19, 1693
PAPAL	Cardinal Tournon. Received in audience December 31, 1705
RUSSIAN	Ismailoff (with John Bell as embassy physician). November 18, 1720, to March 2, 1721
PAPAL	Cardinal Mezzabarba. December 15, 1720, to March 24, 1721
	(Chinese embassy to St. Petersburg, 1733)
PORTUGUESE	Pacheco. May 1 to June 8, 1753
BRITISH	Lord Macartney. August 21 to October 7, 1793
DUTCH	Van Braam and Titsingh. January 9 to February 15, 1795
BRITISH	Lord Amherst. August 28–29, 1816
	(Treaty of Nanking, August 29, 1842—the first "unequal treaty")

CHAPTER ELEVEN

THE GRAND ALLIANCE

Ferdinand Verbiest and the Emperor K'ang-hsi

AT THE time of Baikoff and Nieuhof in Peking, Adam Schall was in high favor with the Chinese court. In 1645 he had been appointed chief of the committee for correction of the calendar. But the wind of such a Western mind blowing through the ranks of the official class was to reap a jealous Eastern whirlwind. Schall's correct prediction of the eclipse of that year, when the indigenous astronomers failed, had confirmed him in his position, but at the same time had added force to the approaching storm. The emperor Shun-chih, however, ennobled Schall's ancestors, and even provided an imperial inscription for Ricci's old church, the Nan T'ang. The learned but circuitous fathers Buglio and Magalhaens, whom we saw attempting bribery with the emperor's vests, now managed to get permission for other Jesuits to settle in provincial China. On the surface all went well.

But with the death of the Shun-chih emperor in 1662, the Confucian literati who controlled the six boards of the government put their heads together and seized on a tract written years before by one of the Chinese Moslem astronomers in Peking which had raised the old bogey of Jesuit intent to bring Portuguese armies into China. Buglio and his friend wrote a refutation. But, arraigned before the Board of Rites, the aged and paralyzed Schall (now seventy-four) was unable to appear to offer his defense; and the pleading of Ferdinand Verbiest, a Jesuit new to Peking, failed to stem the tide of anti-Jesuit, anti-Western feeling. The Christian religion was condemned as immoral (by strict Confucian standards it probably was), tending to rebellion (which it might have done had it got a grip) and the invasion of China (which at the time was unproven but was later proved to the hilt there and elsewhere). Schall was sentenced to strangulation—not an unusual punishment in China at the time. Jesuit science was pronounced untrue and, in effect, un-Chinese. Verbiest himself, though imprisoned, was not deeply incriminated, because he had not yet reached China when Schall altered the calendar, and he escaped without penalty.

In this reverse there occurred what probably seemed to the Jesuits an intervention of Divine Providence. The earthquake of 1665 threw the court into a panic and led the empress dowager, who always had a soft spot for the Jesuits anyway, to arrange for an amnesty. Schall, now completely broken, returned from prison to his church, Nan T'ang, and died the following year. That same year a decree proscribed Christianity, ordered all priests to the capital, and then sent them to captivity in Canton before banishing them to Macao. Of the thirty-eight Jesuits then in China, only four were allowed to stay in Peking.

But times were changing. Shun-chih's successor, K'ang-hsi, was a boy of eight under a regency. He was one of those children apparently marked by fate not only to rule, but to rule magnificently, to rule wisely and long. In 1667, at the age of thirteen, he took over the government of the empire. It is said that he dismissed the regents and, with the moral support of the court, began to rule, largely because of this persecution of his old tutor Schall, which he is said to have resented. Since the fall of Schall, the president of the Bureau of Astronomy in Peking had been his old opponent, the anti-Jesuit Yang Kuang-hsien, who was a Chinese Moslem. In all his ignorance and traditionalism he was ensconced in Schall's house with Schall's instruments and large scientific library. From there, in 1668, he was foolish enough to publish a calendar for the following year. A copy was transmitted to Ferdinand Verbiest on Christmas day. He at once challenged it, writing to

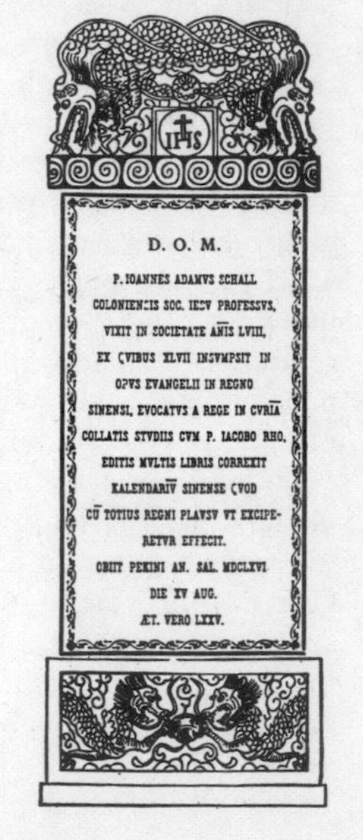

THE STELE OF ADAM SCHALL erected near his grave in the Jesuit cemetery near Peking. Summing up his long years of work in China, the inscription says: "He edited many books and his correction of the Chinese calendar was praised and adopted throughout the kingdom." (From Favier)

THE YOUNG EMPEROR K'ANG-HSI.
(From Favier)

K'ang-hsi with the request that Yang prove his points experimentally. Imperial decree gave Verbiest a month to substantiate his own points.

We must pause a moment to salute the courage of Verbiest. His letter of January, 1669, to the emperor reveals by its wording the thin ice on which he knew he skated: "Ferdinand Verbiest, your servant from the Far West, addresses you respectfully to reply to your command . . . and presents you with a pamphlet which he has drawn up in conformity with your order to examine and check the calendar of the Moor, [i.e., the Moslem Yang]. . . . Your servant, far from his country and alone in a strange land, in recognition of the emperor's personal interest in him, concentrated all his abilities and all his heart . . . to assist you in your strong desire to give to your peoples a calendar precisely in conformity with the laws of astronomy. . . . I find the calendar contains grave errors. Your servant learned in his childhood to follow the path of virtue; his words are not intended to criticize anyone. But, having been charged by the emperor to verify the calendar . . . he does not dare but to state the truth frankly and clearly, fearing otherwise to be guilty of a lack of sincerity. I respectfully present you with this pamphlet containing my observations, sending you back also the two volumes of the [Moor's] calendar, so that you may see them both yourself. I submit myself with confidence to your august decision. Your servant protests his profound sentiments of respectful humility."[1]

1. Quoted by Josson and Willaert.

Yang was deposed and Verbiest took his place. Not perhaps since Ricci had there been such a man as the portly and erudite and valiant Verbiest. And it was with him, in his long friendship with K'ang-hsi, that the status and fortunes of the Jesuit mission and of Western science once more rose phenomenally.

Verbiest was a Fleming born in a small town on Belgium's western border.[2] He studied at Courtrai and Bruges, and took philosophy at the University of Louvain. At eighteen he began his novitiate. Later he studied theology at Rome. Twice as a young man he traveled from Belgium to the Iberian Peninsula, intending to embark for the Americas as a missionary. But in the end he made the hazardous trip by sea to Macao; and so to Peking. Once there in 1660, a man of thirty-seven, he virtually never left the capital again.

He could hardly have come at a worse time. No sooner was he settled in Peking than Shun-chih died and was replaced by the council of anti-Jesuit regents who, in 1664, cast Schall, Buglio, Magalhaens, and Verbiest into prison, manacled and loaded with chains. The earthquake, as we have seen, secured their release, Verbiest refusing his freedom until Schall was set free too. Even after the young K'ang-hsi took over the reins of government, life was difficult for the Jesuit fathers. At first, even in his qualified approval, the lusty young K'ang-hsi had suspicions of them. Unable to imagine a group of grown men celibate and apparently without sex life, he suspected the worst. He sent "a young well-built Manchu" to live in their house on the pretext of learning philosophy, but really, as the Jesuit writer Le Comte delicately puts it, "to discover the most secret things about them, and to be, himself, it would seem, a subject of scandal." The handsome Manchu stayed a year in the Jesuit house, reporting every detail of the activities therein, and especially what happened to him personally. Unmolested at the expiry of this time, his information reassured K'ang-hsi that all was at least seemly, if not comprehensible to him. Le Comte quotes the emperor as saying: "These men are teaching us nothing that they do not practise themselves; and they are just as chaste as they appear on the surface." The hanging query now cleared from his mind, the emperor's respect grew. Very gradually he made a friend, almost a confidant, of Ferdinand Verbiest. Perhaps a father figure in the person of Verbiest, who was thirty-one years his senior, was no bad thing for the youthful K'ang-hsi.

The life of the new emperor is characterized by a vigor, an open-mindedness, and a lively appreciation of life's possibilities. Such an outlook was so unusual as to be all but revolutionary in either a Chinese or a Manchu of the period. But in K'ang-hsi, no less an emperor than any other of China's rulers, those qualities were necessarily limited by the bounds of his enforced conformity to rigid tradition. Still, within the fairly spacious cage of ideas

2. Brief biography in Pfister.

FERDINAND VERBIEST wearing the sumptuous robes of a high Chinese official. Beside him are a sextant and a celestial globe. This Japanese print by Utagawa Kuniyoshi (1797–1861) succeeds well in conveying something of the gravity of the great scholar and confidant of the emperor K'ang-hsi. The inscription praises him as a military hero, doubtless alluding to his role as a maker of cannon for the Manchu. (From *Revue Coloniale Belge,* No. 6, 1951)

and immutable laws of great antiquity, the young emperor stretched quite un-Oriental intellectual wings. The learned and manly Verbiest was perhaps K'ang-hsi's ideal intellectual complement. Almost the sole aspect of life they could not share was the life of sexual freedom taken as the due of any Chinese or Manchu.

In the *Astronomia Perpetua,* one of forty or so books which Verbiest wrote or translated in China, he says: "I used to go to the palace at break of day, and did not quit it until three or four in the afternoon; and during this time I remained alone with the emperor reading and explaining. Very often he would keep me to dinner, and entertain me with most dainty dishes, served on gold plate. To appreciate fully these marks of friendship shown me by the emperor, a European must remember that the sovereign is revered as a divinity and is scarcely ever seen by anyone, especially not by foreigners. Those who come from most distant courts, as ambassadors, consider themselves fortunate if they are admitted but once to a private audience, and even then the emperor is only seen by them at a considerable distance. . . . The ministers and even his nearest relations, appear before him in silence, and with manifestations of the most profound respect, and when they have occasion to speak to him they always kneel."[3]

It was science, that sharpening spearhead of Western culture, which constituted the Jesuits' distinguished strength in China. And, little though

3. Quoted by Huc.

they knew it in the reign of K'ang-hsi, it was that same preoccupation with science, the prying finger which was supposed to introduce the later hand of Christianity, which was in the end to constitute the sole employment of the Jesuits at the Manchu court. But that was some distance ahead. In the time of Verbiest, K'ang-hsi's fascination with science makes a story at once absorbing in its depiction of an intelligence equal at least to that of any European youth, and baffling in its demonstration of the hard core of thus-far-and-no-further. No further than the hallowed and sacrosanct precincts of the imperial Mandate of Heaven.

With Schall, Verbiest was casting cannon for the Manchu to use against the Ming (and Schall has the doubtful distinction of having also cast them for the Ming to use against the Manchu!). Later Verbiest was to cast many more. Oddly enough, it is this Jesuit armament industry more than any other activity of Verbiest which seems to have caught the imagination of the Chinese. As late as 1901, when a memorial was sent to the throne on the subject of learning from Western techniques, its authors remarked: "Emperor K'ang-hsi used the Westerner Nan Huai-jen [Verbiest] to cast the red-barbarian cannon, and up to the present on these cannon [modeled on those cast by Verbiest] the name of Nan Huai-jen is still inscribed."[4] Verbiest himself, with a touch of pious whimsy, had the names of male and female saints cut on his own cannon. K'ang-hsi rewarded him for these efforts, notably by a personal visit to the Jesuit house (a positively astounding honor), and at another time by publicly taking off his sable mantle and dragon-embroidered gold tunic and giving them to Verbiest. But Verbiest was severely criticized in Europe for making armaments. Only a decision of Pope Innocent XI absolved him from blame.

During his lifetime Verbiest never ceased to be on excellent terms with the emperor. As adviser, interpreter, even arbiter, he assisted K'ang-hsi in a semiofficial capacity each time an embassy appeared from the West to trouble the waters of official Peking. As a highly placed official—of the second of the nine high grades, first as president of the Bureau of Astronomy and later as vice-president of the Board of Works as well—he was aware of the whole climate of opinion, of the crosscurrents of ambition, the curtains of distortion and secrecy and special pleading, that surrounded the throne, the councils of state, and the boards of the civil service. He was one of the "Great Mandareens" (as contemporary English writers called them) of the greatest and most populous and most sophisticated empire of the world—no mean feat for a boy from a Belgian country place. Never before and never after was Jesuit influence in China so deep. This came about because of the conjunction of a brilliant emperor and a brilliant Jesuit. It was a partnership absolutely unmatched in the history of Sino-Western relations.

4. Quoted by Teng and Fairbank.

Verbiest's days were not wholly occupied with state affairs. A spate of books, original and in translation into Chinese and Manchu, poured from his pen, on science, morality, aspects of Christianity. He it was who refitted the observatory on Peking's eastern wall, leaving it in much the state in which one finds it today: his are most of the instruments, replacements of those Yuan instruments there which he called with his usual tact the products of a "ruder muse." He it was who introduced the thermometer, as well as a use of steam power which the Chinese had not thought of themselves. He built into a small cart (and later into a small paddle boat) a mechanism which ejected steam. The steam jet played on the vanes of a wheel, which in turn drove the wheels of the cart (or the paddles of the boat). This primitive fixed-jet turbine, doubtless not his own invention since

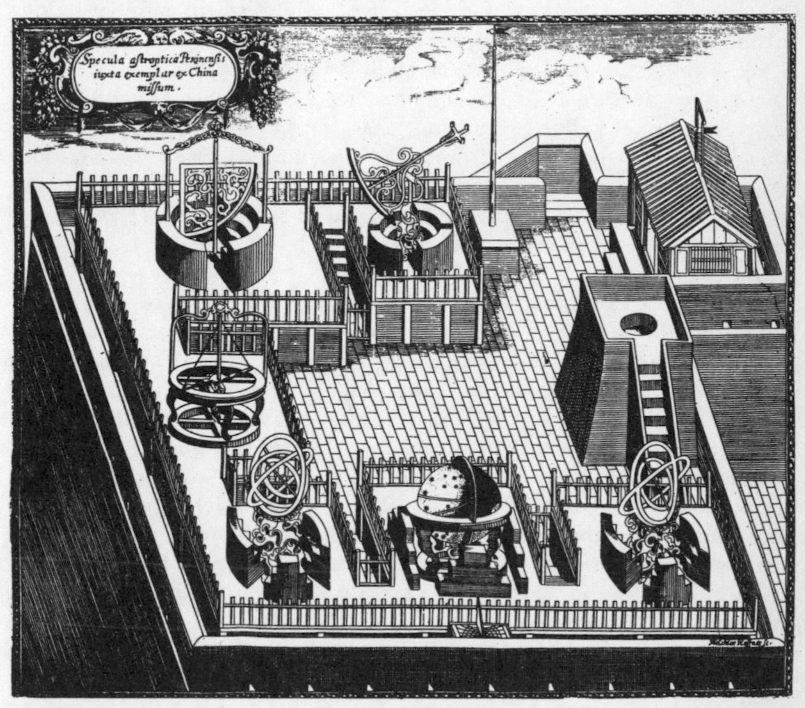

Verbiest's Observatory on the eastern wall of Peking. One of the achievements of Verbiest's career in the capital was the refitting of this observatory. Reading counterclockwise from the top center, the instruments are: sextant, quadrant, horizontal circle, ecliptic armillary sphere, celestial globe, and equatorial armillary sphere. The observatory is still on the Peking wall. (From Verbiest's *Astronomica europaea sub imperatore tartaro-sinico Cam-Hy appellato,* etc., published by Rencard at Silligen, 1687)

it had been described by Giovanni Branca in 1629, was peculiarly suited, one might think, to Chinese taste. For the Chinese had had manually operated paddle boats since at least the T'ang dynasty (eight or nine hundred years previously), and also knew of the power of steam. What they had never attempted was to harness steam to paddles or to wheels—probably because their abundant manpower made this irrelevant to them. But, like many an invention of their own, and others from outside, this of Verbiest was regarded as an amusing toy, diverting in the palace, and was then allowed to drop out of mind.

We may ponder a moment in passing what it was about the structure of Chinese life that caused this lack of urge in the sciences, and the lack of urge to utilize what the sciences discovered. The question is bound up with the unique Chinese political and ideological system, which endured so long unchanged: the answers, as yet, are as many as the investigators.

Nor were Verbiest's days even then full. He and his brother priests continually pushed on with their attempts to spread Christianity, to convert the emperor and court. But to interrupt that stream of divine power (as the Chinese saw it) which flowed (and had ever flowed) down from Heaven through the demi-divine person of the emperor, through the complex hierarchy of ministers and officials to the Seed of the Dragon—the Chinese people at large—was as impossible as to melt the Great Pyramid with prayers.

Apart from this fundamental impediment, there were others more particular. Outstanding among these were, first, polygamy and concubinage and, second, that troublesome question of the Chinese rites. The respectable custom of having a wife and as many concubines as one could afford was often the last hurdle over which the Chinese near-convert stumbled and fell. And the Chinese rites,[5] codified and sanctified by Confucian orthodoxy, deeply ingrained as normal in the heart of every Chinese, consisted of paying respects to one's ancestors by prayers, kowtows, offerings. Ricci, as we have seen, viewed the rites as purely secular, mere reverence to great men or antecedents. In this he took his stand on the writings of Confucius himself. Confucianism, including the rites, was not an impediment to the acceptance of Christianity—this was his conclusion. On the whole most succeeding Jesuits agreed that the rites should not be a bar to Christian conversion, believing moreover that the rites and their practice would wither away with the convert's increasing immersion in the new religion. But an opposing view was gaining increasing support, both among other Catholic missionaries in China and in high councils of the Roman church, that the rites were to be condemned as gross superstition and idolatry, that no Christian could practice them. Such a view—which was shortly to be given

5. The Chinese Rites Controversy was extremely complex and is only lightly touched on here. For a good brief outline see Cary-Elwes.

papal sanction, with the dire consequences we shall see—and such a public argument among the missionaries themselves, inevitably proved a stumbling block to the conversion of the Chinese.

For the moment, however, these were only minor problems, and the Jesuits found cause for optimism. During Verbiest's life, although edicts forbidding the preaching of Christianity remained in effect, such was his personal influence that provincial governors and viceroys dared not enforce them. The expelled priests were allowed to return to China.

Down the years Verbiest's alliance with the emperor continued. In 1682, when the emperor was twenty-eight and Verbiest fifty-nine, and again in the following year, K'ang-hsi took the Jesuit on safari in the wild Manchurian terrain of his ancestors. The purpose of these trips seems to have been chiefly to see for himself what was going on there and to toughen his army commanders and men who, traditionally tribal cavaliers, were now translated into courtiers and administrators in the world's most brilliant and wealthy capital, and were showing signs of becoming soft. Verbiest traveled with K'ang-hsi's father-in-law, living in his tent, and took with him his scientific and astronomical instruments, equipped thus to record the "disposition of the heavens, the elevation of the pole, the declination of each country . . . the height of mountains, and the distance of places."[6] He rode on the emperor's horses and was one of the inner circle of K'ang-hsi's entourage.

"The emperour took with him his eldest son, a young prince of ten years . . . heir of the empire . . . the three chief queens . . . every one of them in a gilt chariot, as likewise the principal Kings who compose the empire . . . all the great men of the court, and the most considerable mandarins of all the orders; who having a numerous retinue and splendid equipage, made in all an attendance . . . of above threescore and ten thousand souls."

Verbiest gives a spirited description of this veritable migration of people—seventy thousand of them swarming over mountains, crossing rivers by bridges thrown over by the engineers at their head, encamping in thousands of tents, bringing with them not only the baggage trains containing three months' provisions, but thousands of spare horses and every conceivable item of impedimenta that might be needed on the long journey. In the path of this imperial horde, the villagers and townsmen swept and smoothed the streets. "Christians are not so careful to sweep the streets . . . through which the Holy Sacrament is to pass . . . as are these infidels to clean the way by which their kings . . . are to go. . . . The emperour hardly ever kept to that highway, spending his time for the most part a-hunting. . . . He marched most commonly at the head of that . . . army. . . . When he came up with the queens, he kept along the side [of the prepared road]

6. The Verbiest letters concerning the journey are in his book *A Journey of the Emperour*.

lest the great number of horses that followed him might spoil it. . . . And though that vast number of men, horses and flocks, kept away pretty distant from the emperour's road, yet they raised such a terrible dust that we seemed to march in a cloud, and had much ado to discern those that were fifteen or twenty steps before us."

Verbiest enjoyed the imperial hunting. Huge numbers of beaters were deployed in a wide circle, closing in over the hills and through the forests until thousands of hares, deer, bears, wild boar, and tigers were driven toward the position of the huntsmen and slaughtered. The emperor, also, had great pleasure in the chase, as we know from a letter he wrote on another such occasion to one of his sons in Peking: "The hares of the Ordos region have an exquisite flavor; everything here has more savor than the best that Peking can provide."[7]

"It was the emperour's pleasure," Verbiest continues, "that I should be present at all these different ways of hunting, and he recommended it to his father-in-law . . . that he should have a special care for me, and see that I were not exposed to any danger in the hunting of tigers and other fierce beasts. Of all the mandarins I was the only person near the emperor without arms. Though I had been inured to fatigue from the time we set out on our progress, yet I was so weary every evening when I returned to my tent, that I had much ado to stand upon my legs, and many times I would have spared myself the labor of following the emperour, if . . . I had not been afraid he would have taken it ill."

Now rains began to pour steadily and bridges to collapse. The emperor's hopes of fishing were disappointed. "And as we were returning up the river, the bark wherein I was with the emperor's father-in-law was so beaten by the waves that we were forced to go ashore and to get into a cart drawn by an ox." Which brought them very late to join the party. "At night . . . discoursing with the emperour about that adventure, he said laughing: 'The fish have made fools of us.' " Verbiest describes the immense struggles of animals and men in the ensuing days as they went on remorselessly across the inundated and mud-clogged countryside. "The emperour himself, his son, and all the lords of the court, were oftener than once forced to cross over mires and fens on foot, fearing that they might be exposed to greater danger if they attempted to pass them on horseback."

One evening, unable to ford a swollen stream, the emperor went over in a tiny boat; but, reaching the other side, he at once returned, asking where Verbiest was. " 'Let him come into the boat,' K'ang-hsi said, 'and come over with us.' " The next day much the same thing happened. "When we were over, the emperour sat down by the waterside and made me sit down by

7. Quoted by Grousset.

him. . . . The night being clear, and the sky very serene, he would have me name to him all the constellations that at that time appeared above the horizon, and he himself named first of all those which he knew already; then unfolding a little map of the heavens, which some years before I had presented him with, he fell a-searching for the hour of the night by the Star of the Meridian; delighting to show to all the skill he had in those sciences. All these and the like favors which he showed to me often enough . . . were so public and extraordinary that the emperor's two uncles . . . said upon their return to Peking, that when the emperour was out of humor, or appeared melancholick, he resumed his usual cheerfulness so soon as he saw me."

And so they returned after many months to Peking. At the end of the second journey, Verbiest gives his opinion of his emperor: "With wonderful equity he punishes the great as well as the small, he turns them out of their places, and degrades them from their dignity, always proportioning the punishment to the heinousness of their crime. He himself takes cognizance of the affairs that are handled in the Royal Council, and in other tribunals, requiring an exact account of the judgements and sentences that have been passed. . . . In a word he disposes of all. . . . That absolute authority . . . is the cause that the greatest lords of the court, and those of the highest quality in the empire, even the Princes of the Blood themselves, never appear in his presence but with profound respect and reverence. . . .

"I have spoken elsewhere of the fruit that religion may reap from our journey. Let it suffice in this place to say that the emperour, whose will and pleasure we cannot in the least resist without exposing this mission to manifest danger, commanded us to follow him. I have nevertheless spoken to that Court-Lord who is our particular friend, that he might get us excused from such long journeys, and especially myself, who am not now of an age fit for them. . . ."

A few years later, in Peking in 1688, Ferdinand Verbiest died, aged sixty-five. He was given what amounted to a state funeral attended, at the emperor's command, by some of the highest dignitaries of the court and administration.[8] He was laid to rest near Ricci in that ground given by the former Ming emperor Wan-li, his name honored and remembered, the state of those church affairs to which he was so great an ornament seemingly in a healthy condition, and his brother priests hopeful of continued imperial favors.

Verbiest, like Ricci before him, left his work in good order—in better order than he knew. For, a few days after his death, there arrived in Peking a party of Jesuits (for whom, again like Ricci, he had often asked) which

8. The impressive funeral is well described by Father de Fontenoy in *Lettres Edifiantes*, vol. XVII, edited by Querbeuf. Fontenoy was one of the Jesuits who arrived at Peking just after Verbiest's death.

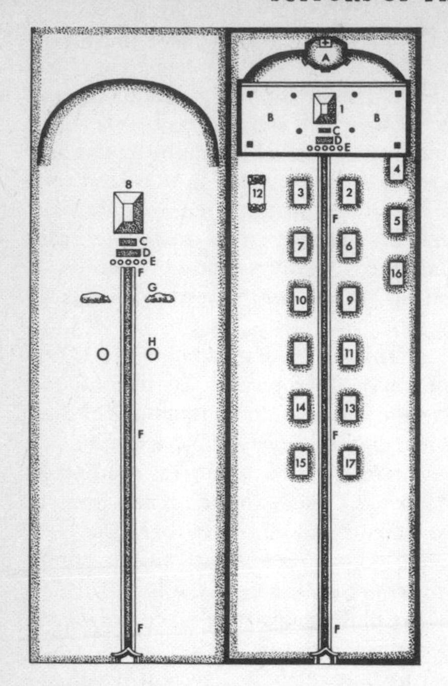

PLAN OF THE JESUIT CEMETERY NEAR PEKING. The tombs of Ricci (1) and Schall (8) occupy the most prominent positions, with that of Verbiest (12), as befits his importance, also larger than the others. *Graves and dates of death:* 1) Matteo Ricci, 1610. 2) J. Terrenz (Schreck), 1630. 3) J. Rho, 1638. 4) F. Christopher, 1640. 5) F. Mendez, 1640. 6) Nicolas Longobardi, 1654. 7) D. Coronatus, 1666. 8) Adam Schall, 1666. 9) De Sequeira, 1673. 10) G. de Magalhaens, 1677. 11) L. Buglio, 1682. 12) Ferdinand Verbiest, 1688. 13) F. Simois, 1694. 14) C. Dolze, 1710. 15) L. Pernon, 1702. 16) P. Frapperie, 1703. 17) C. de Boissia, 1704. *Structures:* a) Hexagonal chapel. b) Courtyard. c) Stone stele. d) Stone altar table. e) Incense burners. f) Paved walks. g) Carvings of two recumbent horses. h) Carvings of two officials, much like those at the approach of an imperial tomb, but here on a smaller scale. The arrangement in front of Ricci's and Schall's tombs of stele, altar, and incense burners is of course also typically Chinese. Indeed, the entire complex is quite Chinese in feeling. (Redrawn from Favier)

included the great Father Gerbillon. The newcomer was later to step into the shoes of Verbiest, the shoes of K'ang-hsi's friendship, and to continue the long slow struggle of Verbiest. There is no question but that Verbiest's life and association with the emperor paved the way for Gerbillon, Ripa, and Attiret, and for others. And no question either but that it led to the edict of toleration of Christianity of 1692:

"We have seriously considered this question of the Europeans. . . . Since they have been living among us they have merited our esteem and gratitude by the great services they have rendered us in civil and foreign wars, by their diligence in composing useful and curious books, their integrity and their sincere regard for the public welfare. . . . The Europeans are very quiet; they do not excite disturbances in the provinces, they do no harm to anyone . . . and their doctrine has nothing in common with that of the false sects in the empire, nor has it any tendency to excite sedition.

"Since, then we do not hinder either the Lamas of Tartary or the bonzes of China from building temples . . . much less can we forbid these Europeans, who teach only good laws, from having also their churches and preaching their religion publicly in them."[9]

Such is the story of Ferdinand Verbiest, one of the most potent influences ever brought to bear by the West on the court and intelligentsia of China. He broadened and solidified the foundations laid by Ricci, and built on them an edifice as strong as anyone could. He ushered in an age of Jesuit and Western influence that endured for about a hundred years. But we cannot leave him there. What we have seen in outline is the story commonly written of the great Jesuit and the great emperor. It is true. But there is another side to the coin. For in the tale of another brilliant European visitor to the court of K'ang-hsi we see also another Verbiest, through the eyes of a Russian.

9. Quoted by Huc.

CHAPTER TWELVE

THE DEPTHS OF DIPLOMACY

Spathary, Verbiest, and K'ang-hsi

IT MUST have been a considerable shock for the Manchu officials in Peking when they encountered for the first time the Russian ambassador Nikolai Gavrilovich Spathary. Never before had they met a Western ambassador who was anywhere near their match in argument. Nor had they been required to deal with one who was in a position of some strength when he arrived in the capital. In fact, the embassy of Spathary is probably more important than any other that ever came to the capital. In a manner which is best described as definitive, it displays the crux of all the others, the reasons underlying the frustrations of them all. In the carefully written, precise, subtle, and lengthy dispatch of Spathary,[1] the whole irreconcilable ambivalence of Chinese and the Western viewpoints emerges with sharp, even alarming clarity.

Spathary was a Greek, with a Greek's logical and flexible intellect. His family had settled in Moldavia and risen to the rank of minor nobility in a country then under Turkish domination. Educated in Constantinople, he had at his command ancient and modern Greek, Russian, Turkish, Arabic, and later Latin and Italian. He had also studied theology, philosophy, history and literature, mathematics, and natural sciences. A learned friend once described him as "replete with universal knowledge." He served two successive kings of Moldavia, and other nobility, in the guise of diplomatic henchman, a role admirably suited to his sinuous character. His misfortune was that at one time he attempted to usurp the position of one of his masters, and failed. For this failure he was punished by having his nose either cut off or split, it is not clear which.[2] Fleeing to Constantinople, he was involved

1. Quoted by Baddeley, whom we follow in this chapter.

2. It seems probable that somewhere in the Chinese archives at Peking there may still exist a portrait of Spathary painted at K'ang-hsi's order while the ambassador was there.

A RUSSIAN AMBASSADOR AND HIS SUITE, 1627. Unfortunately, there is no picture of Spathary and his suite as they appeared in China, but doubtless they looked fairly similar to this embassy to the Holy Roman Emperor at Regensburg. The ambassador (wearing a wide fur hat) has his three principal aides around him and is followed by his secretary (holding a dispatch case) and by others carrying sable skins as gifts, just as did the members of the Spathary embassy to Peking. (The Ratisbon Print, from Baddeley)

in another betrayal and took flight to Germany, where he had a cosmetic operation on his nose. By 1671 we find him in Moscow, squirming his way upward in favor until his talents procured for him an appointment as ambassador on the mission sent by the tsar to China in 1675.

Spathary was too clever for the people he served. Consequently he was a tricky individual, yet one whose loyalty was fierce while it lasted. Then, loyalty failing, his antagonism was equally ferocious. In Peking he was loyal to the tsar. Intellectually he was a good representative of the European culture of the times; morally he was not so reprehensible, in those same times, as he would appear today. Nothing like him from the West had ever before confronted the officials in Peking.

Twenty-nine days' journey distant from Peking, Spathary and his suite with their baggage and their load of gifts for the emperor, and an even more cumbersome load of goods for trading in China, were met according to routine by Manchu officials. The journey onward was apparently amicable, but an interesting incident occurred. A high official was seen approaching, bearing an imperial letter slung on his back in its casing of imperial yellow silk. Spathary's escort asked him to dismount and kowtow by the wayside as the imperial letter passed. Spathary refused. The official then dropped behind a good distance. The letter passed Spathary. The official, far behind, kowtowed as it passed and later caught up with his ambassadorial charge. Nothing more was said. Thus, in a perfectly Chinese gesture of accommodation, the problem was circumvented. In Peking, on both sides, others were similarly approached, with close argument but with some courtesy. The incident set the tone of all the ensuing negotiations.

The Russian party was lodged according to custom in the Palace of the Four Barbarians in the manner of all tributary envoys.[3] The traditional Chinese formulae were then applied, but with a detectable nervousness on the part of the Manchu officials. For Russia was not only a neighbor (unlike any other Western country), but also for some time had been a troublesome one. The problems along the Amur River which we noted in the time of Baikoff had by no means diminished since that irresolute envoy stumbled out of Peking in disgrace. Moreover, the Manchu government was in trouble with its own rebellious south at the same time. The Jesuits were busily engaged in the armament factory and their Portuguese-style cannons were rushed as soon as completed to one or other of the several fronts. This situation, in essence, was known to Spathary; and what he did not know was soon told to him by the Jesuits. The situation was his strength and the Manchu weakness. Hence the nervous tenor of his reception in Peking.

The tattered manuscript read in the Russian archives before the Revolution and translated by Baddeley unfortunately lacks the initial pages of what happened in Peking. But their contents can be reconstructed from what follows. The officials of the Board of Rites arrived to question Spathary according to form, to ask him for the tsar's letter to the emperor, and to list the presents. Spathary evidently demurred at once. The manuscript begins:

"We know very well what sort of monarchs are the Dalai Lama, and the Kalmuk, Mongol, Bukharan and other chiefs, because they are our neighbors; also the Portuguese and the Hollanders; but if you gathered together all the dominions of all the kings you have mentioned . . . and added to these many more—you would still not have so mighty a kingdom as His Majesty the Tzar, by the grace of God, governs with absolute power; and it is even especially unfriendly on the part of the Bogdikhan [the emperor][4] to compare with the Portuguese, Dutch and others, the credentials and representative of His Majesty. If there should come ambassadors to the Chinese emperor from the Caesar of Rome, or the Sultan of Turkey . . . it might be said of them that they are the great sovereigns of the earth, but even they cannot compare with our Sovereign Lord in importance [to you] seeing that we are neighbors; and one neighbor cannot be indifferent to another." This,

3. Spathary complains of the poor quarters and restrictions on his movements while there. But in Moscow in the sixteenth and seventeenth centuries various ambassadors also complain of being "kept, as it were, in honorable confinement and allowed neither to go out nor to receive visitors—in large rooms without any trace of furniture."

4. The word "Emperor" has been inserted in most places after this point, since it is less confusing for the reader than Spathary's "Bogdikhan" or "Kaan." Similarly, I have called his various Manchu officials simply that, in place of the profusion of terms Spathary uses.

of course was putting his finger exactly on the point. The Chinese could not afford to be indifferent to the ungauged power of neighbor Russia.

"Furthermore, the Emperor invited His Majesty the Tzar to be friends in his letter of some years back . . . and if that letter had not been sent, His Majesty would never have dispatched this embassy, after learning from Baikoff your arrogant ways and the reception you accord to ambassadors—dishonoring, and contrary to established custom in all other countries. Even if the Emperor thinks himself greater than all other sovereigns, nevertheless, for the sake of the friendship and love manifested by His Majesty, he might still receive the Tzar's letters from me in person. We ask nothing but what, from the beginning of the world, has been done by all monarchs. . . . When it is seen that we proceed openly through the city, to his Chinese Majesty, with the letters and gifts of our Sovereign Lord the Tzar . . . then all the Emperor's friends will rejoice and his enemies tremble, seeing that an embassy comes from so glorious and great a King and neighbor, in friendship and with gifts. For we are well aware that you have enemies as well as friends. . . . I am ready even now to discuss with you the affairs of His Majesty, but not to go to the Board with His Majesty's letter. . . . If I go to the [Emperor] I am ready now. But it is a remarkable thing . . . that ambassadors are received by the Emperor, and not the letters of their master."

The official of the board replied: "You must not be surprised that ambassadors are received by the emperor, and not letters, for that custom dates from long ago. . . ." And he went on to say that once upon a time an ambassador was allowed to present such a letter in person, and that it contained insulting terms. "So from that time it was established in China that

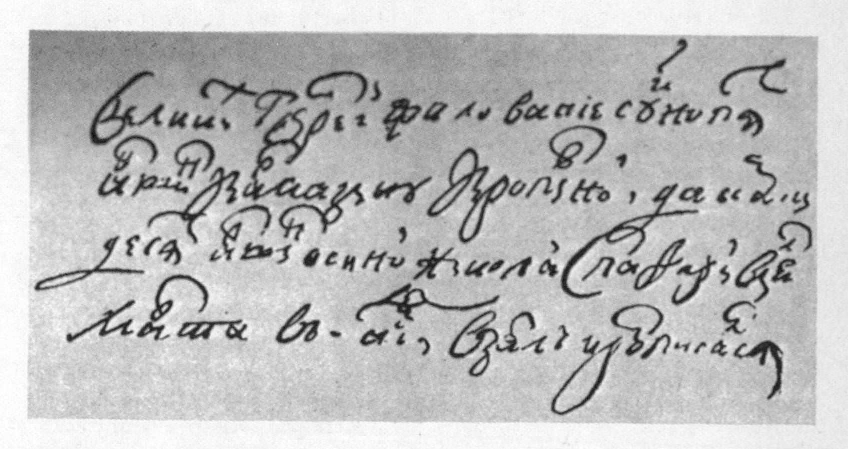

SPATHARY'S HANDWRITING, from the report he wrote on returning to Moscow from Peking. (From Baddeley)

. . . all missives . . . should be taken and perused, to see in what style they were written. . . . Yet out of friendship and regard to the dignity of His Majesty, [the Emperor] has commanded, contrary to custom, that the Tzar's letter should be received by those two great officers of State." This was a big concession on the Chinese part. "But as to your being taken before the Emperor with your letter you need not dream of it!" The official went on to cite the case of Baikoff, in which this concession was not given, and to recount how the Dutch and Portugese letters were taken by minor officials. "If the Emperor received you now in person [bearing the letter], all the sovereigns of neighboring states would leave off calling him Emperor; seeing that he had, contrary to custom, taken the Tzar's letter himself; by which he would suffer much dishonor."

To which Spathary, never at a loss, replied tartly that if by taking in person the tsar's letter, the dignity at the emperor would suffer, so would the tsar's honor and fame "be much lessened through the fact that, having sent an embassy with so much friendliness and such gifts" to the emperor, the Chinese appeared to despise that friendship and thereby to count the tsar as an inferior. "Nor could those Kings who . . . sent their representatives here . . . put themselves in comparison with the Tzar, or complain that [Spathary] was given preference—for each knows very well his relative standing." Spathary then assured the official that the contents of the tsar's letter could do nothing to offend. He professed to be astounded that he was not told when met far from Peking that the letter would be received by the board and not personally by the emperor. For "I should have declined to continue my journey, realising that no proper respect would be paid to our Sovereign Lord the Tzar."

The official replied coolly: "I would not assert that I actually told you so, but my orders from the emperor were these: to find out from your talk and the Tzar's letters whether any threats or insults were intended, and if so to thrust you back at once. I was then to collect all the troops I could, march on Nerchinsk . . . and destroy it and its foundations; for we know exactly how many men you have. . . ! It is enough to say now that if you refuse to take the Tzar's letters to the Board you will go back as Baikoff did!"

The ambassador, says Spathary (who usually, except in moments of extreme emotion, refers to himself in the third person), replied that it was unnecessary to bring up the subject of how the Chinese had destroyed Russian forts. They, the Russians, did not brag about *their* successes, desiring to live in peace. "That is why he has sent me; if His Majesty had ordered me to declare war, I should have done so long ago. As to your saying that I shall be sent back—that is as your master wishes." Spathary then insinuated that the Russian garrisons in the disputed areas were not perhaps as puny as the Chinese might think. "Does the Emperor wish to maintain love and friendship with His Majesty, or not?"

"If the Tzar wishes to dwell in love and friendship with the Emperor, then so does the latter with him,—and if not, not," said the official.

Spathary then called on God to witness that the tsar only wished good relations. It was not with any thought of dishonoring the emperor that he had come to present his credentials in person to the emperor, but because ambassadors always personally present their credentials to the tsar, who would, were these ambassadors Chinese, ask after the emperor's health. He offered to put this in writing and threatened that if he was not allowed to present his letter in person, the tsar would retaliate in like manner.

The official retorted that even if the emperor sent the highest-ranking ambassadors to the tsar, they would "do as they were told—go where they were bidden to go, sit where they were told to sit. . . . How otherwise, seeing that each country has its own customs?" The Chinese in Peking could not alter theirs. "But there has been talk enough. Hand over the Tzar's letter; if not you will have had your trouble for nothing, like Baikoff."

Spathary responded that they could do with him what they wished, even take the letter by force; but he would never of his own free will surender it to the board.

"One of those two things will certainly happen, and soon; unless you give up the Tzar's letter whilst we are still friendly." Saying which, adds Spathary, the official went away.

Such was the first of more than thirty discussions Spathary had with the officials.

But at noon the same day the official was back—with Ferdinand Verbiest, and the president of the Board of Rites, no less. Verbiest wore Manchu dress and spoke Latin. The ambassador thanked the officials for bringing the Jesuit. Thereafter almost all the interpretation was done by Verbiest.

Now they embarked on a recapitulation of the morning's interview, and raised also many questions of the history of the disputed territories in Siberia with immense detail and complicated arguments and counterarguments. Once more Spathary was assured that special arrangements had been made to receive the tsar's letter in the most unusual and honorable manner at the board—a procedure quite different from the reception of the letters from other nations' heads of state.

But, said Spathary, "if the letters are the more honorable, why present the bearer to the emperor and not the letter?" To which there were replies about ancient custom, the emperor's dignity, and "thirdly, that so long as they have only the ambassador's word of mouth, and have not . . . seen that all is written . . . in terms of honor, they could not be sure that this was the case." But on pressure the officials agreed to listen to Spathary's report of what the tsar's letter contained and consult with the emperor on the matter. They said he must not be put out, for customs differed so much. They had heard that in the tsar's presence no one was allowed to wear a hat when making his bow. Now in China "God forbid that anyone should commit so great an insult as to bow *without* a hat on!" Still, they, the Chinese, would certainly take off their hats in the tsar's presence.

Finally it was agreed that Verbiest should write at Spathary's dictation the contents of the tsar's letter and give the memorandum to the officials. While he was doing this, a young man entered "who came from the Emperor, one of his chief intimates. . . . The Jesuit looked at the letter as if he were reading it and said to the ambassador that the young man was the Emperor's chief favorite," who had been sent to see if Verbiest could really converse with Spathary. But, Verbiest went on, Spathary must dissemble, so that the favorite should not know that he was recognized by the Russian. The favorite inquired what language they spoke, was assured that it was Latin, wrote this on a piece of paper in case he forgot the unfamiliar word on the way back to his master, and departed.

Spathary's next statement is important and deserves the italics I have added: "Between the conversations the Jesuit told the ambassador that he was glad to serve the Tzar as best he might, for Christianity's sake. . . . But he regretted that an embassy should have arrived from so glorious a monarch, *seeing that the Chinese were barbarians, who rendered honor to no ambassadors . . . and answer in their letters as a master would to his servant; and other contemptuous treatment there is of which he would tell me another time. . . . And he made the ambassador swear before the ikon to tell no one what he had said 'Nor to write it down until you leave China.'* "

It is with a shock of surprise that one reads this first confidential remark of Verbiest's—the professed intimate and friend of K'ang-hsi. One hopes that there is some misattribution in the original. But the following day, when the officials said they had not yet consulted the emperor on the dictated version of the letter, Verbiest, in Latin and surreptitiously, "managed to say that [this] was a lie; for already, in his presence, they had made their report—and the Emperor had commanded to search in the ancient books" for a precedent; but that the officials had held out against doing so.

On about twenty occasions in the following weeks Verbiest seized his chance to talk in Latin, which was incomprehensible to the officials. Later on this same day, Verbiest added a few more words for the private Russian ear. Spathary had been complaining that he and his suite were incarcerated in the Palace of the Four Barbarians. The official promised he would speak to his superiors about it. "At this moment, the Jesuit again made pretense of reading [a document], but told the ambassador that they were keeping back the answer on purpose to force him to give way." Verbiest had recently been taken aside by the emperor, he said, "and he asked after you, and the negotiations . . . and I could judge . . . that he was inclined to receive you; but certain not very important persons take an arrogant line in the Council. The Emperor, though young—23 years old [actually he was just a few weeks past his twenty-first birthday at the time]—leans to the good in everything, but the ministers . . . are very proud and obstinate."

Here Verbiest seems to take an almost paternal attitude to K'ang-hsi, and one does not feel any betrayal is involved. A few days afterward, however, Verbiest was plotting again. He said that one of the officials had taken him aside and told him privately that some of the mandarins would not object to the scheme Spathary had suggested, that the letter and presents should be handed over in the presence of, though not directly to, the emperor. But that "others were against it." Indeed, it was a miracle that they were going to concede anything at all, "for they were arrogant people."

At last, after the most tedious negotiations, a formula for the reception of the tsar's letter in the presence of the emperor—who was to be invisible behind a screen—was agreed upon. Then came the ancient problem—the kowtow. One can see it looming in everyone's thoughts. Even the urbane Spathary breaks into the first person in his narrative; "On that a violent quarrel arose and lasted a long time. They insisted on an answer. I refused to give one, saying, 'It seems you want to compel me by force to follow your customs in everything. . . . And you delay your answer from day to day to wear us out.' "

Spathary refused even to entertain the idea of kowtowing until "I see what respect is shown to the Tzar's letters. . . . As to the presents—I refuse altogether to give them up in advance—they and the letter must go together. . . . We have heard your people call the presents tribute! That, we repudiate; for our Sovereign Lord and Master takes tribute, indeed, from many countries, but renders it to none." The officials demurred, suggested all sorts of variations, and eventually agreed on a way to do it. Only then did Spathary agree to do the kowtow, but in fact no one really knew whether, at the last moment, he would do it or not.

Spathary had his way, more or less. On June 5 he took the tsar's letter and presents to the court, having been assured that the emperor would be present though screened from view. "The letters were carried by the clerks, and the gifts were carried on tables by the Emperor's people." He went "through many gates in the Court, and came to a hall where a square place had been prepared covered in yellow, a color no one else dares to have. . . . And on the left-hand stood the intimate Counsellors; and the ambassador, taking . . . the letter of His Majesty the Tzar, stood with them; and the [highest mandarin of the court] took them and laid them on that four-cornered place; and the gifts in other places and Nikolai (Spathary), saying never a word, nor the counsellors to him, walked back . . . and from there rode home to the courtyard."

It was a curious, eerie scene he had forced the Chinese to enact, one all but devoid of normal court procedures; and Spathary's own absence of comment makes one feel the achievement was a hollow triumph. The honors of winning the first round, however, must be allowed to go to the Russians; those of courtesy to the Chinese; and those of flexibility to both.

A cloud of problems now gathered around the negotiators' heads. The question of an audience for Spathary, which he requested; the kowtow, which the Chinese demanded; the form of the emperor's reply to the tsar's letter. Would the reply be as from one equal to another?—for nothing less would be acceptable. Would its contents even be made known to Spathary?—for unless they were, he would refuse to accept it.

The Manchu officials attempted to cut through this swarm of queries by asking, quite simply, if the Russian had any oral messages to give. He had—a number of requests: 1) That a former letter from the emperor to the tsar be translated from Chinese into a language comprehensible to the Russians. 2) That a language be agreed upon for future use in letters, etc. 3) That the titles of Tsar and Emperor be agreed upon and model letters be drafted in those forms. 4) That the emperor send an ambassador to Russia with Spathary. 5) That merchants be allowed to trade freely on either side. 6) That Russian prisoners, if any, be set free. 7) That, yearly, forty thousand pounds' weight of silver be sent from China to Moscow in exchange for Russian goods. 8) That agreement be reached on trading routes. And, finally, 9) that these and a few other points be accepted in love and friendship.

At this point again there are passages missing from the manuscript. But it recommences with Spathary attempting, *through Verbiest,* to bribe the high officials in order to procure better treatment at audience. Verbiest in person took the bribes to the officials.

In a day or two, fifteen horses were sent to bring the Russians to audience. Spathary gives one of the best factual descriptions of the scenes as he passed through successive courtyards of the Forbidden City—closely documented, balanced, unemotional. In such details, which can be checked for their accuracy, one can judge of his general accuracy. And any doubts that he might have been twisting Verbiest's character for his own ends disappear when, with his known accuracy for detail, we ask ourselves: Would any useful purpose have been served in falsifying the record in regard to Verbiest? The answer is: Probably none.

Reaching the great courtyard before the Pavilion of Purple Light, "the sun was already up . . . and . . . we saw the mandarins sitting there, each according to his rank; and at the near end . . . a place had been prepared." Here they waited, seated, for the ceremony to begin. Without doubt the scene he then goes on to describe was one of the most extraordinary court spectacles in the world. Round the central area of the square, where the greatest men of the country are seated waiting, are six pavilions, the most elaborate being the Pavilion of Purple Light approached by marble stairs "all of the most wonderful art, very beautiful to gaze at. . . . In the midst of the courtyard as you go . . . towards the Emperor's throne, there is a road paved with white marble on which none but the Emperor ever walks. . . . On each side . . . stood people . . . holding in their hands banners; and on each side were forty-eight banners of taffeta . . . inscribed with Chinese characters in gold, the tops of the banner staves being fashioned into spear-heads, with, just beneath the blades, dyed horse-hair tufts. . . . Besides these . . . they held up on either side twenty-five others of carved wood. . . . These banners are not military, but part of the decoration specially devoted to the reception of ambassadors. Behind, on either side, stood five white horses, having yellow saddles, and behind them again, thirty men in yellow raiment. Still further back . . . were yellow umbrellas and twenty men on either side, standing near the platform surrounding the hall, but outside the doors, clad in cherry-colored silk embroidered in gold, bearing for arms . . . great two-handed gilt swords . . . and others with huge spears, with long leopards' tails by way of streamers."

Here, 980 feet from the pavilion and throne and still unable to see the emperor, Spathary perfunctorily kowtowed, "bending neither low nor slowly" (to the annoyance of the officials). The officials then told the ambassador and his party to go up nearer to the emperor, at the run as was the custom. Spathary naturally refused to run, and proceeded at a dignified pace to a spot fifty-six feet from the emperor. "The emperor sat . . . a young man with a pock-marked face" surrounded by his brothers and relatives. Tea was served in "large, yellow wooden cups, and boiled together with butter and milk, in Tartar, not Chinese, fashion." After tea the emperor left and the company went away.

During the ensuing few days Spathary was given by Verbiest a very detailed account of the military and financial problems afflicting the Manchu government in China. Verbiest was helped out in his treacherous explanations by Buglio and Magalhaens, ever ready, it seems, to intrigue against the Manchu regime. The Jesuits themselves were surprised to learn that Russia was so extensive. They were "sorry that so great a sovereign [the tsar] had sent a mission to these people [the Manchu] . . . as they were barbarians, who knew not how to render honor where honor was due. . . . This was why the Dutch and the Portuguese, after sending embassies more than once, seeing their insolence, did so no longer. . . . *Would to God they might see just once His Majesty the Tzar's armed forces and be made to understand the difference between him and them!*"[5] Later still, Verbiest, only slightly pressed by Spathary, divulged other secret information, "but it must be kept absolutely secret or he would lose his head."

The hundred pages which Spathary's report occupies in translation give a full account of those long negotiations with the Chinese government. The whole question of Sino-Russian relations, the problems arising from pris-

5. The italics here and in the second paragraph below are mine.

oners taken, from territory which until recently had hardly been given a thought by either side—all those aspects are argued with skill. And, one must admit, with forbearance on both sides. The incidents make interesting reading and are by far the most clearly recorded account of a Western envoy's reception by the Chinese court and administration that we have while China was still an independent power. One or two points relating to Verbiest, however, should be retold here.

According to Spathary's report, in return for an ikon which Verbiest begged of Spathary for the church in Peking, the Jesuit "told the ambassador under oath of secrecy, that the emperor intended [if the Russians did not surrender one particular prisoner] to go to war for him; he meant also to capture the forts of Albazin and Nerchinsk, for the Russians had become formidable in their eyes, especially since they had learned . . . that in truth the Russians were there by the Tzar's orders and not . . . as lawless people such as formerly infested the Amur river, and could be destroyed whenever necessary. 'They know [continued Verbiest] that at present the garrisons in those places are not numerous and that Moscow is far away, while they are comparatively near; but their plan is to wait until the number of their troops on the frontier is augmented. . . . If [the tsar's] intention were to refuse to give up [the prisoner] *troops in large numbers should be sent without delay to defend these forts;* for the Manchus themselves wondered how we dare dwell in such small numbers in proximity to so mighty an empire.' *They, the Jesuits, were glad to serve the Tzar as they serve God, for they love not the Manchus, as they did the Chinese.*"

Despite these pieces of vital information about the enemy's intentions, the Russians did not act, and they lost the Amur in 1689. But, as Baddeley exclaims, the Jesuit betrayal of K'ang-hsi seems complete. Throughout the Russian stay in Peking, Verbiest carried the secrets of the administration to Spathary's ears very promptly. There appears to be no doubt that he sided entirely with the Russians, while at the same time posing as the friend of the Manchu and of the emperor himself. Yet in reality he seems to have been genuinely fond of K'ang-hsi. But, more importantly, Verbiest appears to have been working for a return of the Ming or some other Chinese administration in China. We may wonder from this how much more sabotage the Jesuits in Peking and elsewhere in China at this time managed to perpetrate under cover of their advisory position in the Manchu administration.

The embassy of Spathary dragged on, for over three months, until the beginning of September. Spathary finally achieved a more or less private audience with the emperor—the interpreter being, of course, Verbiest—and was treated with courtesy. Interminable arguments went on about the manner in which Spathary would accept the emperor's presents for the tsar, and after a fiasco in which Spathary appears, even in his own account, a stiff and almost ludicrous figure of pomposity, he finally gained his point and was given the presents without kowtowing. But the emperor's letter to the tsar presented even greater problems. Spathary refused categorically to accept a letter which did not address the tsar as from one sovereign to an equal. The emperor and his administration refused to write any such (from their point of view) humiliating missive. For, in the whole history of relations between China and all the other states of the world no such letter had ever been written. The tsar's letter to the emperor remained unanswered, as did Spathary's numerous verbal requests.

The embassy prepared to depart on its journey back to Russia. But before quitting the Palace of the Four Barbarians, Spathary completed one piece of business which had occupied him now and then in the past weeks. In the intervals of his diplomatic activity, Spathary had bargained with merchants for a jewel of great size and beauty—a ruby—and now he bought it. With the help of Verbiest, the merchants, who saw the ambassador about to depart, sold it very cheaply to him almost as he rode off for good. The ruby eventually found its way to the crown of Empress Anne of Russia, now on display in the Kremlin museum.[6]

Thus, with thinly veiled acrimony, with a ruby and the goods they had been unable to sell in Peking, and with some rather miserable presents for the tsar, the ambassador and his party left Peking in poor order with insufficient carts and animals to transport themselves. And that was the end of a highly unsatisfactory but revealing East-West encounter.

Spathary's narrative of his negotiations in Peking is probably the most important single document from a Western source on Sino-European relations until we come to those relating to negotiations preceding the Treaty of Nanking in 1842. Of all the Western envoys until that time Spathary was the only one who spoke from a position of comparative strength, yet the basic problem encountered by Spathary in dealing with the Chinese state was the same as that of every other envoy from the West—the rigidly held opinion of the Chinese that they were the real leaders of the world, that theirs was the absolute power because theirs was the absolute and ultimate civilization, and that all other peoples could only ask for favors and be granted what seemed, to the Chinese, good for them. Spathary was, of course, quite unable to accept this attitude, and in return was stung into exaggerating the power and magnificence and influence of his sovereign. Neither side could admit that to accept each other's customs would not damage the honor of their respective rulers.

So it was impossible to get any further at the time, despite the fact that

6. This huge jewel had been offered to Spathary earlier in his stay. The mandarin who owned it at first asked a very high price but was gradually beaten down. The negotiations were conducted in a most obsequious manner by Verbiest, who perhaps felt that the secret information he had supplied was not enough in exchange for the ikon Spathary had given for his church.

both the Russians and the Manchu administration would have liked to reach some sort of settlement of the Siberian question. The problem was eventually settled fourteen years later, while China was still under the rule of K'ang-hsi, by the Treaty of Nerchinsk, a document partly negotiated by Verbiest's successor, Father Gerbillon. This was the first treaty ever signed by China with a Western state, and in general may be described as a recognition by China and Russia that their countries had a common frontier, and that this frontier would be respected by both. It was a treaty as between equals. In it one sees the wise and moderate hand of the older K'ang-hsi. Verbiest and others had failed to set Russia at China's throat.

Reading through the pages of Spathary's *stateini spisok,* one can hardly suppress gasps of regret as the perfidy of Verbiest unfolds sentence by sentence. The standard picture of the Jesuit that we followed in the preceding chapter is suddenly seen to be superficial. The reader must judge for himself what accommodation to make between the light and the dark of Ferdinand Verbiest. But, at least partly, through a long look at Verbiest and Spathary and K'ang-hsi we may understand a little better the perpetual Chinese distrust of the Jesuits and their activity in China—a distrust that never receded further than to lurk behind temporary acceptance of them. The fears recounted in the edicts limiting and prohibiting Christianity in China at various times constitute the real Chinese attitude to the religion and its priests (not by any means all of whom were Jesuits). There was a foundation for those fears that they would eventually incite or permit sedition and encourage the rebellious sections of Chinese life. The great Verbiest, for one, is no mean example of the reason behind those Chinese opinions.

CHAPTER THIRTEEN

RIPA THE NEAPOLITAN AND HONEST JOHN BELL

IN THE year 1840, Dr. Karl August Mayer, a German, published a book called *Naples and the Neapolitans* in which he gives the following description:[1] "Close to the Ponte della Sanitá, to the northwest on a neighbouring declivity, stands the Chinese College; for this is the name given to a religious institution . . . which educates young Chinese as teachers and missionaries for their native land. . . . We went into the capacious hall. . . . A servant pointed out to us in the Refectory the portraits of Matteo Ripa, the founder of the college . . . and of several young Chinese, whose names and the dates of whose existence were appended to the pictures. After a while the rector appeared. . . . The conversation then turned on Matteo Ripa. He was an Italian and a missionary who . . . preached Christianity in China, where he had been appointed court painter. We heard the following anecdote touching the pictures: — As soon as the young Chinese are sufficiently instructed to understand their business . . . they return to China; and the portrait of each youth is then taken. . . . Should one happen to die in Naples, he is painted either before or immediately after his death. Some of the faces from this reason have death strongly marked upon them. . . . The number of pupils at present amounts to eight, of whom six are Chinese. . . . The instruction is given in Latin; but the pupils have picked up Italian in their intercourse with the servants. The rector himself does not understand Chinese, and the newcomers can only follow his lessons after they have learned some Latin from their fellow-countrymen.

"We were then conducted into another room, and a few Chinese made their appearance, clad in long priest's robes. . . . They greeted us in the most friendly manner, and plenty of time was given to us to observe their ways, and to talk with them in Italian. The colour of their faces is yellow,

1. The English translation is from Prandi's rendering of Ripa's *Memoirs*.

but not disagreeably so, and their shining black hair lies straight and smooth over their low foreheads: their small, strange, half-closed eyes are jet black, and full of vivacity, and are placed, turning upwards, towards the temples. . . . The form of the face is oval and flat, the nose flat and short, so that they have scarcely any profile. When they laugh, and this they do incessantly, owing to their childish good humour, it is with a grin which shows all their teeth. Their heavy, monotonous way of moving suits well their round, short, and diminutive bodies. . . . One might almost lay these Chinese down and roll them like barrels.

"They showed us a map of the Celestial Empire . . . also a charming little model of the famous porcelain tower at Nankin, and they gave us the necessary explanations with evident joy. . . . One of them read some passages out of the New Testament translated into Chinese, which sounded strangely enough. . . . Another opened his mouth awfully wide, and sang us a national song to a most barbarian tune.

"The rector then took us over the beautiful terrace of the house, from whence we over-looked all the northeast portion of the city. . . . We enquired of the ecclesiastic whether he was satisfied with the progress his pupils made: their memory, he replied, was exceedingly good, and one of them showed a pleasure in and great aptitude for the sciences. . . ."

To find the beginnings of this unlikely Chinese College, we must descend the hill, cross the Ponte della Sanità, and take our way into the city to the Palazzo Reale. Here, says Ripa in his book,[2] "in the year 1700, as I was strolling one day about the streets of Naples, I came to the open space before [the Palazzo] just at the moment when a Franciscan friar, mounted on a bench, began to address the people. I was only eighteen; but though so young, I was then leading a life which I could scarcely describe without shocking the reader." The friar preached on a text showing how there were a certain number of sins which God would forgive, but that after that number there was no salvation. "It was with a gleam of heavenly light that I perceived the dangerous path I was treading. . . ." From that day it was only five years till he was ordained at Salerno. Returning to Naples, he went to see a priest whom he knew, a member of the order of Pious Laborers. "The moment he saw me, he said, 'Good morning to you, good man; prepare for China.' I was surprised . . . for I had never heard anything about China." And thus it was that Ripa too became a Pious Laborer.

Ripa and his several companions took a British East India Company ship from London for the voyage east. At this time the company had strict rules that no ecclesiastics should be carried, and the priests had recourse to disguise in order to obtain passage. Held up at the mouth of the Thames for months, Ripa slept on a straw mattress under the beam of the rudder. "I

2. This and all other quotations from Ripa come from Prandi's abridged translations of his *Memoirs*.

MATTEO RIPA IN YOUTH AND MIDDLE AGE. The story of a lifetime's difficult endeavor is summed up in these two portraits: the almost cherubic young man with his bright, clear eyes changes to the more than middle-aged priest with many years of work in China—the eyes dulled, the face sharpened by trouble. (Church in the Salita Cinesi, Naples; photos by the author)

could have borne this and other miseries . . . but that which was insufferable to me was that, close to my bed, were the berths of three officers, who, during the four months we remained in the river, were frequently visited by their wives: those who know what liberties English women allow themselves, may understand what a poor missionary must endure in being obliged to remain day and night with such company. One of the women was so barefaced in her actions, that no sooner was her husband out of sight than she behaved in the most infamous manner."[3]

The ship at last sailed for India. Ripa's spirits, filled with indescribable hopes, expanded almost painfully. The ardent young man was setting out, at what risk he neither knew nor cared, to conquer the Eastern world, the China of his still totally European imagination. "I fancied I breathed more freely, as though I had been relieved of an oppressive burden. . . . I felt like a bird which, freed from the constraint of the cage, can spread its wings and rove where it pleases. . . . I had just begun to be an apostle . . . clad in a ragged cassock [and] was about to wander through the vast regions of China, preaching to those blind pagans the Holy Word of God."

Written more than thirty years afterward, Ripa's book captures the freshness of his hopes, unconsciously tracing the way in which his naïve ardor changed in China, over the frustrating years in K'ang-hsi's service, to an ideal that slowly formed in his mind. The ardor itself turns into the very real compassion which now and then flames in his recollection of Chinese scenes and horrors; but the missionary zeal turns—as we will see—into something slightly different.

At Calcutta, the young priest transshipped and sailed to Manila, then under Spanish control. Out of Manila he took a vessel piloted by the farcical priest Don Pedrini, who fancied himself in the role of navigator. "His inexperience in nautical matters nearly cost us our lives two or three times." Once, in fact, he steered them onto a rock. Seeing the danger, Ripa rushed to the bows, blessed the sea with holy water, took a candle given him by Pope Innocent XI and cast it on the violent waves. "Very soon after . . . we were out of danger." On the night of January 2, 1710, they reached Macao.[4]

The immediate reason for the voyage of Ripa and the others was to convey a cardinal's hat conferred by the pope on his legate De Tournon, who had reached Peking in 1705. De Tournon had come loaded with papal gifts for K'ang-hsi and with papal decisions as unedifying to the monarch as the

3. In another passage he relates that, on his return to England, while traveling in a coach for London, his party was in the company of "a colonel and the wife of a merchant, the women of England being indulged with such freedom owing to the entire absence of restraint which prevails in their island."

4. In the fourteenth century, as we have seen, Odoric had raised a fair wind with a few martyr's bones. But by now such relics were evidently in short supply.

gifts were pleasing. His mission was the result of misunderstanding in Europe (in part due to slow communication with China) of the situation in regard to Christianity there. De Tournon himself was one of the worst envoys anyone ever sent to China. His peevish, foot-stamping attitude to the opinions of priests in China, his incivility to K'ang-hsi (who was patient with him), his childish tearing up of petitions sent on serious matters, his storms of tears —all combine to deny him the sympathy one might feel when he was finally shuttled down to Macao and kept under arrest by the Portuguese. It is hardly cause for grief that he died before the cardinal's hat could be placed on his head.

But—alas for that century of Jesuit and other Catholic effort from Matteo Ricci to Matteo Ripa—the damage was done. The sinister Rites Controversy had now crystallized into papal condemnation of the Chinese practice of making offerings to their ancestors and kowtowing before their shrines. Almost at the same time came a statement from K'ang-hsi, made for the benefit of the pope and the Jesuits in China, that the rites were of a purely secular nature. This vindication of the views of Ricci came too late. Instead of being congratulated, the Jesuit fathers in China were accused in Europe of bowing to secular authority in the person of K'ang-hsi on a question of Christian dogma; and the papal bull on the subject flatly contradicted the fiat of the emperor in his own domain of Chinese belief.

Naturally enough, the emperor resented the affront. He was accustomed to have the final say on matters Chinese. When he finally learned from the amateur navigator Pedrini the contents of the bull, which set at naught everything more judicious priests in China had advised and decided in conformity with his own views, the emperor was very angry. Pedrini completely failed to understand the implications and, with a blindness characteristic of him, imagined he was now K'ang-hsi's favorite. K'ang-hsi, wise old man that he was, later cast Pedrini into prison for a while, then let him out—just to show who was who at court.

Ripa, charitable almost to the point of blindness, skirts round this seepage of Jesuit dissension and contents himself with a denunciation of the conduct of missionaries in China in general terms. "If our missionaries . . . would conduct themselves with less ostentation, and accommodate their manners to persons of all ranks . . . the number of converts would be immensely increased; for the Chinese possess excellent natural abilities, and are both prudent and docile. . . . Unfortunately our missionaries have adopted the lofty and pompous manner. . . . Their garments are made of the richest materials; they go nowhere on foot, but always in sedans, on horseback, or in boats, and with numerous attendants. . . . With a few honorable exceptions, all . . . live in this manner. . . . The diffusion of our holy religion . . . has been almost entirely owing to the catechists who are in their service. . . . Scarcely a single missionary . . . can boast of having made a convert by his own preaching, for they merely baptise those who have been already converted by others."

It is not hard to imagine what Chinese now felt about Christians, Christianity, and everything pertaining to Europeans in general. For here—among these men who proposed to teach the only right way of life—were as many opinions and as many quarrels as there were priests. To Chinese the spectacle of Catholic disarray over the rites question was a demonstration of what, in their hearts, they had long suspected. It confirmed their own faith in Chinese ways of thought—for those at least had a monolithic unanimity, a certain dignity, a sure foundation. In vivid contrast was this Western religion whose priests seemed unable even to agree on a common attitude.

Such, in brief, was the climate of Peking when Matteo Ripa arrived. He was twenty-nine years old. Verbiest and Gerbillon (Verbiest's successor in the friendship of the emperor) were long dead. The emperor himself was now fifty-seven, secure in his power, an experienced, sagacious, and wily ruler. Christians he found useful, sometimes amusing, at his court. He tolerated their quarrels for this reason. The final blow was not to come for another nine years, until the arrival of another papal legate in 1720.

"Being safely arrived in Peking," Ripa says, "we were by [K'ang-hsi's] command immediately conducted to the palace, without being permitted to see any of the Europeans." Met there by officials and eunuchs, Ripa was asked "whether we had come prepared to serve the emperor even unto death; and we replied that such was exactly our wish." They were taken to see K'ang-hsi in his private apartments, where they found him seated cross-legged in Manchu fashion on a divan covered in velvet, a table with books and writing materials before him. In the presence of some other missionaries, Ripa performed the full three kneelings and nine head-knockings of the kowtow. (But, he remarks, he did not have to do it again except upon important ceremonial occasions.) K'ang-hsi proceeded to question the newcomers, wondering, evidently, to what uses he could best put them. He told Pedrini to play some music, asked Fabri about mathematics. He talked to Ripa about painting. "The emperor now commanded me to answer the next question in Chinese. . . . He addressed me very slowly, employing many synonymous words, in order that I might understand him; and was very patient with me, making me repeat the words until at length he made out what I meant. . . ." Later, "I was informed . . . that it was his Majesty's pleasure that I should go to the palace to paint; and, accordingly, I entered upon my duty on the following day."[5]

In those few words, without telling anything of what he felt, Ripa accepts his lot in Peking. The zealot turns court painter, engraver, artistic handy-

5. Ripa makes no mention of his ability to paint nor of how and where he acquired it, until he reaches Macao. And then he mentions it in what amounts to an aside, simply stating he painted two pictures.

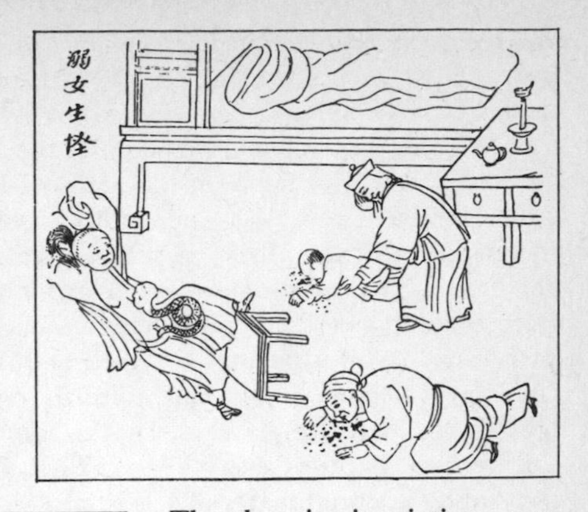

INFANTICIDE. Ripa and other Westerners had much to say about the custom of disposing of girl children at birth. The Chinese also officially discouraged the practice, as indicated by these two popular prints of the Ch'ing dynasty. The woman is giving birth to a snake with a human head as the penalty for having drowned her daughters.

The laconic inscription says: "Drowned a girl, gave birth to a monster." In the other picture, the venerable goat-bearded gentleman is surveying his rejuvenated countenance—a reward for having saved a baby girl from death. Between them, these two prints tell their cautionary tale. (From Matignon)

man to the emperor. Only later in his book does he murmur ruefully about being an "honorable galley-slave." And only then does he turn from descriptive matter on Peking to what he was able to do for his faith, and to the ideal which had formed in his mind.

By far the larger part of Ripa's writing on his China days deals with secular Peking, of which he constructs an interesting, highly idiosyncratic picture. Like that of Marco Polo, his description is often as remarkable for what is omitted as for what is said. But there are many passages of perception and sensitivity.

Early on he discovers the custom of abandoning infants, and after a tender scene in which he rescues one of them, only to see it die, he launches into an explanation: "There is nothing unusual in seeing children thus abandoned; it occurs daily throughout this vast empire. When mothers are poor, and have large families, or observe any defect upon the body of an infant . . . they cast away the little creatures without remorse. This cruel custom is also generally practised by unmarried women who have children. . . . The poor infants are secretly thrown into a river, or left near the public road, in the hope that some passenger may take pity on them and carry them home. This sometimes happens, but generally the unfortunate beings are devoured by wild beasts. Not far from the walls of Peking, I myself saw one infant under the paws of a dog, and another between the teeth of a hog. By a charitable provision of the emperor, carts are sent round the walls of his immense capital every morning in order to collect castaway infants and carry them to a certain temple where a number of women are employed to nurse them at his expense. . . . Well acquainted with this . . . the Jesuits have appointed a Chinese Christian to baptise all the infants that are brought to that temple. . . . In this manner not less than three thousand children are baptised every year." This, incidentally, is a fact one must remember when estimates of Christians in China are put forward by the missionaries.[6]

Ripa gives some account of the city itself, a better one of the palaces, with which he was most familiar, and of court life. His interests range from palace intrigue, through architecture, gastronomic delicacies such as deer sinews, dog meat, and bird's-nest soup, to footbinding.

"At the tender age of three months female infants have their feet bound so tightly that the growth of this part of the body is entirely stopped, and they cannot walk without hobbling or limping; and if upon any occasion they endeavor to quicken their pace, they are in danger of falling. . . . Even when walking at a slow pace, they . . . are obliged to walk like ducks, waddling about from left to right. In the case of marriage, the parties not being able to see each other, it is customary to send the exact dimensions of the lady's foot to her intended, instead of sending him her portrait as we do in Europe. In this particular . . . their taste is perverted to such an extraordinary degree, that I know a physician who lived with a woman with whom he had no other intercourse than that of viewing and fondling her feet." Unlike most Europeans, who thought foot binding served the Chinese aesthetic sense, Ripa hits on the correct, sexual, significance of the custom.[7] His later comments on a scene in the emperor's garden (see page 273) confirm this.

Others have done better than Ripa in topographical description, but none from this period has bettered him in recounting the byways of Chinese life. At a time when European medicine was hardly out of its infancy he tells of his treatment at the hands of a Manchu physician. Riding after the imperial

6. A point also made by John Bell, and by historians in our own time.

7. The swaying walk of women with bound feet is a subject much extolled in Chinese literature. The motion was held to be as delightful as, for example, the swaying of a willow tree in the breeze. But the sexual basis is unquestioned, a point that has been fully treated in Levy's recent book.

entourage on the road to Jehol, a hundred or so miles north of Peking, Ripa was thrown from his horse, "receiving frightful wounds in my head and other parts. . . . As my companions did not dare stop, they recommended me to the care of two heathens, and left me fainting on the street. . . . When I recovered my senses I found myself in a house, but everything appeared dark and indistinct, and I felt as if I had fallen from my horse two months before. The emperor sent me a Tartar surgeon. . . . This surgeon made me sit up in my bed, placing near me a large basin filled with water in which he put . . . ice. . . . Then stripping me to the waist, he bade me stretch my neck over the basin, and, with a cup, he continued . . . to put water on my neck. The pain caused by this operation . . . was so great and insufferable, that it seemed to me unequalled." But the surgeon said it would staunch the flow of blood and restore Ripa to his senses. To the latter's great surprise this actually happened. "He next bound my head with a band, drawn tight by two men who held the ends, while he struck the intermediate part vigorously with a piece of wood," which produced severe pain; but "after this second operation my head felt more free. A third operation was now performed, during which he made me, still stripped to the waist, walk in the open air supported by two persons; and, while thus walking, he unexpectedly threw a bowl of freezing . . . water over my breast." This caused a violent intake of breath and "the surgeon informed me that if any rib had been dislocated . . . this . . . would restore it to its natural position." After some further similar treatments "he ordered that I should continue to walk much . . . that I should not sit long. . . . He assured me that these walks in the open air, while fasting, would prevent the blood from settling on my chest. . . . These remedies were barbarous and excruciating; but I am bound in truth to confess that in seven days I was so completely restored as to be able to resume my journey into Tartary."

The physiological approach to the treatment of traumatic injury, the muscles and blood stream being encouraged to function normally while the patient is prohibited from languishing in bed, may sound normal to us now, but in an early-eighteenth-century context it was far in advance of anything being attempted in Europe.

Such are some of the passing interests of Ripa's book. He never came as close to K'ang-hsi as had Verbiest. The temper of the times did not run with the Christians; nor did Ripa's limited intellect appeal to the erudite emperor as had that of the Belgian. But in the narrative of his time in Peking, Ripa gives here and there a fact, a scene, to illuminate the more private side of K'ang-hsi. Those great autumn hunts described by Verbiest are better described by Ripa. There is an unrivaled description of the celebrations attending the sixtieth birthday of K'ang-hsi on April 4, 1713:

"The whole city of Peking wore an appearance of festivity. All were habited in gala dresses, banquets were given without end, fireworks discharged.

. . . But that which above all . . . struck me with astonishment was the spectacle exhibited upon the royal road from Chan-choon-yuen [a palace outside Peking] . . . which is about three miles in length. This road was adorned on both sides with an artificial wall . . . of mats . . . entirely covered with silks of the most beautiful workmanship, while at certain distances were erected fanciful houses, temples, altars, triumphal arches, and theaters in which musical dramas were presented." Down this three-mile corridor of pure silk, preceded by three thousand caparisoned cavalry and followed by the princes of the blood in their robes and by the officials in their ceremonial dress, rode the emperor K'ang-hsi, wearing his imperial mantle of yellow brocade covered with the five-clawed dragons magnificently blazing in gold embroidery. It is not surprising, as Ripa remarks at this point, that the emperor was frequently referred to as the living Fo, or Buddha.[8]

There is much more on the subject—the banquet K'ang-hsi gave in the palace for a thousand old men from the provinces of China, at which the food was served by his numerous sons. And there are more personal things—the emperor's questions about strawberries and asparagus and the planting of these in the imperial gardens; the curing of a boil on the imperial face by a European priest, and the emperor's vexation that three instead of two hairs of his beard were cut away in the process;[9] K'ang-hsi's amusement at a "mechanical remedy" for constipation suggested by Ripa;[10] his reputation as a miser; the flattery which surrounded him—so that when he merely touched a note of the spinet it "was enough . . . to throw the bystanders into ecstasies of admiration."

And there are more unusual scenes too. "During the stay of the Russian embassy in Peking, Dr. Volta, a Milanese priest and physician arrived . . . and I was summoned to accompany him . . . to the emperor." K'ang-hsi, ever in doubt of the credentials of Europeans reaching his court, commanded the doctor to take his pulse. Dr. Volta did so and said that to form a correct opinion of his majesty's health he must repeat the process in the evening and again the following morning. Having set the ball rolling, the emperor had to agree. "I observed," says Ripa, "that his Majesty's bed was wide enough to contain five or six persons, and had no sheets. The upper part of the mattress, as well as the under part of the quilt, was lined with lambs' skin, and the emperor slept between these, without wearing any night-clothes. As it seldom happens that an emperor is seen in bed by strangers, he said to us, 'You are foreigners and yet you see me in bed. . . . I

8. The last empress dowager, whose malign reign lasted into the present century, was commonly called Old Buddha by Peking people.

9. Even today it is quite common in China to see men whose faces are innocent of beard but who have a few hairs growing from a mole. Such hairs are allowed to grow many inches long since it is thought inauspicious to cut them.

10. Perhaps some form of enema, but this is a guess.

consider you as members of my own house, and very near relatives.'" And that, in fact, seems a good enough definition of Ripa's place in the scheme of things in Peking. The unique glimpse of a Chinese emperor in bed, however, reveals that K'ang-hsi's sleeping habits were still those of a Manchu chieftain rather than of a Chinese emperor.

Another, and most charming, sight of K'ang-hsi, the intellectual who was also a great sensualist, is caught for us by Ripa when, with another priest, he was installed in the apartment of the emperor's mother, who had recently died. The little house was surrounded by a garden and situated on a small promontory commanding an artificial lake in the grounds of the Jehol palace. "On the other side of the lake was a cottage . . . whither his Majesty often repaired to study, accompanied by some of his concubines." Ripa here instructs his readers that windows in China are filled with paper instead of glass and goes on to say how, through the holes in his, "I saw the emperor employed reading and writing, while these wretched women remained sitting upon cushions, as silent as novices. Through these holes I also observed the eunuchs while they were engaged in various ways of fishing. His Majesty would then sit in a superb little boat, with five or six concubines at his feet, some Tartar [Manchu] and others Chinese; all dressed in their national costumes. The boat was always followed by many others," which—in a nice phrase—were "loaded with ladies."

Thus K'ang-hsi disported himself until business called, "when he went by water; and, as he necessarily passed under my window, I also saw him . . . in a boat with some concubines. . . . On reaching the spot where, by a secret door, he entered the room where he held audience, he left the concubines behind in charge of the eunuchs. I saw him several times about the gardens, but never on foot. He was always carried in a sedan chair, surrounded by a crowd of concubines, all walking and smiling. Sometimes he sat upon a high seat in the form of a throne, with . . . eunuchs standing around him; and, watching a favorable moment, he suddenly threw among his ladies, grouped before him on carpets of felt, artificial snakes, toads, and other loathesome animals, in order to enjoy the pleasure of seeing them scamper away with their crippled feet." Then, some days, he would encourage the ladies to clamber on a small hill to gather nuts for him. "He urged on the poor lame creatures . . . until some of them fell to the ground, when he indulged in a loud and hearty laugh."

But, having enjoyed telling the tale, Ripa rounds it off soberly. It shows "a manner of life, which, in my opinion, is one of the most wretched, though the worldly consider it as the height of happiness."

Grateful as we must be for those little revelations, through paper windows, of imperial domesticity, it is a thousand pities that the eye which registered them was not a poet's eye. The visual impact, extraordinary as it is, needs the flash of imagination that only the poet's spirit can give. But even Ripa,

REMAINS OF THE JEHOL SUMMER PALACE in the Manchu country to the north of Peking, a favorite retreat for the imperial families of the Ch'ing dynasty. Many Westerners had occasion to visit the place, including Verbiest, Ripa, and, later, Lord Macartney. The main building is in the Tibetan style, reminiscent of the Potala at Lhasa. (Photo by Hedda Morrison)

one senses, was momentarily shattered by the beauty of a young woman, one of the emperor's principal concubines—"clothed in a scarlet mantle, with a splendid headdress of jewels similar to those . . . upon the heads of Chinese goddesses" whom he saw one day in the garden through that same paper window. There she stood in all her Chinese perfection, a vision on the shores of the artificial lake, with her small son on his knees at her feet. The peculiar magic of those Chinese paintings with which Ripa the painter was familiar fills out the scene with a weeping willow, a languorous group of distant attendants, a bird or two, a boat moored on the gray-silk water of the lake. Ripa, the prosaic (perhaps we may even suggest, the inhibited), could not or dared not let himself go and enjoy the moment. We have to do that for ourselves.

Here, for a page or two, we can conveniently leave him and take a look at the Peking scene in those closing years of the K'ang-hsi reign through the eyes of a very different man—the Scotsman John Bell. He and Ripa met on many occasions in the three months spanning the end of 1720 and the beginning of 1721.

Honest John Bell, as he was to be called in his native Scotland much later in life, makes a refreshing change from Ripa. He was more than honest: he was a genial, intelligent, moderately cultured man, with an observant eye for what he saw, and a mind trained in the discipline of medicine. In his book, written long after, he was under no compulsion to shock, to persuade, to titillate. He had no special ax to grind. Rightly, he understood that what he had to tell about his long journeys in parts of Asia was more important than the manner of the telling—provided he told the truth to the best of his ability. When the book came out in 1763 it aroused in Britain a certain stir of interest in China, which was much more than the publication of Ripa's history of the Chinese College did. Although, like the writing of William of Rubruck, it was quite as interesting as many another volume on the subject of China and Asia of its times, John Bell's work never achieved that long popularity which it deserved. On its publication, the authoritative *Quarterly Review* called it "the best model . . . for travel writing in the English language."[11]

John Bell was born in Stirlingshire, Scotland, in 1691. He studied medicine in Glasgow. "I had, from my early youth, a strong inclination to visit the Eastern parts of the world," he says in his book. So strong, it seems, that he very soon managed to obtain letters of recommendation to a fellow Scot, Dr. Areskine (a Russianization of Erskine), who was privy councilor and physician to Peter the Great, tsar of Russia. To judge by the predilection Peter had for the Scots, of whom he had a number in his immediate service, Bell was sure of a welcome in Moscow before he left London at the age of twenty-three for the Russian capital. And, only a year later, he was setting out from there on a Russian embassy to Persia. His adventures had begun.

In 1718 he returned to Moscow, only to leave again on the embassy sent to Peking in the following year. Under its ambassador, Ismailoff, and after a

11. As this chapter was being written, Stevenson's newly edited version of the Chinese section of Bell's book appeared, thus bringing it within easy reach of the public after many years of unmerited neglect. Unless otherwise stated, the quotations that follow are taken from Bell.

"tedious" journey of sixteen months across the monotony of the Asian steppes, the party arrived in Peking in mid-November, 1720. Like Matteo Ripa on his arrival years before, John Bell was a young man of twenty-nine as he entered the city, his ambition already substantially on the way to fulfillment.

The ambassadorial party made a splendid entry into Peking, one which was unique in the annals of foreign embassies to China. It was led by a mounted officer with drawn sword, who was followed by soldiers, footmen, a drummer, several pages, three interpreters, and the steward. Then came the ambassador himself, flanked by high-ranking Chinese officials and followed by his suite of gentlemen, servants, and other attendants. They approached "over a fine road, through a cloud of dust and multitudes of spectators . . . and entered the city at the great north gate; which opened into a spacious street, perfectly straight as far as the eye could reach. We found it all sprinkled with water, very refreshing after the dust. . . . A guard of five hundred Chinese horsemen was appointed to clear the way; notwithstanding which, we found it very difficult to get through the crowd. One would have imagined all the people of Pekin were assembled to see us. . . ."

Ripa, who was an eyewitness, fills out the picture: "Count Ismailoff . . . was on horseback, and had a man of gigantic height on one side of him, and a dwarf on the other. . . . [He] had a fine person and a noble expression of countenance." It seems odd that Bell does not mention the giant and the dwarf (whom Ripa is not likely to have invented). In passing, we may note that one of the interesting aspects of this period in Peking is that we do not have to rely on Bell and Ripa alone for a record of its events and scenes. There are the Jesuit De Mailla's comments, and the entries in the journal of De Lange, who was also with Ismailoff's party; and there are, for some aspects, the letters of, and book on, the papal legate Mezzabarba, and of other Catholic priests in the religious nexus at Peking.[12]

When we recall the bitter complaints of former ambassadorial parties about the inadequacy of the Palace of the Four Barbarians, it is surprising to read Bell's enthusiastic account of the place. Perhaps the buildings had been refurbished. It would not be unlikely to find this was in fact the case, for the basic Sino-Russian problems had been long settled by the Treaty of Nerchinsk, and Ismailoff was received in Peking with every honor and courtesy.

He had, of course, arrived with the usual encumbrance of instructions, but by virtue of the treaty, his task was a simpler one than that of any previous ambassador to Peking. He was to try to solve problems related to trade caravans passing between the two countries, to obtain permission to build a Russian church in Peking, and to obtain leave to establish a permanent

12. De Mailla's work in thirteen volumes was published at Paris in 1777–85 under the title *Histoire générale de la Chine*. For Mezzabarba, see Servitá. De Lange's journal is included in Bell.

Russian trade commissioner in the capital. He was also to sue for "extra-territoriality" rights. In all this, except the last point, Ismailoff succeeded. The unusual harmony that marked the Russian stay in Peking is demonstrated by the fact that Ismailoff had no fewer than twelve audiences (eleven of them private) with K'ang-hsi. He made only a token show of resistance to kowtowing, and went through the motions with good enough grace.

The first audience took place at Ch'ang Ch'un Yuan, the country palace not far from Peking. Bell, fortunately, was no stranger to courts and ceremony by the time he arrived there, so his description of the place and the scene, while it shows his obvious delight and fascination, is no starry-eyed report. At heart a countryman, he notices things which hardly anyone else bothers to record.

About ten in the morning, having ridden from Peking, the ambassador and party arrived at the palace "where we alighted at the gate, which was guarded by a strong party of soldiers. The commanding officers conducted us into a large room, where we drank tea. . . ." After half an hour "we entered a spacious court, enclosed with high brick walls, and regularly planted with several rows of forest trees . . . which I took to be limes. The walks are spread with small gravel; and the great walk is terminated by the hall of audience. . . . On each side . . . are fine flower-pots and canals. . . ." In the courtyard they found all the ministers and mandarins seated on fur cushions before the open hall where the throne stood. "In this situation we remained, in a cold frosty morning, till the emperor came into the hall."

Bell says they waited "about a quarter of an hour" before the emperor arrived and took his seat on the throne; while Ripa says they were "kept waiting for a good while." Ripa was one of the court's official interpreters and doubtless felt he had a vested interest in the Chinese point of view. It is a pity that he never mentions John Bell by name—and even more of a pity that John Bell fails to give any estimate of Ripa's character.

"The master of ceremonies," Bell continues, "now desired the ambassador . . . to walk into the hall, and conducted him by one hand while he held the credentials in the other. . . . The letter [from the tsar] was laid on a table placed for that purpose, as had been previously agreed; but the emperor beckoned to the ambassador, and directed him to approach; which he no sooner perceived, than he took up the credentials, and . . . walked to the throne." Kneeling, he laid the letter before the emperor, who touched it with his hand and enquired after the tsar's health, saying it was on account of his regard for the tsar that the usual ceremony of reception for credentials had been dispensed with. The ambassador then retired back a little to the body of his suite, who were standing. The master of ceremonies "ordered all the company to kneel and make obeisance nine times to the emperor. . . . Great pains were taken to avoid this piece of homage, but without success." In fact Ismailoff's instructions said if necessary he was to kowtow; and he did, the master of ceremonies calling out the Manchu words for bow and stand—"two words which I cannot soon forget," adds John Bell, his Scottish pride not a little outraged.

But Ripa gives another version of the scene: "Count Ismaïloff then entered, and immediately prostrated himself before the table, holding the Tzar's letter with both hands. The emperor, who had at first behaved graciously to Imsailoff, now thought proper to mortify him by making him remain some time in this . . . posture. The proud Russian was indignant . . . and gave unequivocal signs of resentment by certain motions of his mouth, and by turning his head aside, which under the circumstances was very unseemly. Hereupon his Majesty prudently requested that the ambassador should take up the letter to him." Ismailoff did so, kneeling; and K'ang-hsi received it with his own hands, "thus giving him another mark of regard, and granting what he had previously refused." Without entirely disbelieving Bell, who was not close to the emperor's throne, we must probably believe Ripa in this case, for he was right on the spot interpreting.

The audience lasted for two hours or more, dinner being served. With his own hands K'ang-hsi offered Ismailoff "a gold cup full of warm tarassum; a sweet fermented liquor, made of various sorts of grain, as pure and strong as Canary wine, of a disagreeable smell, though not unpleasant to the taste." K'ang-hsi observed "that this liquor would warm us that cold morning. His Majesty found many faults with our dress, as improper for a cold climate; and, I must confess, I thought him in the right."

Bell gives a sober description of the setting, the participants, and such events as the musical entertainment kept up as they all ate in the crowded hall. "What is surprising, there was not the least noise, hurry, or confusion. Everyone perfectly knows his business; and the thick paper soles of the Chinese boots prevent any noise from their walking. . . . By these means everything goes on with great regularity; but at the same time with wonderful quickness. In short the characteristic of the court of Pekin is order and decency, rather than grandeur and magnificence."

In John Bell's pages we are far from the neuroses of celibate and pious priests, close to the rather canny outlook of a Scot accustomed to great people and great events, but at heart a plain man. It is pleasant to attend a banquet given by a mandarin in a palatial residence with John Bell as guide. "At the entry were placed two large China-cisterns, filled with pure water, in which played some scores of small fishes, catching crumbs of bread thrown into the water. The fishes are about the size of a minnow, but of a different shape, and beautifully varied with red, white, and yellow spots; and are therefore called the gold and silver fish. I never saw any of them out of this country; though, I imagine, they might easily be brought to Europe. . . . I had about twenty of them standing in a window at my lodgings; . . . after a frosty night I found all the water frozen, most of the fishes stiff, and seemingly dead; but, on putting them into fresh cold water, they all recovered, except two or three." He was right about goldfish. They reached Europe later in that century, though not America, where their arrival seems to have been as late as 1874.

Their mandarin host was a collector. Conducted through his house, the party "saw a noble collection of many curiosities . . . particularly a large quantity of old porcelain or China-ware." The secret of porcelain manufacture had in fact only been known in Europe for a few years at this time, and only in Germany; so, naturally enough, Bell was amazed at the quantity and the antiquity of the mandarin's pots. He gives a charming picture of the collector describing and fondling his pieces and regretting (ironically, since he lived in the great porcelain age of K'ang-hsi) that the Chinese had lost the "art of making porcelain in that perfection they did in former times." (But perhaps the mandarin's taste was for T'ang and Sung, or Ming blue-and-white.)

Not at this banquet, but at another, given by the emperor's ninth son, Bell saw a performance of Chinese opera, which puzzled him since he knew virtually no Chinese. But after the opera came a series of farces, in one of which "the last character that appeared on the stage was a European gentleman, completely dressed, having all his cloaths bedawbed with gold and silver lace. He pulled off his hat and made a profound reverence to all that passed him. . . . This scene was interrupted, and the performers dismissed by the master of the feast, from a suspicion that his guests might take offence." Presumably such satires on the obsequious priests and trade-seeking Europeans commonly seen at court in Peking were normal palace entertainment, but this is the only reference to them.[13]

Fortunately for posterity, Bell seems to have had very little in the way of doctoring to do while he stayed in Peking. Instead, he spent a great deal of his time making Chinese friends and exploring the city. Fortunately, too, the Ismailoff embassy was in better odor with the court, or, like Spathary, he would have been prevented. Bell gives us an intimate and homely picture of markets, shops, teahouses, private houses, streets, temples, gardens; of the suburbs that sprawled outside each of the many city gates; of his walks and rides round the walls of Peking; and of many another thing left undescribed by most missionaries on whose accounts we have to depend so largely. Bell was still a young man enjoying immensely this chance of seeing a remote and fabled land. He observes Peking with an almost participant eye, not reciting marvels, nor condemning out of hand, but bringing his experiences within the compass of ordinary life. Evidently he went to no

13. In the last few years Peking opera itself has turned to contemporary themes, in some of which satire on Westerners and their governments is to be seen. Now, as in the days of K'ang-hsi, the Chinese again feel themselves strong and able to confront the outside world.

little trouble on the spot to check his facts. The result is mostly lively and frank.

"The 6th, while walking through the streets, I observed an old beggar picking vermin from his tattered cloaths, and putting them into his mouth; a practice which . . . is very common among this class of people. When a Chinese and a Tartar [Manchu] are angry with each other, the Tartar, in reproach, calls the Chinese a louse-eater; and the latter, in return, calls the other a fish-skin coat; because the Mantzur Tartars who live near the river Amoor, subsist by fishing, and in summer, wear coats made of the skins of fishes."

Giving an account of the celebrations marking the sixtieth year of the K'ang-hsi reign, he remarks on the "cheerfulness" of the emperor. "At taking leave, the emperor told the ambassador that he liked his conversation. He desired to be excused for sending for him in such cold weather, and smiling said, he knew the Russians were not afraid of cold. I cannot omit taking notice of the good nature and affability of this ancient monarch, on all occasions. . . . He still retained a sound judgement, and senses intire; and, to me, seemed more sprightly than many of the Princes his sons."

On one of his walks: "I went to see the market where the provisions were sold. It was a spacious oblong, spread with gravel, very neat and clean. The butchers had their shops in a shade, running quite round the place. I saw little beef, but a great deal of mutton. In the middle was a great store of poultry, wild-fowl, and venison; but what surprised me not a little was to find about a dozen dead badgers exposed for sale. The Chinese, it seems, are very fond of these animals. . . . All the Chinese merchants have the art of exposing their goods to sale dressed up in the most advantageous manner; and, even in purchasing any trifling thing, whatever the case be that holds it, it is half the cost, and often exceeds it in value."

With Ripa, Bell notices the abandoned babies. But he also tells how K'ang-hsi had some knowledge of Old Testament stories, of which Ripa gives not a hint. He recounts the emperor's entirely erroneous story of the antiquity of the Chinese compass. He gives animated descriptions of Chinese acrobats and jugglers, and a paragraph on the brothel quarter; and a charming picture of the emperor's grandchildren exercising with bows and arrows to improve their chest expansion. He discusses Chinese medicine; Chinese women, whom he obviously found attractive; and Chinese painting—"the chief study of their painters seems to be landskip-painting; and I have seen some of their performances, in this way, very natural." His description of the Great Wall is accurate and his opinion of its construction interesting: "I am of the opinion that no nation in the world was able for such an undertaking, except the Chinese. For, though some other kingdom might have furnished a sufficient number of workmen . . . none but the ingenious, sober, and parsimonious Chinese could have preserved order amidst such

K'ANG-HSI IN OLD AGE, an unusually perceptive portrait of a shrewd but kindly old man. (From Hibbert)

labour. . . ." The emperor who built the wall deserves much fame, being superior to him who built the pyramids of Egypt: "a performance of real use excells a work of vanity." A thoroughly Scottish moral remark.

Just as Bell appears to have made no impression on Ripa, and vice versa, so the papal legate Cardinal Mezzabarba appears in Bell's narrative in the character of a stage extra, representing, if anything, the Chinese Rites Controversy, of which Bell gives the essence in a paragraph. Perhaps we should have heard more of him had he been in the "French convent" when Bell went there to call. But there was no one at home except "an old gentleman," Father Bouvet, who was one of the mathematicians sent by Louis XIV of France in 1687, and now getting on in years. Being a Protestant, disinterested in Catholic affairs, Bell did not realize that with the advent of Mezzabarba the shroud was about to be wrapped round the body of Jesuit and other Catholic missionary activity, preparatory to its being placed in the Chinese coffin ready to receive it.

Mezzabarba brought with him the fateful constitution *Ex illa die,* which commanded missionaries under pain of excommunication to abide by the decision of the Holy See on the Chinese rites. In this document the pope decreed that Chinese worship at family shrines, and other allied traditional practices, were repugnant to Christianity. Therefore no Chinese was to be baptized if he persisted in such ceremonies. K'ang-hsi had already been informed of this by the sneaking Pedrini. The emperor had already confirmed, but stayed execution of, the decrees of 1717 in which missionaries were to be banished, their churches destroyed, and the Christian faith put down. Despite this, he received Mezzabarba with courtesy.

Finally K'ang-hsi saw the text of the constitution Mezzabarba had brought. He returned it with the comment: "All that can be said about this decree is that one asks oneself how the Europeans, ignorant and contemptible as they are, presume to deliver judgement on the lofty teachings of the Chinese, seeing that they [in Europe] know neither their manners, their customs, nor their letters. . . . [The decree] teaches a doctrine similar to that of the impious sects of Hoxans . . . who tear one another with pitiless cruelty. It is not advisable to allow the Europeans to proclaim their law in China. They must be forbidden to speak of it; and in this way many difficulties and embarrassments will be avoided."[14]

When Mezzabarba put forward a list of "concessions," the emperor remained adamant. But, with the mildness toward the Christians that he generally showed, he did not permit his decrees to be acted on while he lived. Persecution under his decrees did not begin until after his death in 1722, and was to continue for a hundred years. K'ang-hsi's basic attitude did not vary. It is well summed up in a paragraph from one of his edicts: "As for the doctrine of the Occident which exalts T'ien Chu [Lord of the Sky], it is . . . contrary to the orthodoxy [of the Chinese Classics], and it is only because its apostles have a thorough knowledge of the mathematical sciences that the State uses them—beware lest perhaps you forget that."[15]

Mezzabarba left Peking three weeks after John Bell and the Russian embassy, having completed, unintentionally and with the unintentional assistance of the pope, the ruin of the Catholic mission in China. The work of Ricci and five hundred other priests (the estimate is Matteo Ripa's) was at an end. Apart from its scientific aspects, Catholic influence in China had not been devoid of good. Its charity, its humanity, its singular capacity, exercised through the often distinguished intellectual qualities of its priests, to light an entirely new sort of civilized beacon amid the ancient glow of the Confucian way of life, should not be denigrated. Like that Chinese way of life, the Christian effort also had its darker sides, redeemed, if it was redeemed, by the intelligent dedication of many a Jesuit and other priest. In the end, however, it was to leave on China almost no mark of its inner significance. What it did achieve was to establish in official Peking the fact that European science was now superior in many ways to that of China. If emperors from this time onward were mostly to use Christian priests and others solely for their skill in attending to the mechanical playthings of the imperial leisure, they also continued to depend on those Westerners for higher astronomy and the regulation of the vital Chinese calendar. Not until Europe, by armed force, brought China to her knees more than a century later did the West and its ebullient post-Renaissance, post-Industrial-Revolution civilization begin to make a more decisive imprint on Chinese thought and Chinese life.

Neither John Bell nor Matteo Ripa (nor, for that matter, the pope, nor Tsar Peter the Great, and certainly not Mezzabarba) was in a position to com-

14. Quoted by Pastor.
15. Quoted by Cordier.

GIFTS FROM THE TSAR TO JOHN BELL in recognition of his services on the Peking and other embassies. The crystal is engraved with the tsar's monogram and the imperial arms. These last surviving mementos of the great Scottish traveler are in the possession of Bell's collateral descendants in London. (Photo by the author)

prehend the wider consequences of the last act in the drama being played out in Peking in the first half of the eighteenth century. Bell and the Russians finished their business there, leaving De Lange, Ismailoff's secretary, as the tsar's representative (the first ever to be accredited to a Chinese court), and returned to Moscow, well enough pleased with their mission.

For his services Bell was given a gold-headed cane and some fine engraved crystal by Peter the Great. He left Russia, but eventually returned there as secretary to the British minister in St. Petersburg in 1734. Later he was in the Middle East, and later still returned to Russia again for a time. Turning merchant, he again went to Constantinople; and at the age of fifty-five he married Marie Peters (probably a Russian) and retired to spend his remaining years as a country gentleman in Scotland. His country mansion, Antermony House, was graced by an avenue of lime trees which flourished with a fine improbability in the Scottish air, and perhaps reminded him of those upon which he remarked at Peking. His charity in the neighborhood is noted in the local chronicles, as is the fact that he was sometimes to be seen riding about the countryside of Stirlingshire in his "oriental costume." While it is amusing to visualize him surrealistically attired in silk mandarin robes, exotic in the undemonstrative landscape of Campsie parish, it would probably be nearer the truth to see him in the long blue or gray padded gown worn in North China winters until only recently. When we remember that K'ang-hsi in person had recommended something warmer than European clothes, and recall the inclement nature of many Scottish days, this is likely to be the picture of John Bell we ought to conjure up.

It was not until 1763 and after many doubts about his literary ability, that he finally published his book of travels.[16] The book was an immediate success. With the shrewdness we have noticed in his writing, he sold Antermony House subject to his own life-tenancy, and lived on there to the ripe old age of eighty-nine. Antermony House was demolished during the First World War. From a photograph preserved by the tenant of its lodge, it was no great thing of beauty; particularly after the addition of a Victorian conservatory. The avenue of lime trees has gone too. And in the overgrown churchyard of the Clachan of Campsie not far away, the broken tomb slab of Honest John Bell lies forgotten beside that of his wife Marie Peters. His only tangible mementos are the crystal of Peter the Great and a fine gold watch which possibly went to Peking with him. The curios he bought in Peking—recollected to have been in a glass cabinet in Antermony House during the childhood of a relative—have disappeared.

From Scotland we must now return, if only for a short time, to Peking—where Matteo Ripa was increasingly troubled and unhappy at developments. His spiritual ardor began to find another outlet. Ripa established a school for Chinese boys in the capital to teach them the faith with a view to making priests of them.

It would be helpful in assessing Ripa's character to have some studies of the psychology of religious celibates. Ripa's only affections seem to have been turned toward his Chinese pupils, youths in whose lives he became very involved. But how much his drive to establish a school for them, in order to convert them into priests, was a product of his affections and how much of his reason or of his faith, there is no means of deciding.

In Peking his efforts to form and later to sustain his small school for Chinese boys were dogged by the persecutions of resentful colleagues and distrustful officials, by whom he was accused of "abominable practices" with the boys. "They stated . . . that it was highly discreditable to the Chinese mission that one of its members should travel about in the suite of the emperor with a carriage full of boys, just as some of the chief courtiers were doing, to the great scandal of the nation." Repudiating such accusations, Ripa persevered, even when his favorite pupil, a highly emotional youth called John In, was taken away from him by his parents. But Ripa eventually got the youth back.

With the death of K'ang-hsi, just before Christmas in 1722, Ripa begins to

16. Perhaps his doubts were finally dispelled by a learned gentleman's telling him that if he took *Gulliver's Travels* for a model, he could not go wrong.

talk of a "new state of things." A host of restrictions descended on the Christians. Ripa determined to return to Naples. "Considering . . . how little I could effect in China for the propagation of Christianity, and how repeatedly I was exposed to the danger either of participating in idolatrous practices or of perishing, in obedience to the Holy Word,—'But when they persecute you in this city, then flee ye into another,' I resolved to return.

"On the 15th November, 1723, I at last left that Babylon, Peking, with my four pupils and their Chinese master, myself on a litter, the two youngest boys on another, the other three and two servants on horseback. The wind blew so furiously that it upset our litters several times. . . . It seemed as if the Evil one, foreseeing the great good which at some future time would arise from my little flock of Chinese, had mustered all his forces to drive us back to that capital of his dominions." Ripa is one of the very few Europeans on record as detesting Peking—but he had been sorely tried. And even then, he carried away not only his young Chinese, but a firm respect for the Chinese people at large.

At Canton they took passage on an English ship. The English took a dislike to the Chinese, partly because one of them contracted a skin condition which the crew declared to be leprosy, and because the youngest "dirtied the cabin." The ship's surgeon threatened to give the Chinese with the skin complaint "a powerful dose that would carry him off," and suggested throwing all the youths overboard, one by one. So the voyage was a nightmare of insults from the crew, as it was also of foul weather. At the Cape of Good Hope, Ripa took his Chinese ashore to refresh them, but, so unusual a sight were they that they were "teased in every way." Somewhere ashore Ripa discovered that some English and Dutch residents, for the amusement of the company at large, had been "pushing the landlord's daughter against the youth [John In], who, weeping and trembling with dread from such temptation, had at last crept under the bed."

At length they reached London. King George I was interested enough to invite Ripa and his five Chinese to the palace to dine. And then to Italy. In Naples the road to the formation of Ripa's heart's desire, the Chinese College, was a long and tortuous one, pitted with ecclesiastical jealousy and financial problems. But with the opening of the institution in 1732, Ripa at last reached his goal. The college endured for over a hundred and fifty years and trained one hundred and six Chinese priests before it was suppressed by the Italian government in 1888.

Today, traveling out of the town of Naples from the square in front of the Palazzo Reale, where Ripa was converted from his profligate ways, it is a long way through the uproarious traffic of the Via Roma, and farther on, before you come to the Ponte della Sanità. The bridge itself has changed from a rustic arch to a major viaduct, and the old Church of Santa Maria della Sanità, which is almost the only landmark Ripa would recognize, is

DOORWAY OF THE CAPPELINA of the Chinese College and Congregation of the Holy Family, built by Matteo Ripa in Naples. (Photo by the author)

crowded round with haphazard building. The Chinese College is gone, demolished forty years ago to make way for a hospital for the poor. What was field and stream and trees between the college and the city in the eighteenth century is now house and shop and commerce.

Turning off the main road after the bridge you can find your way on foot to a short narrow street whose houses look out toward Naples—the Salita Cinesi, the Street of the Chinese. The people who live here, with an unconscious historical memory, call it Salita dei Piccoli Cinesi, the Street of the Little Chinese. Here, returning from walks in the countryside, the Chinese pupils of Matteo Ripa's college used to climb to the front door of their temporary home. Now, when you reach the top, there is a door, and through it a small part of that terrace on which Dr. Mayer in the nineteenth century looked over the city of Naples after his tour of the college. A hideous bust of Ripa has recently been placed there, its back to the sky and the fingers of the television antennae poking over the wall. But the view is still a fine one. All that remains of the buildings raised by Ripa are the church and the cappelina, whose doors open onto the remnants of the terrace, that of the cappelina being inscribed "Chinese College and Congregation of the Holy Family."

In the church itself, on the floor, is a marble slab over the tomb of Matteo

Ripa with his date of death, 1756. In the small, dark cappelina there is a death mask, now falling to pieces, little resembling the Ripa of the two oil portraits there. One of the paintings shows him as a young man, perhaps soon after he went to Peking, round faced, soft lipped, the eyes large and speaking, the brow high and domed. The other, damaged and poorly restored, presents him in his old age, recognizably the same man, the eyes now deep and strained, almost tortured.

There are no small Chinese now to question, to hear recite the scriptures in their curious tongue. Even the paintings of those who died have vanished. The present "rector," as Dr. Mayer called the priest he met, is a Neapolitan also, but neither "oily" nor tall. A charming, helpful little round man, he looks after the spiritual needs of the patients of the hospital as well as presiding in the church. He led the way to the right of the altar, pulling aside a curtain to reveal a small door. With his assistance we lit candles and descended some narrow broken steps into the vaults beneath the church. There, in the dry, dead, cell-like rooms, the walls have niches, in each of which is a roughly made seat only wide enough for the pelvis. On each of those was placed in death one of the small Chinese who succumbed before he could be returned to his native land. There, and sometimes in conventional coffins, the bodies were left to desiccate, to mingle with the fine volcanic dust of the Neapolitan earth, only their small bones remaining, now friable as chalk. The foot, carelessly placed, crushes a femur, a stray Chinese finger bone. That is all that remains of the life and work and hope of Matteo Ripa so long ago.

Death Mask of Ripa, in very poor repair and likely to disintegrate. No photographs of this mask seem to have been published before. (Church in the Salita Cinesi, Naples; photo by the author)

CHAPTER FOURTEEN

FOREIGN LORDS AND FOREIGN MUD

IN THE spring of 1792, Sir George Leonard Staunton, lately knighted by George III for his services in India, set out from London for Naples with his son, George Thomas Staunton, aged eleven. The baronet had just been appointed secretary to Lord Macartney, his chief in a former Indian campaign, whom the king had charged with an embassy to China. The immediate reason for the journey of father and son to Naples was quite simple, if surprising; in the whole of England at that time, not a single man could be found who knew anything of the Chinese language, and, as the embassy would certainly require interpreters of its own if it were not to depend on the Chinese to provide them, they must be found in Europe. In the words of a writer in the following century, the Protestant English "were somewhat discreditably reduced to the necessity of engaging the services of two Romish priests to aid the important objects of the mission in the quality of interpreters."[1]

The destination of the noble lord and his son in Naples was the "Popish institution" of Matteo Ripa's Chinese College on the hill overlooking the city. "It was in Naples that I first saw any of the natives of China," the younger Staunton was to recall many years later. "And it was in the course of the journey home [from Italy to England] . . . and on the outward-bound voyage to China, which soon followed, that my ears were first familiarized to the sounds of a language in which, during the next five-and-twenty years of my life, I had so much exercise."

Before we follow the "page," as young Staunton was generally styled, to Peking, we must look briefly at the two sides of the world which, once more, were about to make contact on an official plane. For since the days of Ripa much had happened in Europe that was relevant to China; and in the Middle Kingdom itself a certain attitude had finally been struck that was pertinent to Europe.

In France, the eighteenth century saw the development of what is termed

1. John F. Davis.

the Enlightenment, of which Voltaire was a central figure, and with it the cult of chinoiserie. In his biography Brumfitt tells how Voltaire remarked that China was "better known in our times than many a province of Europe." The Enlightenment and the cult of chinoiserie were, of course, almost entirely separate movements. Nor were they confined to France. The repercussions spread over most of Europe, making the era one of the most fascinating in European cultural history. Within seven or eight decades the settled basis of traditional Western civilization came up for questioning, its Christian mainspring under heavy fire. Pure reason and pure reasonableness were raised to great heights. Unfortunately, the advance of science in that age still lagged behind such intellectual aggression and could not adequately support its conclusions.

With doubts about the perfection of Europe's Christian-based way of life, sincere attempts were made to assess the civilizations of the East as alternative cultural forms, and therefore as possible substitutes for the traditional outlook of Europe. Drastically revised was the former European concept (almost Chinese in its hermetic exclusiveness) that other parts of the world were barbarous in varying degree, and in need of European civilization to ameliorate the malady of their false gods and even falser prophets. The pendulum swung very far in an Eastward direction. Those gods, and especially those prophets Mohammed and Confucius, were intensively investigated by leading European minds, who came to some surprising conclusions and took up some even more surprising postures. At one point, as Brumfitt again remarks, the great Voltaire was saying that "the human spirit cannot imagine a better government" than that of China. And a portrait of Confucius hung on the wall in his house at Ferney—Confucius, whose moral code of "tolerant deism," as opposed to Christian mysticism, "consisted in being just."

The starting point of all these speculations was the writings of such lay and priestly travelers as we have been following in previous chapters. There were, of course, many others who sent back their stories—travelers not only to China but to the Arab world, to India, and elsewhere. Initially in Europe it was Arabia which awakened the learned professors of Paris, Cambridge, and Oxford. For the first time Moslem texts were studied with a certain open-mindedness. The Arab world emerged in a new light. Paul Hazard sums up these opinions ably: "So vast a section of the human race as the Arabs could never have followed in the footsteps of Mohammed if he had been no more than a dreamer and an epileptic." Such a conclusion, unthinkable in the previous century, sent the learned men of Europe scurrying to the original texts of the thought of the East. There were not many of those texts to be found, and the searchers had in general to rely on travelers' reports; so the exaggeration of their consequent opinions is understandable.

The mystic teachings of Mohammed were soon discarded as being in a sense too similar to Christianity. But the discovery of the reasonable maxims of the Chinese sage Confucius caused a wave of enthusiasm for Chinese thought to flood through the salons and studies of Europe. The frenzy of this new interest in the Chinese mind, in what was thought to be a philosophy of inspired common sense, is nowadays hard to grasp, but it was very real in the eighteenth century. The Confucian code was seen at first as one of "natural" as opposed to European "revealed" religion. With the intellectual dissatisfaction and drive characteristic of Europe as opposed to the East, and never questioning the accuracy of the reports on which their knowledge of China was based, the best minds of Europe singled out the traditional thought of China (as it was then understood) and placed it on a pedestal. All Europe turned with a kind of delight and relief to China.

Astonishing things were said by great minds.[2] "China offers an enchanting picture of what the whole world might become, if the laws of that empire were to become the laws of all nations. Go to Peking! Gaze upon the mightiest of Mortals; he is the true and perfect image of Heaven!" Thus Poivre, in France, saw fit to eulogize the Chinese essence, perhaps echoing the Jesuit Parennin's description of the emperor K'ang-hsi as "one of the most extraordinary of men, such as one finds only once in several centuries." And Voltaire, in a passage which Hudson says might be the motto of the Enlightenment, said: "The Chinese have perfected moral science, and this is the first of the sciences. . . . [Confucius] spoke only in wisdom, and never in prophecy." The philosopher Leibnitz even went so far as to suggest that "Chinese missionaries should be sent to us to teach us the aim and practice of natural theology, as we send missionaries to them to instruct them in revealed theology." And Voltaire again: "The organization of the [Chinese] empire is in truth the best that the world has ever seen."

Some of the attitudes of the Enlightenment now appear a trifle ludicrous. Even the brilliant and cantankerous Dr. Johnson in England was not immune to the pandemic sinophilia, although with reservations. Having written: "Let Observation with extensive view, Survey Mankind from China to Peru," he did, however, urge caution in assessing the results of the survey.[3] "It is difficult to avoid praising too little or too much. The boundless panegyricks, which have been lavished on Chinese learning, policy, and arts, show with what power novelty attracts regard, and how naturally esteem swells into admiration. I am far from desiring to be numbered among the exaggerators of Chinese excellence. I consider them as great, or wise, only in comparison with the nations that surround them; and have no intention to place them in competition with the antients, or with the moderns of this part of the world; yet they must be allowed to claim our notice as a distinct and very

2. The following examples are quoted by the historian of Sino-Western relations G. F. Hudson.

3. All the quotations from Johnson are taken from Fan Tseng-chung.

singular race of men . . . who have formed their own manners, and invented their own arts, without the assistance of example."

In thus defining his opinion of China and the Chinese, Johnson the Augustan intends simply to reassert that the taste of Europe, derived from classical European civilization, is superior to anything of a similar nature in China. The challenge of Chinese philosophy was probably a very real one in his mind at the time; for the content of that philosophy was disturbingly rational, disturbingly attractive. However flimsy the basis of Europe's knowledge of China, Johnson apparently saw clearly that in the thought of the Eastern world there was a nucleus of strong sense.

Johnson's caution makes him urge his listeners several times to return to the most reasonable sources on China—to John Bell, for example, as he said to Boswell. But Johnson too could be wide of the mark in criticism, as were others in hyperbole. The Chinese, he said, had not been able to invent an alphabet. "They have not been able to form what all other nations have formed." Their written language of an immense number of characters "is only more difficult from its rudeness; as there is more labour in hewing down a tree with a stone than with an axe." Fine Johnsonian stuff, but—like the conclusions of all the thinkers of the time—it suffers from the fundamental lack of accurate information.

It was just this partial nature and distortion of information available to Europe from Jesuits and others (each with an ax to grind or, as Johnson might have said, each with an ax peculiarly fitted to fell his own particular species of tree) that allowed the wildness of the eighteenth-century-intellectual China cult. It was this same information that eventually, when its faults were recognized, caused a strong reaction in Europe and led to the eclipse of the Chinese luminary there. When actual conditions in China came under the scrutiny of European intellectual honesty, the vast erudition that had been expended on constructing a Chinese logic-pyramid of "natural religion" and common sense quickly turned on itself and bit its tail in exasperation.

That renowned "justice" of Confucianism, which had been supposed to be embodied in emperor and state, was now found not to exert the pan-Chinese influence attributed to it. The whole idealized concept, the image of an empire run by pure reason, turned quickly sour. No one paused to examine the possibility that this European reaction was in part as invalid as the earlier European concept of Chinese philosophy; nor did anyone pause to consider that there might even be something in Chinese philosophy of real value after all.

Turning from the philosophical to the aesthetic aspects of European China-appreciation at the time, we find that even the skeptic Dr. Johnson was taken in. The passage quoted above (page 290) comes from Johnson's introduction to a book called *Designs of Chinese Buildings, Furniture, Dresses,*

CHINOISERIE AND ITS CHINESE COUNTERPART. The first illustration shows a superb wool and silk tapestry woven at Beauvais around 1725–30 and entitled "The Audience." In a delicately florid pavilion that owes nothing to China and everything to European fancy, the emperor of China sits in unexampled splendor as tributaries kowtow before him. (Designed by G. L. Vernasal, J. B. de Fontenay, and J. J. Dummons; size 3.58×5.29 meters; Residenzmuseum, Munich.)

The third illustration shows an early example of the newly discovered art of porcelain making in Europe. It is called "The Chinese Tiger Hunt" and was made about 1765 at the Nymphenburg Factory in Germany. Obviously modeled from

Machines, and Utensils, to which is annexed a Description of their Temples, Houses, Gardens, etc., by William Chambers. If in this introduction Johnson was cautious on Chinese philosophy, elsewhere he was enchanted by what he supposed to be Chinese aesthetics. In fact, what he—and the rest of Europe—admired was not Chinese art and craft, but that peculiarly European interpretation of them known as chinoiserie.[4]

4. The most valuable and readable book on chinoiserie is that recently published by Hugh Honour.

some imported Chinese example, like all chinoiserie it succeeds in expressing an entirely European and totally un-Chinese spirit. (Bavarian National Museum, Munich)

The same comment is relevant to the second illustration, a Chinese painting of a madonna and child attributed to Tang Yin. The eighteenth-century artist has been quite unable to discard the ancient Chinese representation of Kuan-yin, Goddess of Mercy, from his mind. And indeed the two concepts are entirely consonant in graphic terms. The result is a charming example of chinoiserie's Chinese equivalent. (British Museum, London)

Acknowledging the actual Chinese origins of chinoiserie in general (Chambers's ideas of Chinese gardens, for example, were based in part on his own visits to South China) we must recognize that the result was a European and not in any way a Chinese form of art. People thought that chinoiserie consisted of imitations of Chinese style. In fact, until very recently European folklore contained a picture composed of varying elements such as delicate pagodas, mad little bridges, pig-tailed men, women with bound feet, dragon-embroidered silks, long fingernails, tea in blue-and-white cups without handles, and a pervading placidity of Chinese mien amounting to an incomprehensible and rather vague "inscrutability." How far from truth this was and is, we have seen in part in the foregoing chapters; just as we have seen how Confucian morality—while it was an intellectual ideal most of the time in China—was never more than partially realized in practice, either in government or in society.

But turning from the chinoiserie world of the genteelly whimsical, we must recapitulate a little. As suggested earlier, the story of Europe's knowledge of China was one of total ignorance, followed by disregard of the sources which became available; then came a commercial, followed by an intellectual, interest; then intellectual disillusion, when sources were fully examined, coincidentally with continuing commercial interest. To understand what followed—the crucial stages of the long contact—we must take a glance at China herself in the eighteenth century, at what she thought of the West as that time of brilliance in both hemispheres wore toward its close.

The most significant matter is the Chinese attitude to Western visitors, their culture (as far as that could be understood in China) and its meaning for Chinese. The answers China arrived at are simple, flat, dogmatic, unreal, and, unfortunately, final (as Lord Macartney and, later, Lord Amherst were to discover). In the treatment met by Westerners who came to Peking we have seen a sampling of the evolution of the Chinese attitude. The Russians, so far, had come off best (even though the Treaty of Nerchinsk ran in China's favor). The probable reason for this was that Russia was China's immediate neighbor and had to be neutralized in one way or another. As it was, Russia was held at the tips of the imperial fingers, with a Chinese smile.

As for the rest of the West—it was to be excluded whenever and wherever and however possible. It was to be dealt with only at Canton. No diplomatic contact was needful, none in theory allowed: certainly no formal agreements. Not without a degree of justification, the Chinese had come to the conclusion that dealings with the West were generally not to China's advantage and caused problems which did not fall within the competence of the machinery regulating contact with the outer world. Therefore contact was to be purely commercial, confined to Canton, officially discouraged. What the government did not really reckon with (or was from time to time bribed to close its eyes to) was that Canton merchants, governors, and the viceroys controlling the area found it comforting to the pocket to permit more than that. The Europeans were not loath to expand their trade, nor the Cantonese either. But official China was, and knew herself to be, intellectually, spiritually, and commercially self-sufficient. This was a dangerous idea. For what China did not understand was that to remain a world unto herself, she would henceforth have to *defend* that world against the West.

As it approached China in 1793 the British embassy under Lord Ma-

THE EARL OF MACARTNEY, an engraving. (Museum and Art Gallery, City Hall, Hong Kong)

cartney, to whom young Staunton was page, seemed, and indeed was in itself, peaceful enough.[5] It was also unquestionably the most magnificent ever to arrive from the West. No expense or refinement had been spared by the British cabinet in order to make it a success—such was the importance attached to improving trade conditions with the Celestial Empire. Lord Macartney himself was a man of noble birth, a successful diplomat whose qualities had already been proven. He was cultured, urbane, a brilliant conversationalist, highly personable, intelligent, and had that facility of the highly born and cosmopolitan of making himself acceptable in all elevated society. His party included such able men as Staunton, the father of the page, and others who in later years proved themselves men of parts. There were two Scottish doctors aboard the *Lion* as she sailed up the China coast from Macao, where the embassy had been given the unusual permission to approach Peking by sea. With them were a botanist, two Chinese interpreters from Ripa's college at Naples, and two artists. One of the artists, Alexander, was a considerable painter through whose eyes the landscape, townscapes, people, and personalities of China, as well as those of the embassy, come alive in a manner unprecedented before that date. (The

5. In the following account of the embassy we rely for many details and a number of quotations on the invaluable journals of Lord Macartney, which he kept assiduously and later edited and published. Cranmer-Byng's recent edition of the journals is the best and most readily available guide to the subject.

Jesuit Castiglione, a little before this time in Peking, and Alexander are probably the best European draughtsmen ever to work in China.) No expense had been spared by the British in gathering together a superb collection of presents for the emperor that would indicate the splendor and superiority of British culture.

As the *Lion* reached the mouth of the Pei River, a few days' journey from Peking, the party was in a state of high expectation. Young Staunton had completed the first part of his diary (now lost) and started the second. Twelve years old—a mere boy who wrote, in a poorly formed hand, a badly spelled and thoroughly schoolboy account of the trip to Peking (see illustration on page 303)—he had nevertheless managed to acquire from the Chinese interpreters during the sea voyage sufficient spoken Chinese to enable him to converse in the language. Even more remarkable, he seems to have learned enough of written Chinese to be useful in copying Chinese translations of letters from the ambassador to the Chinese during their stay. He is also said to have known French and German, as well as Latin and Greek. While he was apparently something of a prodigy, it is difficult to square this with his featureless and deservedly unpublished diary.[6]

Apart from negotiations on improving trading conditions at Canton, the purpose of the embassy was to attempt to establish at Peking a permanent representative of Great Britain in the manner of world diplomatic usage. Macartney's instructions were intelligent. He was left wide freedom to conduct his business in the manner that seemed best on the spot, and was told not to be punctilious over matters of no great import. Despite the flood of knowledge about China that had coursed through Europe in the preceding decades of the eighteenth century and the high tide of interest in the country, the British cabinet was still unsure how best to tackle what it recognized as a thorny problem—Chinese insularity and *amour propre.* Hence their choice of the sophisticated and practiced Macartney as a match for the equally sophisticated and wily emperor Ch'ien-lung, who, through the Jesuits, had learned much of the situation in cultured Europe of his day. The character of the emperor, one of efflorescent brilliance, prolix rather than profound, was known in essence in Europe; but his capacities in the arts were overrated, just as it was wrongly thought that his intellectual capacities would make him more receptive to European opinion on the subject of international trade and diplomatic relations.

Macartney carried a letter to Ch'ien-lung from King George III. This letter, couched in cautious terms, going as far as it could toward what was felt to be suitable terminology for the ears of a Confucian emperor, reveals the uncertainty of Britain on the best way to proceed. George writes as a friendly equal, recommending his ambassador as his "trusty and well-

6. The MS is in the Duke University Library, Durham, North Carolina.

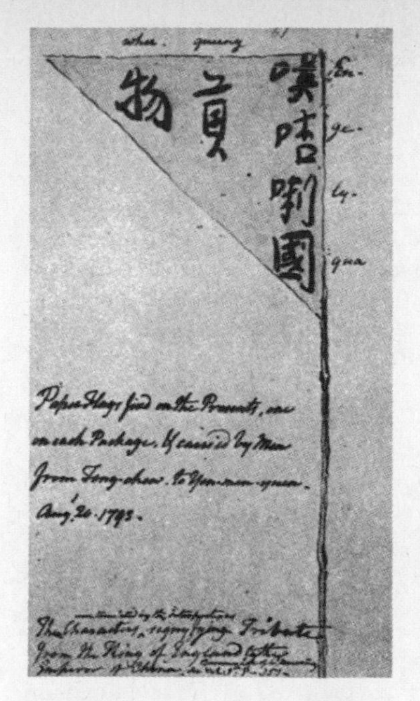

A Tribute Flag. A sketch by Alexander of one of the paper flags attached to the presents Macartney brought for the emperor. The four characters next to the staff mean "England," and the two larger characters, "Tribute." (British Museum, London)

beloved Cousin and Counsellor . . . a Nobleman of high rank and quality, of great virtue, wisdom, and ability . . . sent to obtain such information of Your Celebrated institutions as will enable him to enlighten Our People on his return." He adds the hope that as he and Ch'ien-lung are "Brethren in Sovereignty, so may a Brotherly affection ever subsist between Us"; and he signs himself "Brother and Friend."

Arriving at the mouth of the Pei River, the embassy with its mountain of baggage and its presents for the emperor transferred from the *Lion* to Chinese barges for the river journey toward Peking, ninety miles distant. Apart from the fact (noted by Macartney, but tactfully overlooked by him) that the barges flew pennants with an inscription informing everyone that a tribute embassy was on its way to the emperor, the journey was smooth. The winds of diplomacy seemed, from the Western side, set reasonably fair.

From the Chinese point of view, however, the outlook was in some ways less cut and dried, in others more so. For, as Macartney approached Peking it was almost exactly twelve years since the first consignment of opium—foreign mud, as the Chinese called it—had been sent by the British from India and arrived at Canton. And by 1792, the year preceding the Macartney mission, the revenue from opium accounted for more than one quarter of the total value of all shipments from India to China. The British passion for China tea was largely responsible for this trade into which the East India Company and the British government were drawn. It is easy to see why the opium trade began, less easy to gloss over its continuance against the express decrees the Chinese government promulgated in 1792 and 1796, and against the better impulses of the British people as a whole. The fleets of ships sailing away from Canton to the West were heavy laden with cases of expensive tea, paid for in silver, the only acceptable currency in China. But on their return they had formerly arrived in China almost empty, or with cargoes of low-priced goods; for relatively few articles produced in the West were desired by the Chinese, whose way of life and whose indigenous industries supplied almost all the Chinese appeared to want. Consequently more and more silver poured into China from the coffers of the East India Company.

The answer to this alarming state of financial affairs was soon found—opium. While the company forbade its carriage in their own ships, they freely permitted its sale in India to others, including English vessels registered in India, which carried the drug to Canton. Worse, the opium was grown for the most part in territory under the jurisdiction of the company in India. A little later than the date of which we are talking, the English-language *Chinese Recorder,* published in Canton, remarked: "In all the territories belonging to the Company, the cultivation of the poppy, the preparation of the drug, and the traffic in it until it is sold at auction for exportation [from India] are under a strict monopoly. . . . The cultivation of the plant is compulsory. . . . Vast tracts of the very best land in Benares, Bahar and elsewhere in the northern and central parts of India are now covered with poppies. . . ." The ships bringing opium to Canton were, by a legal fiction, not regarded as belonging to the company, though, as the historian Fitzgerald says, "all Indian commerce was under the authority of the Company, the sovereign power. The trade was important to the Company, and served to finance the China investment."

One naturally asks oneself why the Chinese took to opium so avidly. The subject is a complex one and involves some consideration of the conditions of social misery in those days of corrupt and faltering government. But perhaps some understanding of the story of opium in China may be gained by reflecting how, in eighteenth-century England, when cheap Dutch gin flooded into a realm no more and no less addicted to excessive alcohol intake than any other, the populace—certainly sections of the more miserably paid and housed—abandoned themselves to its intoxicating embrace.

While the English were keen to sell the drug in China against Chinese decrees forbidding such commerce, it must be acknowledged that the Chinese merchants and others in Canton were no less backward in flouting those same decrees. To increase the sale was no problem, only a matter of time. The officials (tax collectors, the governor, other high officials of

Canton, and even the viceroy of the area himself) found financial advantage from bribes in turning a blind eye to the transactions. In turn they bribed even higher officials in the Peking government itself, who grew rich on the proceeds. But by the time of Macartney in the capital, informed circles were beginning to be worried about the matter. China, however, was still earning more silver from the sale of tea than was lost in buying opium.

This background is essential to an understanding of the mixture of cordiality and brilliantly conducted evasiveness which met Macartney in Peking. The old distrust of foreigners was at work too. But the opium trade, with its tangle of intrigue, its illegality, certainly did not cause Chinese officials in Peking to deal any more accommodatingly with the embassy. Naturally they were careful never to allow opium to emerge into the open.

Leaving the state barges near Peking, the embassy changed, according to the rank of its members, to palanquins, horseback, or springless Peking carts. They entered Peking by one of the eastern gates, preceded, says Marcartney, "by a great number of soldiers, brandishing long whips . . . in order to keep off the enormous crowds which incessantly thronged about us and obstructed the passage. We were fifteen minutes from entering the east suburb to the east gate. We were above two hours in our progress through the city, fifteen minutes from the west gate to the end of the west suburb, and two hours from thence to Yuan-ming Yuan." The emperor was not in Peking but in his summer residence at Jehol, a hundred miles away to the north in the hills of Tartary, where the court went to escape the Peking heat and dust.

Macartney omits any description of Peking as he crosses it, but the artist Alexander, in his unpublished diaries (preserved in the British Museum), supplies one: "our attention was . . . attracted by the splendid triumphal arches which we passed under in the Peking streets. These have dragons at the several angles and appropriate inscriptions, in letters of red and gold, but from the various colors employed have no simplicity, but on the contrary are tawdry and of bad taste." Accustomed to the European taste of chinoiserie, Alexander found here and elsewhere the taste of Chinese decoration hard to take. "Part of our way was by the . . . wall of the imperial palace. This is of brick painted red and is covered with a roof of glazed tiles of a rich yellow color. This extended little short of a mile in length. None of the houses have more than one story. At the edge of the footway before the houses of the tradesmen etc., poles are erected of framework on which boards are affixed with the names of articles for sale depicted in letters of green, red and gold. Lanterns (some of enormous size) and streamers of silk are also suspended from them. . . . The variety of rich colors in paint, and silks, exclusive of the garments worn by the upper class, makes altogether a scene of uncommon brilliancy, such as can hardly be conceived, and cannot possibly be described."

A CHINESE OPIUM DEN. (From *China in a Series of Views Displaying the Scenery, Architecture, and Social Habits of That Ancient Empire;* drawn, "from original and authentic sketches," by Thomas Allom, Esq., with historical and descriptive notices by the Rev. G. N. Wright; published in four volumes by M. A. Fisher, Son, and Co., London and Paris, 1843)

That same day, August 21, 1793, the embassy reached its residence near the Yuan-ming Yuan summer palace just outside Peking. By the following week Macartney was preparing his proposals on how to avoid the kowtow. These had to be transcribed into Chinese and, as Macartney remarks, "little Staunton was able to supply my wants . . . for . . . he had learned to write the characters with great neatness and celerity, so that he was of material use to me . . . as he had already been in transcribing the catalogue of the presents." The presents were set out during the next days in a section of the summer palace. Together with numerous scientific instruments, there were other European luxury articles, of which the Chinese court, although

MACARTNEY'S GIFTS FOR THE EMPEROR. A Chinese tapestry showing British sailors carrying the embassy's gifts for display in the grounds of the summer palace near Peking. (National Maritime Museum, Greenwich, England)

Macartney did not know it, already had an all but wholesale supply. The scientific part of the embassy had brought along as a demonstration piece a hot-air balloon, ascents in this device being all the rage in Europe. But the Chinese, feigning utter disinterest in this as in other machines, refused permission for its ascent into the sacrosanct Peking airs. And the capital had to wait until 1900 before a contrivance more modern than the ancient Chinese whistling kite took flight over the medieval city.

On September 2, Macartney was on his way to Jehol. That morning, he probably awoke to "the heart-stirring drum and ear-piercing pipe" which, Alexander records in his diary, "aroused all from their beds to prepare for the journey to Tartary." Young Staunton remarks in his diary: "Early this morning we set out for Jehol, and Lord Macartney and I went in our English carriage, but my papa went in a Chinese sedan . . . having a little of the gout." Alexander was upset: "It was determined that Mr. Hickey [the "artist" to whom he was junior despite the fact that Hickey produced hardly a scribble during the whole embassy] . . . and myself should remain at Pekin. This to me was a most severe decision, to have been within fifty miles of the famous Great Wall, that stupendous monument of human labor, and not to have seen that which might have been the boast of a man's grandson, as Dr. Johnson has said, I have to regret forever."[7]

After seven days on the road, the embassy made its entry into Jehol. Assuredly it was the finest and most noble ambassadorial cortege ever to approach the Son of Heaven. Through the wild countryside, toward the Tibetan-style palace on its slopes, the English were preceded by a hundred mandarins on horseback. The commander of the British cavalry escort followed with four rows of four Light Dragoons each. Then came a lieutenant, a drum and a fife, and two ranks of artillerymen with their corporal. Then another lieutenant and four ranks of infantry, four abreast, and their sergeant, followed by a file of eight servants in two's and by two couriers, all in "rich green and gold livery." Four musicians led the suite of six gentlemen of the embassy in "uniform of scarlet embroidered with gold." The ambassador followed with Sir George Staunton and his son in a chariot, with a servant in livery. Macartney later notes that he was told the emperor "had seen my entry and procession from one of the heights of his park, and was much pleased . . . and . . . had immediately ordered the First Minister and another Grand Secretary to wait upon me." The emperor was indeed pleased, but not for the reasons Macartney imagined. The size and brilliance of the first British embassy to the court of Peking was, to the Chinese, a fine sign that the most powerful of the barbarian nations had come to pay tribute to the emperor, and to acknowledge his absolute world power.

Not understanding this, Macartney anticipated his first audience with hope and interest. Ch'ien-lung was receiving other tributaries—one from the king of Burma, for example—and the ambassadors were brought very early on the morning of September 14 to one of a number of richly decorated tents in the palace gardens. Soon the emperor approached, borne on a gilded throne by sixteen men through the still-chilly shrubberies of the

7. From Alexander's unpublished diary, now in the British Museum, London. Alexander was doubtless recalling this passage from Boswell's *Life of Samuel Johnson:* "He expressed particular enthusiasm with respect to visiting the wall of China. I catched it for the moment, and said I really believed I should go and see the wall of China had I not children. . . . 'Sir', (said he) 'by doing so, you would do what would be of importance in raising your children to eminence. There would be a lustre reflected upon them from your spirit and curiosity. They would be at all times regarded as the children of a man who had gone to view the wall of China. I am serious, Sir.' "

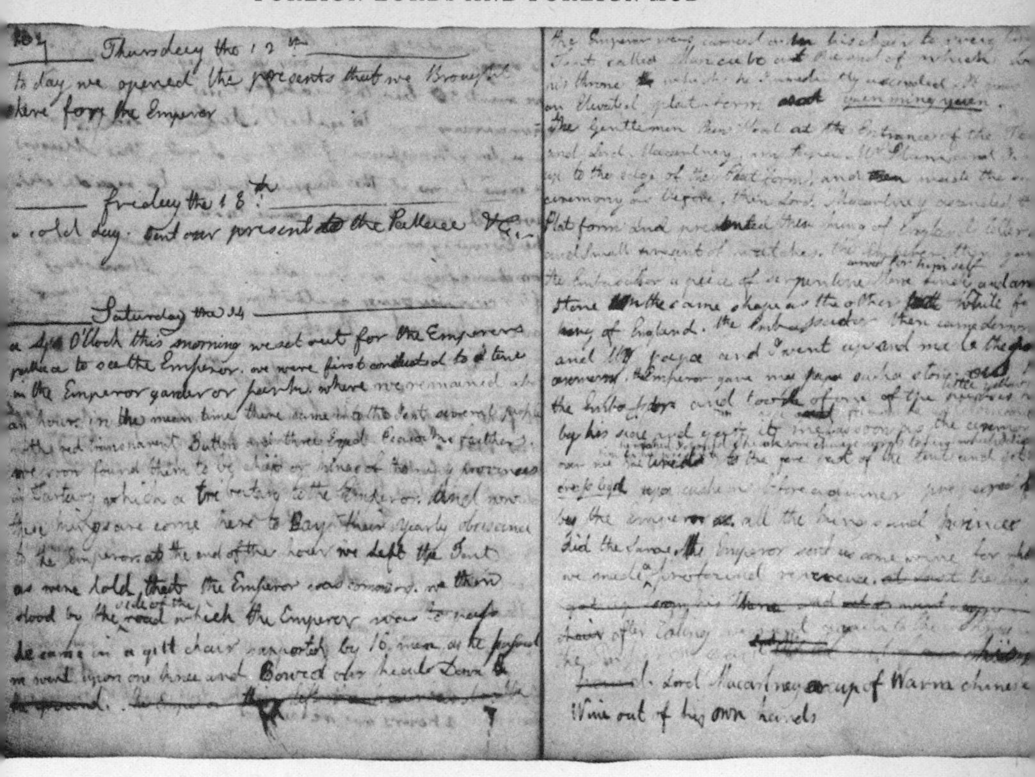

Thursday the 12th
to day we opened the presents that we Brought here for the Emperor

Friday the 13th
a cold day. sent our present to the Pallace &c.

Saturday the 14
at Six O'Clock this morning we set out for the Emperors pallace to see the Emperor. we were first conducted to a tent in the Emperor garden or park. where we remained above an hour. in the mean time there came into the tent several people with red transparent Button and three Eyed Peacock feathers. we soon found them to be chiefs or kings of [illegible] in Tartary which a tributary to the Emperor. And now these kings are come here to Pay their yearly obeisance to the Emperor. at the end of the hour we left the Tent as were told that the Emperor was coming. we then stood by the side of the road which the Emperor was to pass he came in a gilt chair supported by 16 men. as he passed we went upon one knee and Bowed our heads Down to the ground.

PAGES FROM YOUNG STAUNTON'S DIARY. The left-hand page describes the audience with the emperor. Note the deletions at the bottom of the page. (Duke University Library, Durham, North Carolina)

early morning. To this point Macartney's and young Staunton's diaries agree. Macartney continues: "And as he passed we paid him our compliments by kneeling on one knee, whilst all the Chinese made their usual prostrations." Young Staunton, however, remarks frankly in his diary: "We left the tent as we were told the emperor was coming, we stood by the side of the road which the emperor was to pass. He came in a gilt chair supported by 16 men, as he passed we went down on one knee and Bowed our heads Down to the ground." But the word "to" is scribbled over, and "the ground" has a single line drawn through it. Perhaps the elder Staunton, reading the diary afterward, told his son to strike out the revealing remark. None of the historians of the Macartney embassy has remarked on this amusing little sidelight to the noble lord's conduct before the emperor as related by the page and then deleted. Sidelight it is, for the fact remains that Macartney did not actually kowtow. It is impossible to put the head to the ground in that ritual while on one knee. But it seems probable that Macartney did make a more profound reverence with the head than he later saw fit to record.

Set in the imperial park (through which they were later conducted by the notorious Ho-shen), the scene of the audience was one quite different from any we have so far encountered with earlier visitors. Lord Macartney's prose is equal to the occasion: "The emperor's tent or pavilion, which is circular,[8] I should calculate to be about twenty-four or twenty-five yards in diameter, and is supported by a number of pillars, either gilded, painted, or varnished. . . . In the front was an opening of six yards and from it a yellow fly-tent projected so as to lengthen considerably the space between the entrance and the throne.

"The materials and distribution of the furniture within at once displayed grandeur and elegance. The tapestry, the curtains, the carpets, the lanterns, the fringes, the tassels were disposed with such harmony, the colors so artfully varied, the light and shades so judiciously managed, that the whole assemblage filled the eye with delight, and diffused over the mind a pleasing serenity and repose undisturbed by glitter or affected embellishments. The commanding feature of the ceremony was that calm dignity, that sober pomp of Asiatic greatness, which European refinements have not yet attained. . . .

"Thus, then, have I seen 'King Solomon in all his glory.' I use this expression, as the scene recalled perfectly to my memory a puppet show . . . seen in my childhood, and which made so strong an impression on my mind that I then thought it a true representation of the highest pitch of human greatness and felicity. Over rich embroidered velvet I wore the mantle of the Order of the Bath, with the collar, a diamond badge and a diamond star. Sir George Staunton was dressed in a rich embroidered velvet also, and, being a Doctor of Laws in the University of Oxford, wore the habit of his degree, which is of scarlet silk, full and flowing. . . .

"As soon as Ch'ien-lung had ascended his throne I came to the entrance of the tent, and, holding in both my hands a large gold box enriched with diamonds in which was enclosed the King's letter, I walked deliberately up, and ascending the side-steps of the throne, delivered it into the emperor's own hands."[9]

With the whole Chinese court in attendance, in the crisp morning air,

8. Reminiscent of the yurts of Mangu Khan and others, described by William of Rubruck. The Manchu were, after all, cousins to the Mongols.

9. How big an alteration the court had made in its normal ceremonial, and what a large concession was being made to suit his dignity, Macartney did not perhaps fully understand.

A Portfolio of
William Alexander's Sketches

William Alexander (1767–1816) was a young man of twenty-five, already trained as an artist and recommended for his post by the famous Ibbetson, when he joined Lord Macartney's embassy to China. Officially he was the "draughtsman," understudy to Thomas Hickey, the "artist," but since the latter evidently produced not a single drawing on the expedition (or none, at least, that has survived), it is through Alexander's robust and realistic sketches that we know so much of the appearance both of the ambassadorial party and of the Chinese in their country at that time. Alexander must have worked hard during his China sojourn: there are well over a thousand sketches and wash drawings in existence, most of which look like on-the-spot work. He also kept an illustrated diary, still in manuscript in the British Museum.

The impact of his work on the England of his times was considerable. For some decades England, and Europe in general, had been scaling down its concept, derived from the enthusiasm of the Enlightenment, of a China of romantic cloud and mountain, of fragile pagodas and delicate, elfin people stepping out a kind of Oriental quadrille in their never-never land. With the publication of Alexander's sketches and engravings, obviously done from life, the chinoiserie myth of China burst. Suddenly China was seen to be a land of soil and farms, of workaday people of rather chunky aspect and quite solid features. The floating draperies of an elfin land changed to dark blue clothes, coarse and poor in most cases, and the poem of Chinese life was seen very clearly to be Oriental prose. Later history and later artists and even later photographers have amply confirmed the factual content of Alexander's work. The China he portrayed long ago could indeed still be seen until a mere decade ago. His pen and brush were accurate, sympathetic, and not idealizing instruments.

Oddly enough, very few of his drawings have been reproduced. Even his self-portrait (see next page) seems to have rested unknown in the British Museum since is was placed there long ago. The picture of an amiable, alert, rather Schubertian young man is borne out by perusal of his diary, and also by critical appraisal of his drawings. The patch over one eye, however, is unexplained by any known fact, and one might imagine it was added by another, less skillful hand.

On his return to England, Alexander illustrated several books written by members of the embassy and published three of his own. He became the first Keeper of Prints and Drawings of the British Museum and died at the early age of forty-nine of "brain-fever." His grave can still be seen in Kent at Boxley, near his native Maidstone.

With the exception of the self-portrait, all the sketches reproduced are from the India Office Library, London.

SELF-PORTRAIT OF WILLIAM ALEXANDER

MACARTNEY'S FIRST AUDIENCE WITH CH'IEN-LUNG. The ambassador, in a plumed headdress, is on one knee before the emperor on his throne, and is surrounded by his suite. Each Englishman has been numbered by Alexander and the key is at the top left of the sketch. Sir George Staunton wears a dark hat and has his hand on the shoulder of the "page," his son. Alexander, who was not present at the ceremony, based his sketch upon drawings by an officer accompanying Lord Macartney.

MACARTNEY IN COURT DRESS. Something of the elegance and panache of the ambassador and his embassy comes through in this sketch made in China.

A CHINESE MAGISTRATE, together with his clerk and a police officer. Judging by the inscription beneath it, the sketch was a study for a larger work that was to show the culprit as well.

CHINESE VILLAGERS gathered on the banks of the canal outside their village to watch the passage of the embassy barges.

A CHINESE WOMAN smoking a long pipe.

DOMINUS NEAN, a Chinese from Ripa's college at Naples who went with the embassy as one of its interpreters.

with perhaps just the first rays of the autumn sun slanting across the yellow of the fly tent—it must have been a very fine sight indeed. Breakfast was then served. The emperor with his own hands gave a cup of warm Chinese wine to the ambassador, who, like John Bell, was astonished at "the order and regularity in serving and removing the dinner . . . every function performed with such silence and solemnity as in some measure to resemble the celebration of a religious mystery." Here, Macartney was unconsciously very near the truth. The emperor was a semidivine figure. Had the ambassador realized fully that what he was attending was in many respects the Chinese equivalent of a High Mass in the Roman Church, he would have been more prepared for the outcome of his efforts in Peking. He noted, instead, that Ch'ien-lung was dignified and affable in his manner "and in his reception of us has been very gracious and satisfactory. He is a very fine old gentleman, still healthy and vigorous, not having the appearance of a man of more than sixty." The emperor was in fact within ten days of his eighty-second birthday. Despite his sprightly state, Ch'ien-lung's knowledge of and influence on what was taking place in China at this period was probably extremely limited. Ho-shen, a Manchu, was the real ruler, and had the emperor's implicit trust.

Young Staunton, naturally enough, notes his own little moment in his diary: "The ambassador then came down and my papa and I went up to the throne and made the proper ceremony. The emperor gave my papa such a stone as he gave the ambassador and took off one of the little yellow purses hanging by his side and gave it to me as a favor he seldom conferred. He asked I should speak some Chinese words to him which I did thanking him for his present."[10]

Thrice after this audience Macartney saw the emperor again. But never for an instant did Macartney have a chance to converse with him on serious matters. The go-between was Ho-shen, virtual ruler of China, then at the zenith of his power. He was by far the wealthiest man in the world, his personal fortune estimated, when he fell from grace at the death of Ch'ien-lung, at over three hundred million sterling, practically none of it acquired in honest ways. Without question he was involved in the opium traffic at Canton and saw that the trade if left to itself would inevitably increase. His interests would therefore be best served by the *status quo,* as far as the British in Canton were concerned.

Ho-shen was probably the best Chinese foil to Macartney. "I could not help admiring," says the latter, "the address with which the Minister parried all my attempts to speak to him on the business of the day, and how artfully he evaded every opportunity that offered for any particular con-

10. The purse was later given by Staunton to the Royal Asiatic Society, London. But it seems to have disappeared many years ago when some objects were transferred from there to the Victoria and Albert Museum, London.

versation with me." And there was probably yet another reason why Ho-shen wanted no real contact with the British—the current Chinese fears that British power in India constituted a threat to the western borders of the Chinese empire. In this he was astute, for the frontiers of India and China have remained a thorn in the flesh of both sides ever since, for reasons very similar.[11]

Macartney's conduct in the encounters he had with official China was admirably unruffled. After his first talk at Jehol with Ho-shen, as he was shown the park, he had his fears. But for a time it seemed that the panache of his embassy and the brave show it made at court would so impress the Chinese that they would recognize the British were not like other "barbarians" and would grant facilities at Canton and elsewhere. Not a cross word was exchanged between the two sides in the six weeks Macartney was in Jehol and Peking. But no progress toward agreement was made either.

By the end of September, emperor and mandarinate were taking the attitude that the British must not risk staying longer on account of Peking's inclement winter. They smiled with the utmost blandness and the politest disbelief when Macartney replied that this was of no importance as the British also had terrible winter weather. His memoranda remained unanswered, except with civility. Like Ricci before him, Macartney discovered in Peking a "desert of gentility."

Eventually, knowing he had failed, he was summoned to receive the imperial letter in reply to that of George III. There it lay on a yellow silk chair, wrapped in yellow silk, a scroll whose contents he could not guess. Then it was tied to the back of a mandarin prostrated with reverence at the mere presence of the imperial word, and so taken to Macartney's residence. There was nothing to do but to accept. "It is now beyond doubt . . . that the Court wishes us to be gone, and if we don't take the hints already given, they may possibly be imparted to us in a broader and coarser manner." A last attempt, however. A last memorandum of six points outlining the British government's requests—an eminently reasonable note, in European terms. We know the answer before, in Macartney's pages, the note reaches the Court; the answer, when it came, that was brief, negative, and polite. Foreigners traditionally have no rights and are simply granted indulgences from time to time so as to trade.

The letter from Ch'ien-lung to King George III of England is by now a well-known classic of its kind. But we must quote some of its more striking passages, remembering always that from the Chinese point of view the mere suggestion in George's letter that his people might be allowed freely to trade with China sounded as brash and unfeasible as does the emperor's reply:

"You, O King, live beyond the confines of many seas, nevertheless, im-

11. This interesting aspect of Sino-British relations is dealt with by Lamb.

pelled by your humble desire to partake of the benefits of our civilization, you have dispatched a mission respectfully bearing your memorial. Your Envoy crossed the seas and paid his respects at my Court on the anniversary of my birthday. To show your devotion you have also sent offerings of your country's produce.

"I have perused your memorial. The humble terms in which it is couched reveal a respectful humility on your part, which is highly praiseworthy. In consideration of the fact that your Ambassador and his deputy have come a long way . . . I have shown them high favor and have allowed them to be introduced into my presence. . . .

"As to your entreaty to send one of your nationals to be accredited to my Celestial Court and to be in control of your country's trade with China, this request is contrary to all usage of my dynasty and cannot possibly be entertained. . . . You are presumably familiar with our dynastic regulations. Your proposed Envoy . . . could not be placed in a position similar to that of European officials in Peking who are forbidden to leave China [i.e., Christian priests in the Manchu service], nor could he, on the other hand, be allowed liberty of movement and the privilege of communication with his own country; so you would gain nothing by his residence in our midst."

The letter proceeds to say that the empire has vast territories and hundreds of tribute missions—all of whom are dealt with in a fair way by the Department of Tributary States. An envoy would not fit into this picture. And besides, all possible facilities for trade have been provided at Canton already. As to the question of China's sending an ambassador to Europe: "How could you possibly make for him the requisite arrangements? Europe consists of many nations besides your own; if each and all demanded to be represented at our Court, how could we possibly consent? The thing is utterly impracticable.

"If you assert that your reverence for our Celestial dynasty fills you with a desire to acquire our civilization, our ceremonies and code of laws differ so completely from your own that, even if your Envoy were able to acquire the rudiments of our civilization, you could not possibly transplant our manners and customs to your alien soil.

"Swaying the wide world, I have but one aim in view . . . to maintain a perfect governance and to fulfil the duties of the State; strange and costly objects do not interest me." He has accepted the gifts solely out of "consideration for the spirit which prompted" their dispatch from afar. "Our dynasty's majestic virtue has penetrated into every country under Heaven. . . . We possess all things. I set no value on objects strange and ingenious, and have no use for your country's manufactures. . . .

"It behoves you, O King, to respect my sentiments and to display even greater devotion and loyalty in future, so that, by perpetual submission to our Throne," etc. "Take note of my tender goodwill toward you!"

In a later letter, Ch'ien-lung replied in detail to that last memorandum Macartney sent before he departed from Peking. The language is the same, the burden of the reply identical but set out in detail. These and a further letter some years after are almost the last signs we have of that blissfully ignorant Chinese world outlook. It was not long now till the first blow fell, not long before it was never again possible to write from the throne such letters of lofty condescension to a European country. But Macartney did not know this. He was baffled by the form of the letter and by its assumption of a power which he well knew China did not possess. In fact, Ch'ien-lung's letters were couched in the ordinary form of letters to tributary states. They were little more than formal replies to impossible requests, such as one might receive today in England where even prime ministers still sign themselves "Your humble servant," which is hardly less unreal.

Macartney's bafflement is understandable enough. Though he was a man of enlarged understanding (a phrase he would have approved), and in his Chinese way so was Ch'ien-lung, both were men of their own times. Ch'ien-lung's policy in regard to foreigners varied very little from the traditional Chinese outlook that had prevailed since the beginning of the Ming. Macartney's outlook, however, was much more the product of the late eighteenth century than of an unexamined tradition. And the philosophy of the times, in regard to China, was a product of those many records that had been sent from China, of a Europe which had for several decades turned the searchlight of its most ingenious thought and analysis on those records: and was now beginning to be disillusioned. The East-West intellectual honeymoon was nearing the going-home stage and not much but the chinoiserie trousseau was left of that first legitimate rapture. But the Enlightenment itself, which had celebrated the marriage, was a product in part of the science which was beginning to reshape the physical as well as the mental European environment. No such scientific revolution (or intellectual one either) had occurred in China. In this—broadly—lay the absoluteness of the vacuum between them. A sample of Macartney's outlook is to all intents a sample of Britain's. Sailing from Peking down the Grand Canal, subsidized to the tune of about $10,000 a day by the munificence of Ch'ien-lung (or Ho-shen, whichever one pleases), he reflected:

"Whatever taste the Emperor K'ang-hsi might have shown for the sciences . . . his successors have not inherited it with his other great qualities. . . . For it would now seem that the policy of the Court . . . concurred in endeavoring to keep out of sight whatever can manifest our British pre-eminence, which they undoubtedly feel, but have not yet learned to make use of. It is, however, vain to attempt arresting the progress of human knowledge. The human mind is of a soaring nature and having once gained the lower steps of the ascent, struggles incessantly against every difficulty to reach the highest." And he adds the afterthought: "Whatever ought to be will be. The resistance of adamant is insufficient to defeat the insinuation of a fibre. Time is the great wonder worker of our world, the exterminator of prejudice and the touchstone of truth. It is endless to oppose it. Power becomes enervate and efforts ridiculous. The tyranny . . . of a state may stalk abroad in all its terrors and for a while may force a base currency on the timorous multitude, but . . . there is always a certain counteraction fearlessly working in the mind of common sense . . . imperceptibly issuing . . . a standard metal whose intrinsic value soon degrades and baffles every artifice of impure coinage."

This is not only the doctrine of inevitable progress, but of a basic democracy working toward that progress. It is a rootedly European idea. Macartney, naturally, could not see that it did not at all apply in China—that it never had and, as it turns out, looks as though it never will—because the basic concepts on which it stood and grew in the West were there totally absent. He himself, and to some extent his philosophy, were children of expanding European capitalism, and no such economic institution ever developed in China.

It was not unnatural, then, that Macartney, and most Westerners before and long after his time, should be unaware that the European outlook, which seemed to them sweetly reasonable, was to the Chinese unlikable, baffling, presumptuous, sometimes threatening—and quite as outlandish as the Chinese outlook seemed to the Europeans. The Chinese world view itself was unreal, but this is not the point.

The two nuggets of adamant, to follow the noble lord's metaphor, had been in opposition many times before Macartney, as we have seen. On their home ground and often the wiser of the two, the Chinese generally took avoiding action when they saw a collision ahead. But now, with the constant growth in Europe's need to trade, the adamants were approaching dangerously close. The forces narrowing the gap were increasing. Like the field of a magnet which induces outside itself a reaction relative to its own power, the expansion of British trading drove the rootedly insular Chinese to responses they would much rather never have made, responses which were ill calculated to limit that trade, or to preserve the *status quo* in China.

But we are running a little ahead. What was the result of the embassy when it returned from the brilliant ballet enacted in Peking? In terms of what the British government sent Macartney to do—nil. Indirectly, through Macartney's dispatches and through the spate of books published by members of the embassy, England learned how inflexible was the Chinese outlook on the rest of the world. With India at England's feet and the seas dominated by the British flag, it was perhaps natural that Britian's attitude to China hardened. It is a saddening thought how little notice the Manchu government took of sea power. Not all the years of British ships at Canton, which could outsail, outmaneuver, and outshoot every known Chinese vessel, had impressed an iota of their danger on the authorities. Like those forts at the inappropriately named Tiger's Mouth entry to the river down from Canton, whose guns proved unable to prevent the entry of Western ships, and were never altered so as to be able to do so, the Chinese failed to reckon with the outer world.

Six years after the embassy left China, young Staunton, now a youth of eighteen, returned as a clerk to work at the East India Company's "factory" (warehouses) at Canton. He was unhappy, because he was unpopular. His father had obtained the appointment for him by influence, and he was resented. But his knowledge of Chinese and his ponderous ability in his work gradually raised him to a position where it hardly mattered if he was disliked or not. He was soon too useful to be overlooked.

Staunton's account of life in the intensely interesting first twenty years of the nineteenth century in the Canton factories is pompous and self-congratulatory. The British dominated the Canton trade in the two commodities that mattered, export tea and import opium. True, American merchantmen also imported opium (from Turkey), but the quantity was small and the opium of poor quality. It was at this time that the opium trade came into its own and Chinese silver began to flow out of China in increasing quantity to the West. It was a vicious era in Canton. The perfidy and astuteness of the Western merchants were well matched by the same qualities in the Chinese with whom they dealt. It was a roistering era of British and Chinese rogues, of astronomical profits on both sides, of debauched British sailors dead drunk on the grog dispensed in the stews of Hog Lane adjacent to the factories, of brawls between the dregs of East and West.

With opium came its attendants, the Protestant missionaries. Robert Morrison, an Englishman and the first Protestant missionary to China, arrived in Canton in an American ship in 1803. By 1809 his command of Chinese obtained him the post of interpreter to the East India Company. He translated the Bible, and wrote his remarkable dictionary, which, for the first time, put knowledge of Chinese on something like a solid footing. To a colleague of his[12] we owe one of the amusing definitions of the language: "To acquire Chinese is a work for men with bodies of brass, lungs of steel, heads of oak, eyes of eagles, hearts of apostles, memories of angels, lives of Methuselah." Morrison was a sincere and orthodox Christian; yet, being a child of his times, he seems to have had no compunction in interpreting for an organization which derived almost its entire profit from illegally purveying opium.

Morrison's son, John Robert Morrison, was born at Macao and, later, succeeded his father as Chinese secretary and interpreter at Canton. He wrote

12. William Milne, who arrived in China in 1813 and ordained the first Chinese Protestant minister there. The quotation comes from Neill.

a helpful book, published there in 1834, called *A Chinese Commercial Guide*—a volume which transports us back over the centuries to Francis Balducci Pegolotti with his book aimed at merchants on the (legal) trans-Asian caravans. The difference between the two men is that Morrison advised on a trade which he knew to be illegal. "Opium," he says, "is chiefly in demand along coastal South China. . . . To avoid as much as possible the officers of the government, disregarding alike their promises and their threats, is a rule the observance of which is necessary to ensure the least success. The people are generally glad to meet foreigners, and not unwilling to expose themselves to considerable risk. Such being the case a gradual opening of the trade may be looked for. . . ." Thus Morrison "treading in the steps of his honored father . . . devoting his energies to the benefit of China," as his contemporary Karl F. A. Gutzlaff said of him.

Gutzlaff was a Prussian missionary, a much more important man than Morrison. "Prussian buccaneer missionary interpreter," Arthur Waley calls him.[13] And Commissioner Lin, one of the few sane men involved in the opium wars, thought him dangerous: "I hear . . . that they have made a certain Gutzlaff 'prefect' of Ting-hai. He speaks Cantonese, and steps should be taken to thwart his machinations." Another Chinese says of the same period and place: "Here Gutzlaff trains the foreign troops. . . . The woman singer of local airs . . . has two daughters, whom she has given to Gutzlaff as his wives." The youthful Harry Parkes, later to become British consul in Canton, recalls him as a "short squat figure, the clothes that for shape might have been cut in a village of his native Pomerania ages ago: the broad-brimmed straw hat; the great face beneath it, with that sinister eye!" "In short," as Waley says, "a cross between parson and pirate, charlatan and genius, philanthropist and crook." James Hudson Taylor, the founder of the China Mission in the fifties of the century, called Gutzlaff its grandfather, apparently unaware of the doubtful quality of such ancestry. As to Gutzlaff himself, his opinion of the Chinese, on whom he lavished a lifetime's work and a hail of Christian tracts, was far from love, or even respect. "How much," he said, "foreign intercourse has improved Chinese manners at Canton!"

Like Morrison, Gutzlaff was financed by the East India Company as an interpreter (after he ran through the money he got on the death of his first wife). Up and down the China coast he went, often in opium ships, assisting in the sale of opium and handing out his Christian literature. Both temperament and education cast Gutzlaff on the wilder shores of evangelism, led him by what at best we may call eccentric paths toward converting Chinese to his faith. With an opium coating he offered every man's salvation. Both he and Morrison acted, *passim,* as intelligence agents for the

13. All the quotations in this and the following paragraph are from Waley's book on the opium wars.

Karl Gutzlaff in Chinese dress. This is a quite believable portrait in view of contemporary descriptions of this flamboyant man. Why he should be wearing an extremely un-Chinese turban is not easily explained. (Radio Times Hulton Picture Library, London)

"Company" (and therefore at a later date as it turned out, for Britain and her naval power). One of his gang of criminal-fringe Chinese wrote to him, for example, in May, 1838: "Specially to let you know that I now have the maritime chart of the Inner River at Foochow, where the Governor gave permission for cannon to be placed at the point near Lo-hsing Pagoda and knock holes all over your ship. . . ." And the same spy remarks that for a favor from Gutzlaff he will work in his interests "like a horse or dog."

With such willing help, the company expanded its trade along the China coast and at Canton. And its officers, Staunton among them, garnered fortunes from it. There are still trading houses thriving in Hong Kong today whose commercial empires were founded at this time in Canton on Chinese opium silver.

Gutzlaff was in the thick of it during most of his time in China. Among the spectrum of influences he exerted were several of great moment to the future of Sino-British relations. He it was who taught Chinese to Harry Parkes, a man whose character Waley has summed up as "a firm believer in gunboat diplomacy." The founder of the China Inland Mission, James Hudson Taylor was sent to China by the Gutzlaff Association in 1853. Reports of Gutzlaff's work and ill-conceived schemes brought about twenty missionaries from America and England to China in the 1830's. And later Gutzlaff himself organized a squad of young Chinese to rove over the country to teach the Gospels. Setting out from Hong Kong, these men sent back glowing reports of their successes. But "a fearful exposure was to follow," as Bishop Neill, the historian of missions, puts it. It was discovered that, equipped with funds from the pockets of pious European societies, few if any of these young men had gone farther than the opium dens and brothels of Canton, where they had quickly settled in to enjoy the opportunities afforded them. And when Gutzlaff died all that remained of his Christian endeavor was a very putrescent smell in the West's Christian nose: and the fact that it was from his translation of the Bible that the leader of the Taiping Rebellion later acquired his form of Christianity. Add to these activities Gutzlaff's participation in the Opium War, and the picture of the missionary is complete. To Europeans in China at the time it was generally an acceptable one.

To be quite fair to Gutzlaff's complicated character we ought to allow him a word in his own defense: "I am fully aware that I shall be stigmatised as a headstrong enthusiast, an unprincipled rambler, who rashly sallies forth, without waiting for any indications of divine providence, without first seeing the door opened by the hand of the Lord. . . . I have weighed the arguments for and against the course I am endeavoring to pursue, and have formed the resolution to publish the Gospel to the inhabitants of China proper, in all ways, and by all means, which the Lord . . . appoints in this world . . . and rather to be blotted out from the list of mortals, than to behold with indifference the uncontrolled triumph of Satan over the Chinese." While he often spoke against the smoking of opium, he seems nonetheless to have managed to accommodate himself to its sale in China in order to assist his distribution of Gospels. "While representing Christianity as the only effectual means of establishing a friendly intercourse, I would not reject the efforts of commercial enterprise to open a trade with the maritime provinces, but rather regard them as the probable means of introducing that Gospel into a country to which the only access is by sea." What one doubts is his mental balance. Unlike the Jesuits, he was unaware of his own casuistry. That is the kindest interpretation of his deeds in China.

Meanwhile, before a suitable excuse for the Opium War was found, Staunton was off again to Peking with another pompous gentleman, Lord Amherst, on an embassy which, by the time Amherst reached Canton, was patently unnecessary. But this was an age, in China, compounded of follies: of British and Cantonese commercial rapacity, of a Manchu government sick unto death, of missionaries compliant in the role as helpers in poisoning the Chinese they wished to bring to God, of incipient rebellion in South China, of the gradual approach of the Western extortioner. All the time the opium flowed faster and faster, and Chinese silver cascaded into the pockets of the merchant barbarians.

George III, who had sent Macartney to Peking, was now mad. Amherst was sent with a brief letter from the regent to the emperor. Ch'ien-lung too was long gone, and his fifteenth son, Chia-ch'ing, reigned. Amherst was to try once more for better trading conditions at Canton. But the Chinese were even less inclined now than before to open the question. Not only were the illicit profits of the mandarins huge, but there was no change in the old Chinese isolationist policy in regard to Westerners. Anyone in his senses would have listened to Staunton's advice to go no further than Canton. Henry Ellis, who arrived with Amherst, writes: "Those who have perused the accounts of the former Macartney embassy, commenced . . . as it was under better prospects, can scarcely anticipate either public success or private gratification from any events likely to occur." He thought a visit to the Bedouin Arabs would be preferable. The Manchu government, for its part, saw only the kudos that would accrue to it in public should the embassy perform the kowtow as a tributary to the faltering dynasty.

Landing, as had Macartney, at the mouth of the Pei River, Amherst was escorted to Peking. During the journey his ambivalent attitude convinced his escorts that with a little persuasion he would kowtow. They sent word of this ahead to Peking. But when it was too late to recall this letter, Amherst finally stated bluntly that he would not kowtow. The officials were then in a quandary. How could they inform the emperor that after all the ambassador refused? A letter from the emperor then came. Amherst was to be brought to him at once.

OPIUM SHIPS in a South China harbor, 1824. (Museum and Art Gallery, City Hall, Hong Kong)

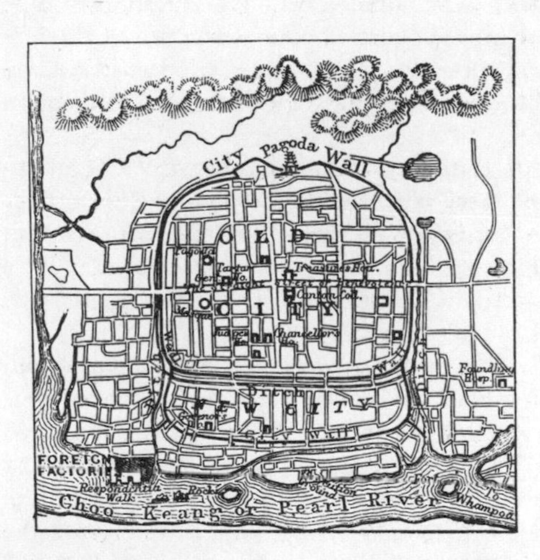

PLAN OF CANTON, 1857. Note the dominating waterfront position of the foreign "factories." (From the *Illustrated London News,* 1857)

The emperor was to hold audience at 6:30 A.M. on August 28, 1816, but postponed it on hearing the ambassadorial party was late. Around this time, then, Amherst tumbled from his carriage, disheveled from a night of entirely sleepless travel, at the gates of the Summer Palace. The whole court with the exception of the emperor was assembled as the disgruntled nobleman was ushered in, unshaven and wearing his crushed traveling clothes. They put him in a chair while elegant courtiers gyrated round him with evident disdain. Chia-ch'ing would not be kept waiting. The Manchu duke responsible took Amherst by the arm in a friendly fashion with the plain intention of leading him directly to the throne room and the kowtow. Amherst, suspecting what was happening, shook him off roughly and refused to go—something unheard of in the palace. The duke, nonplused, rushed back to the emperor and invented the tale that Amherst was transfixed by a bellyache. Chia-ch'ing directed that he be taken to his residence, given medical aid, and brought later. He called for the other members of the embassy to appear at once before him. But the duke said they too had pains in the stomach and could not. Whereupon, the emperor, enraged at this transparent excuse, and imagining the embassy had invented it all, began to dictate a decree expelling them from Peking at once. While he was dictating, the duke returned to Amherst and merely told him to go to his residence, which Amherst did.

Once there, he discovered that the Chinese porters had orders not to unload his baggage. Very soon came the imperial decree expelling him forthwith. Just as the party were about to re-enter their conveyances and be off, hungry and frustrated, a sumptuous breakfast arrived, the gift of the emperor. The English, nothing loath, sat down to a hearty meal. By four in the afternoon they were on their way home. Theirs was the shortest time spent by any embassy to the Chinese court—about ten hours. Amherst's faults were chiefly grandiloquence and the fact that he failed to decide on what to do about the kowtow until too late; and that he had no conception of the rigidity of Chinese custom.

With that abortive embassy we come very near to events which in an important sense are the crux of the story of China and the West. It is not the province of this book to deal extensively with the Opium War as such, but the briefest outline of events which led up to that calamity is necessary if we are to see in perspective the deeds and attitudes of the Europeans who afterward came to China.

Staunton left Canton in 1817. "My public service abroad terminated with Lord Amherst's embassy. Having held the highest place in . . . the service of the East India Company to which I was attached; having accomplished my favorite object, of revisiting Peking in a diplomatic capacity; and having accumulated a competent fortune . . . I gladly abandoned the prospect of increased wealth . . . and I rejoiced to find myself able to terminate the period of my banishment, at the early age of six and thirty." Thus the orotund head of the company's Select Committee at Canton. His fellow members on the committee wrote him an equally fulsome letter of farewell and gave him a "splendid silver-gilt salver valued at six hundred guineas." He later became a member of Parliament, bombarded the British government with advice on China (which went largely unheeded), and eventually founded (with one other) the Royal Asiatic Society, London, where in many a volume on the library shelves his signature is still to be seen.

By 1833, the year before the British parliament ended the monopoly of the East India Company and opened the Canton trade to all, the profits from opium were gigantic. According to the American H. B. Morse, who was later commissioner of Chinese Maritime Customs, the British import of the drug during that year was valued at over twelve million dollars, more than half the value of the total British import to Canton in the same period. American imports to Canton that year were valued at over a quarter of a million dollars.

With the monopoly ended, the Chinese demanded that one man be appointed spokesman for all the traders, in place of the Select Committee. Canton from then onward was a tinderbox ready for the fatal spark. Largely unable to control Canton and the south of China, the Manchu government grew alarmed about the drain of its currency to pay for opium—more worried about the silver than the effects of the opium, one suspects. And it seems from the still-far-from-complete research on this complicated period, that the entire economic structure of South China was in disruption due to the import of opium, and to the effects of foreign trade in general. One of these effects was to encourage the rebellious nature of the south in defying the edicts of the central government, and it can be argued that in later days this was in part responsible for fomenting what turned out to be the Taiping revolt. But, as the American-Chinese scholar Chang Hsin-pao has pointed out, apart from economic factors, the situation involved a "clash between two different cultures." The idea is not unfamiliar to us at this stage in this book. The difference between the past and the present was that now the Western form of culture was backed by great force readily summoned, and by its felonious association with a network of Chinese lawbreakers in its pay. Given the facts of the opium trade as they were, the same writer continues: "Assume . . . that the opium trade was carried on in another country—one with a culture analogous to England's—in obstinate violation of the laws of that country. In such a case could war be avoided?"

In Peking, on the last day of 1838, after agonized debates on how to suppress the opium trade in the south, the Tao-kuang emperor issued his decree appointing Lin Tse-hsu commissioner extraordinary to go to Canton and take action. Lin was a scholar, a proven opium-suppressor elsewhere, an able administrator, a considerable statesman—an excellent choice. He left the capital early in January and arrived in Canton on March 10, carrying the seal authorizing him to take almost any measure in the emperor's name. It was a rarely given power, and probably no one in China at the time was better fitted to use it.[14]

Lin was soon well informed and acted with care and thought, with some caution and a remarkable degree of understanding. He even addressed a letter to Queen Victoria advising her to stop the growing of opium and to prohibit her subjects from bringing it to China—one of the most balanced and reasonable documents penned by anyone on either side during the whole tragic era of conflict.[15] Lin's sincerity is undoubted.

There ensued at Canton a series of discussions, proposals, and counter-proposals, ending in the surrender by the merchants of a staggering quantity of opium—over twenty thousand chests of it. This was destroyed by Commissioner Lin with the approval of the emperor, who congratulated him on his good work.

The reaction of the foreign community was swift. "Almost every Englishman who had an interest in the China trade, from the superintendent in Canton to the financiers and manufacturers in London and Manchester, entreated the British government to intervene," notes Chang. Pressure brought to bear in England was severe and the prime minister, Palmerston, acceded to its demands. An expeditionary force set out, the first vessel arriving at Canton in June, 1840. Others quickly followed. Canton was blockaded and ransomed itself, for the moment, for six million dollars. It was war, the first Anglo-Chinese war. Soon the jubilant London *Times* was exulting: "The British flag waves over a portion of the Chinese empire for the first time. Chusan fell to British hands on Sunday the 5th of July, and one more settlement in the Far East was added to the British crown."

China, without a navy, was at the mercy of the world's strongest naval power. Commissioner Lin failed because he was eventually not supported by the feeble government in Peking. But he would have failed anyway against the Western seaborne invasions of the ports. The time of reckoning had come. One by one those ports fell in 1841 and 1842. The last two, Shanghai and Chinkiang, surrendered in late July. A month later, on August 29, the Treaty of Nanking was signed.

However much the British and other traders had been aided and abetted by sections of the Chinese populace in their illegal trade, that trade was known by them to be illegal; and they persisted in it. The Treaty of Nanking is therefore regarded by Chinese (then and now) as an "unequal

14. Chang Hsin-pao's book is one of the best dissections of Lin's character and his acts in the impossible situation he found at Canton. But see also Waley.

15. The best translation of the letter is in Teng and Fairbank. It is a pity we do not have space to quote it, but the detailed story of Chinese officials in their efforts to counter the commercial and other aggression of the West in China is at least another volume in itself.

A Naval Battle of 1841. The East India Company's ships, including the steamer *Nemesis,* are seen destroying Chinese war junks off the South China coast. (Radio Times Hulton Picture Library, London)

treaty" signed by China under duress. This is undeniable, and constitutes one of the most ineradicable stains on Anglo-Chinese history.

The provisions of the treaty were harsh. Five major ports of China (Canton, Amoy, Foochow, Ningpo, and Shanghai) were opened to British trade. The island of Hong Kong was ceded to the British—who used it as a military and naval springboard, among other things. An enormous indemnity in silver was exacted for the destroyed opium stocks, and opium trading was not prohibited. No more than a five-percent tax was to be levied on any foreign goods imported into China—which led to more imports and a still greater drain of silver. And all British persons were to be subject only to British law while in China, being entirely free of Chinese law—the famous extraterritoriality clause.

The treaty was soon followed by others, made more or less at gunpoint, with other nations, among the first being the United States. The so-called Cushing Treaty was signed in 1844 and granted similar privileges to the Americans. In his *Americans in Eastern Asia* Tyler Dennett remarks that this treaty "became immediately the model for the French treaty" and was "in practice, the smuggler's delight."

For the first time China was almost completely open to the West, and as the century progressed and more treaties were extorted from the effete Manchu dynasty, this became an accomplished fact. It was a long way from the times of Ricci, even from the best days of the great Jesuit scientists, a tremendous step even from Macartney. The West congratulated itself as it raced to trade (if we may dignify that commerce by the name) up and down the coasts of China.

The remainder of the nineteenth century, when Europeans flooded into China and so many went to Peking that they have never been counted, is a radically different story from that we have told in earlier chapters. Every European who came to China after the Treaty of Nanking could consider himself a member of a superior race—the virtual conqueror of the oldest kingdom in the world, Many, alas, did think so. They soon showed all the classical signs of the master race, repeating the cruelties, the extortions, the fatuities, and the good works patiently endured by colonial India. The ancient pride of China had received its deathblow, and now it sickened away in a lingering splendor and a lingering shame. For, whatever unreality (and it was great) there had been in the former Chinese concept of the outer world, the central Chinese concept of themselves was now shattered as well. Every Chinese could see this fact written in the arrogance of the European communities. China was now a semi-colony being divided between a handful of Western nations for their profit. She was to witness their squabbles, their rapacity, their schismatic missionaries, their callousness, and their odd, sporadic humanity; to learn something of their industries, to acquire their armaments; but not yet awhile to understand their civilization. A parallel growth to this colonial power was the degeneration of the colonial communities until each one was a caricature of the Western society from which it originated. The rule of the West in China was quite as corrupt as that of the Manchu—and much more powerful because it was backed by industries, armies, and navies.

Perhaps we need an epitaph for those days of the early nineteenth century in China. There are several good ones. One of the best, the most honest, comes by chance from William B. Langdon's catalogue of the "Chinese Collection now exhibiting at St. George's Place, Hyde Park Corner, London," in the very year of the Treaty of Nanking. The collection had been gathered in Canton by an American named Nathan Dunn, and previously exhibited with éclat in Philadelphia. The annotated catalogue ends with these words, astonishing for their time, and, with a little editing of no great moment, applicable far into the future:

"The Chinese have been, repeatedly, denounced in terms savouring little of Christian forbearance and charity. In their business transactions, they have been presented . . . as a nation of cheats; in their bearing towards foreigners, as scornful and repulsive . . . and in their own social relations, as bereft of every noble sentiment and generous sympathy. The policy, especially of excluding foreign traders from all but a single port of the empire, has been made the subject of the most acrimonious denunciations. Far be it from us to enter the lists in defence of this policy. . . . But Truth and Justice are suitors at the bar, and demand a few words of explanation. . . . We have already seen that this people, at an early day, sought commercial connections with various neighbouring nations; that the Arabians traded freely . . . that the earliest European visitors were received with marked kindness . . . and that the Catholic missionaries had free admission to all parts. . . .

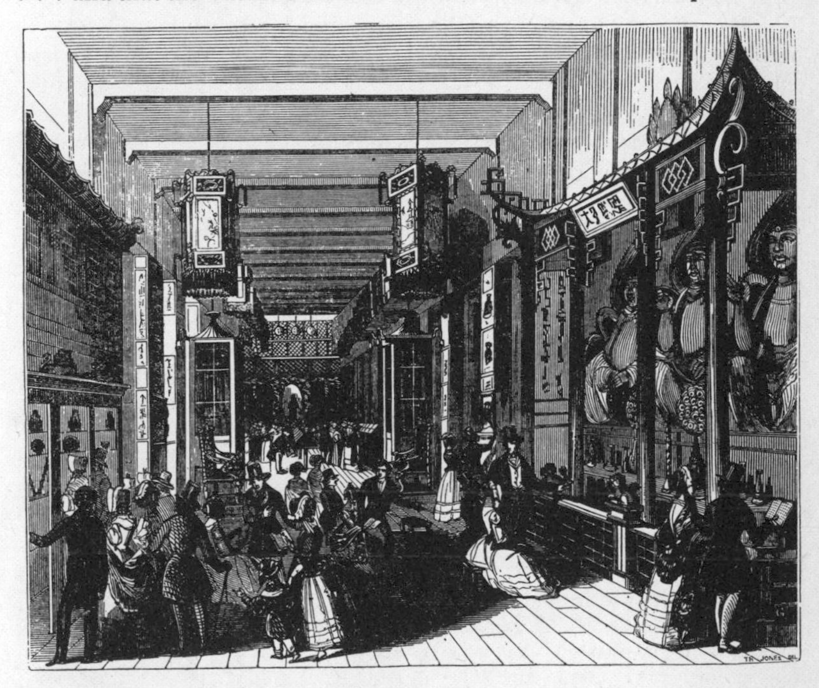

THE CHINESE EXHIBITION, London, 1842. A poem written by a Chinese visitor to the exhibition remarks that in London "the houses are so lofty that you may pluck the stars." An English writer, reviewing the exhibition, remarked that it enabled one "in some sense [to] analyze the mental and moral qualities of the Chinese." (From the *Illustrated London News*, 1842)

These zealous and able sectaries were frequently promoted to the highest dignities of the empire. . . . Hundreds of thousands of Chinese were, nominally at least . . . converted to the Christian faith. They [the Europeans] continued in favour till they indiscreetly began to tamper with government affairs, and attempted to undermine the ancient institutions of the realm. . . ." The writer then turns to the merchants. "But the burning jealousies . . . perpetually kept up between the subjects of the different European governments that sought to share in the rich claims of the China trade . . . inspired no very favourable opinion of their character. . . . These circumstances . . . with various positive abuses of the liberties of the trade . . . caused the government [of China] to commence . . . the work of abridging the privileges of foreigners, and the result appears in the rigid system of restrictions now in force.

"If Europeans and Americans may fairly blame the illiberality of the Chinese, these have certainly just ground of complaint against them in the illegal practices to which their [Western] cupidity tempts them. Fifteen to twenty millions' worth of opium has been for years, in defiance of the laws and known wishes of the Chinese government annually emptied upon the shores of China by Christian merchants!

"Alas for missionary effort, so long as the grasping avarice of the countries whence the missionaries come sets at naught every Christian obligation before the very eyes of the people whom it is sought to convert!"

The exhibition in London was a very great success. Perhaps, since these words appear on the last two pages of a 163-page catalogue, they were not read by many. For at the time such words were not only courageous, but highly unpopular in their sentiments. Perhaps, too, the sudden glimpse of the several thousand Chinese objects displayed so roused the fancy of the visitors that they could afford to overlook the current problems of Christianity and opium on the shores of South China. But as a matter of fact only three years later, in 1845, Madame Tussaud's waxworks exhibited a new model of one Lin, a Chinese commissioner, with his favorite consort. And Tussaud's never yet made a model of anyone who was not a household word. Staunton, either as "page" or as man, certainly never made the grade.

Opium Bales shipped from the growing areas in India to China and stamped with the initials of "Victoria, Empress of India." (From the *Illustrated London News,* 1843)

CHAPTER FIFTEEN

DR. PETER PARKER

AND THE AMERICAN EFFORT

IT HAS been said of Dr. Peter Parker that he opened China to the Gospel at the point of his lancet. Certainly the medical work he did there was the more estimable part, for the rest consisted of an attempt to open China by more usual if less humane means.

Peter Parker's career began much better than it ended. He was the son of farmer parents, born in 1804 at Framingham, Massachusetts. As a child he was more than normally introspective, at first rejecting the narrow life imposed by his pious parents; rejecting too their scriptural instruction. At about the age of fourteen, "I was brought through the goodness of God to a *solemn pause*."[1] He began to feel a deep sense of sin. With it, as violent as his former rejection of Christianity, there came an equally violent acceptance. He read the Bible, resolved never to do wrong again, and promised God to "remember the Sabbath and keep it holy, and also to read religious books."

He decided to prepare for the ministry. By his thirtieth year he had graduated from Yale in medicine and theology, convinced that his life's work lay with the medical missions. A month after graduation he was sent by the American Board of Commissioners for Foreign Missions to Canton, where he arrived three months later—the first Protestant medical missionary to China. Leaving soon after for Singapore to learn Chinese, he returned to Canton in 1835 and on November 4 set up his Ophthalmic Hospital there.

The hospital quickly became an established institution. In a book published only three years later, C. Toogood Downing, a visiting member of the English Royal College of Surgeons, wrote: "There is nothing more interesting . . . in China at the present time than the examination of the Chinese Hospital in Canton." A wild enough statement, since at that time Europeans were unable even to walk outside the "factory" enclave where

1. This and the following quotations are from the biography by Stevens and Marwick.

the hospital stood, far less to examine any other area of China. "During my stay . . . I frequently visited the place and took great pleasure in affording any little assistance . . . to Dr. Parker, the missionary surgeon, whenever any operations were to be performed. . . . Passing along this crowded, dirty thoroughfare [Hog Lane] . . . you find yourself in a large hall paved with stone. . . . The hospital is merely a house. . . . On the first floor is the receiving room. . . . Around the walls are arranged portraits in oil and water colours of some of the most remarkable patients who have been treated here, with their different appearances before and after operation. Whether it be attributed to the skill of the native limners who execute these works of art, and who style themselves over the doors of their shops 'handsome-face painters,' I will not pretend to decide: but certainly many of these men and women appear as good-looking before as after operation, not withstanding the enormous tumours and awkward blemishes which have been removed."

Another contemporary, Henry C. Trumbull, described the doctor himself: "Wherever [he] was he was sure to be recognised and looked up to as a man of nobility and grace. He was of large frame and of imposing . . . presence. Intellect and character and experience showed in his fine face . . . winning the confidence of all to whom he spoke. He was a good illustration of the superiority of the missionary above ordinary men."

This flattering portrait seems to be accurate at least as far as Dr. Parker's medical activities were concerned. His hospital flourished, financially supported by American and British merchants in the city, not, one suspects, for altruistic reasons, but because the patients tended to be drawn from the ranks of the influential Chinese. The treatments that Dr. Parker provided for these people ensured at least praise for Westerners and tended to produce favorable reactions to Western commerce. Those eye cataracts, and other conditions not susceptible to cure by traditional Chinese medicine, that were treated in the Western enclave had their value. And indeed the hospital was a useful and unique institution, continuing long after Dr. Parker had left China.

The official reports of the hospital's work make interesting, sometimes amusing, reading. The sixteenth, for example, dealing with the year 1850–51, gives the number of patients treated since the opening of the hospital fifteen years previously as 42,528. "Every variety of ophthalmic disease, also tumours, hernias, dropsies, dislocations, fractures and stone have presented . . . and have been treated with the usual success. A mere selection from thousands of cases is given, interspersed with the sentiments of thankful patients illustrative of the favourable impression produced upon their grateful hearts, and demonstrating the divine wisdom of medical missionary operations."

Among the testimonials included in the report by Dr. Parker (whose sense

DR. PETER PARKER. (New York Public Library)

of humor appears to have been rudimentary) is the following specimen by one Sie Wan-kwoh: "I have been afflicted for several years with stone, and passing water was each time attended with no ordinary suffering; subsequently I applied to Dr. Parker to cut me open and put me right, and in less than one month, this disease appeared as a thing that is lost. Deep is the sincere gratitude I bear him, and I have composed this couplet [*sic*] as a record of my constant and lasting remembrance.

"By the stream and by the steel,
He can save and he can heal:
Life and health he can impart,
Be he honoured—and his art!"

We do not know the identity of Sie Wan-kwoh, but in the year 1839 the report of the Medical Missionary Society in Canton (founded by Dr. Parker) notes: "Case 6565. Lin Tse-hsu, the Imperial Commissioner." This is the same Commissioner Lin whom we met in a previous chapter attempting to suppress the opium trade at Canton and who, incidentally, requested Dr. Parker to make a translation of a book on international law and also to check the translation of the commissioner's letter to Queen Victoria. Lin suffered from hernia but refused to attend the hospital in person. To his request for medicine, Dr. Parker replied that no medicine would help and that Lin must wear a truss. The appliance was sent and apparently worn, for Parker notes later: "The truss answers tolerably well." Commissioner Lin, whose diary is normally rich in detail, unaccountably makes no reference to the episode.

In those early years in Canton the young doctor was kept fully occupied with his medical work. But he found time to interest himself, as a committee member, in the Society for the Diffusion of Useful Knowledge in China. In the second report of this body, published in 1837, there is an interesting comment, one rare enough in its time: "We have enumerated advantages arising out of such knowledge as we may impart to the Chinese. On the other hand we might also, it is not improbable, were we brought into constant intercourse with intelligent and well-informed natives of this country, derive much practical information, and hence derive considerable direct benefit, even from them. . . ."

Despite the patronizing tone, such a sentence shows a certain openness of mind, in which attitude we must suppose Peter Parker shared. Probably at this time he would have echoed the sentiments expressed by the renowned American missionary Samuel Wells Williams: "As a means, then, to waken the dormant mind of China, may we not place a high value upon medical truth, and seek its introduction with good hope of its becoming the handmaid of religious truth? . . . That enquiry after medical truth may be provoked, there is good reason to expect: for, exclusive as China is in all her systems, she cannot exclude disease nor shut her people . . . from the desire for relief. [And indeed Dr. Parker was at a very early date beginning to train several young Chinese to be doctors]. Does not then the finger of Providence point clearly to one way . . . directing us to seek the introduction of the remedies for sin itself by the same door through which we convey those . . . designed to mitigate or remove its evils? Although medical truths cannot restore the sick . . . to the favor of God, yet perchance the spirit of enquiry about it, once awakened, will not sleep till it enquires about the source of truth. . . . At any rate, this seems the only open door, let us enter it. . . . None can deny that this is a way of Charity that worketh no ill, and our duty to walk in it seems plain and imperative." Though Williams did not know it, medicine and school and university instruction were to prove the only doors to contact with the Chinese. Religion itself was to fail to touch them.

Young Doctor Parker was a success in Canton with Europeans, Americans, and Chinese alike. But the first Anglo-Chinese war, which we have noted in the foregoing chapter, loomed dark on the horizon. Before it came Dr. Parker sailed for the United States on a visit. In America he married a relative of Daniel Webster, who in that year (1840) became secretary of state. Already before he left Canton there had been tinges of political ambition in the doctor's activities, and doubtless his marriage eased his entry into the political field. In June, 1842, he sailed for China with his wife, who became the first foreign woman to reside in Canton. He sailed after a long series of talks with the American president. The Treaty of Nanking, signed between Britain and China in August of that year, was followed two years later by the advent of Caleb Cushing as first American commissioner in China. Cushing, largely through the services of Dr. Parker, who was appointed his secretary and interpreter, managed to extract America's first treaty with the Chinese—the Treaty of Wanghia. Dr. Parker's first operation on China now altered in character. The healing knife was replaced by a blunt instrument—power politics.

The treaty contained the same stipulations giving the same privileges to Americans as had been wrung from the Chinese for the British in the Treaty of Nanking. But it went further. The American document contained thirty-four articles, that of the British only thirteen. There were, in the words of Professor Carrington Goodrich writing in the *Encyclopedia Americana,* "four significant additions" not included in the Treaty of Nanking. "Americans engaging in opium smuggling would receive no protection from the United States; revision of the treaty would be permitted at the end of 12 years; officers and citizens of the United States would have the right to purchase books and to employ Chinese scholars for the study of the languages of the empire; and the article on extra-territorial jurisdiction was more definitely stated." Cushing also managed to insert a clause in which it was agreed that any privileges accorded by the Chinese to any other nation should at once be offered to the United States as well. And, incidentally, it was at this time that America received its definitive Chinese name. Previously the United States had been named the New Nation, and then the Flowery Flag Nation (after the stars on the flag). Cushing was responsible for the name A-mei-li-chia-kuo, which was later contracted to Mei-kuo (Beautiful Country). This form has designated the United States ever since.

Seldom is the life of a man so clearly divisible into two disparate parts as is that of Peter Parker. Before his marriage he was a doctor: after he returned to China he was a diplomat. During the thirteen years until he finally returned to the United States in 1857 he was, according to Tyler Dennett, the sole "element of continuity in the diplomatic relations of the United States with China." Until the last day of 1855, when he was finally appointed commissioner, he served as acting chargé d'affaires no less than five times. He was the only Western commissioner appointed to China who could speak and read and write Chinese.

His own opinion of his part in the Wanghia treaty negotiations marks his changed attitude to his life and to the Chinese. After lamenting perfunctorily the loss of time spent on medical work he continues: "Yet of the path of duty . . . I had not a solitary doubt. The hope was most sanguine that thus a providential opportunity was afforded me of not only serving my country, native and adopted, but of occupying a position in which I might do more in a few months . . . than by all the rest of my life. . . . To the divine praise be it now recorded that my . . . hopes have been exceeded in the results of these negotiations, by which a treaty of peace, amity, and commerce has been concluded between America and China on terms the most honorable and advantageous to both nations."

Perusing certain articles of this treaty of amity, considering how it was extracted as America sheltered under the skirts of the British navy, whose guns had permitted the signature of the Treaty of Nanking—it is obvious that Dr. Parker had graduated to the position of gunboat diplomat. The American historian Tyler Dennett, as we noted some pages back, calls the Wanghia treaty "the smuggler's delight." It was actually more, and better, for Americans, than that. The manner in which it was negotiated and the ease of the operation under virtual British protection were to have far-reaching effects not only on Peter Parker but on the course of American attitudes to, and relations with, China in the future. It was a significant, and ominous, start.

American missionary contact with China had been established for some time. The first Americans to arrive in Canton were Elijah Bridgman and David Abeel. Reaching there in 1830, they were joined three years later by Samuel Wells Williams. Bridgman founded and edited a journal called the *Chinese Repository*, whose pages contain invaluable and unique information on the era in Canton. Williams was a scholar, acted as U.S. chargé d'affaires in China at various periods until 1874, and eventually became the first professor of Chinese at Yale. These ardent Christian men, desirous of converting the Chinese at almost any cost (or so it would seem), thoroughly approved of the British war in China and "demanded that China must bend or break."[2] The American merchants, however, were on the whole not in favor of the hostilities. British action in Canton (to quote the words of one of them) was "one of the most unjust wars ever waged by one nation against another."[3] One may perhaps wonder if they thought this because the war interrupted their trade with the Chinese. Commodore Biddle, the American commissioner at Canton after the departure of Cushing, said in 1846, when the

2. Quoted by Tong from the *Chinese Repository*.
3. Both this and the following quotation are from Tong.

Nanking and Wanghia treaties were safely signed: "God has often made use of the strong arm of civil power to prepare the way for his own kingdom." Presumably he referred to the British power, since at that time America had not sent any aggressive ship to South Chinese waters. And we can presume that Peter Parker concurred in the statement. A successful doctor, he was, however, a very unsuccessful missionary, and ended as a discredited diplomat. His relations with the Chinese after his return from the States were frequently arbitrary and arrogant. With the British he was alternately fawning and quarrelsome.

Mission work in Canton under American aegis singularly failed. By June, 1854, twenty-four years after the landing of the first American missionaries, one of them was to write: "But we dare not say that we have satisfactory evidence that God has renewed a single soul in connection with our labor."[4]

In 1855, much exhausted, Dr. Parker returned to the United States. During his stay he was appointed United States commissioner in China. His arguments and ideas, coming fresh from a practically unknown land, appear to have had a considerable effect in the capital, and the doctor was invested by President Franklin Pierce with full powers to seek the revision of the Wanghia treaty due in the following year. The government asked him to obtain three concessions: first, permission to station a diplomatic officer at Peking; second, facilities for unlimited expansion of American trade in China; third, the removal of every restriction on personal liberty of American citizens in the Celestial Kingdom.

Returning to Canton, Dr. Parker sought in vain for an interview with the Chinese commissioner there. Once more he tried to coerce British and French representatives into taking "joint and energetic action." Among the barrage of proposals he put up at that time were two at least which incurred the ridicule of the European diplomats. One stipulated that the Chinese should reform their system of courts of justice, while the other proposed that China should station ambassadors in foreign capitals. These two ideas were completely lacking in authorization from his own government. They were never, in the outcome, presented to China.

Failing to get an interview at Canton, failing, therefore, to deliver the president's letter addressed to his "Great and Good Friend" the emperor of China, Peter Parker sailed northward along the Chinese coast, the presidential document and others of his own in his hands. He went to Foochow, where he managed to see the governor and to deliver the letter for transmission to Peking. The governor, however, wily man that he was, eventually returned the letter with its seals broken, together with a note to say it should be presented through the commissioner at Canton. Doubtless (though we do not know this) he had had the letter translated, and doubtless the con-

4. Quoted by Varg.

tents were soon known in both Peking and Canton. Dr. Parker regarded the breaking of the seals as a national insult and would have set sail impetuously for Peking had an American ship presented itself to carry him thither. Luckily, none did.

Meanwhile another war was brewing between Britain and China. The commodore of an American ship then in the vicinity foolishly sailed upriver toward Canton, taking soundings near some Chinese forts. He was fired on and at once ordered the fire to be returned, accomplishing the destruction of the forts. This action met with the full approval of British officials and American traders. But the Chinese, seeing the barbarians united against them, climbed down and offered their apologies for the incident.

Parker was jubilant. His dispatch to Washington contains the essence of a project which was to be his crowning folly. "Were the three representatives of England, France and America, on presenting themselves at the Pei-ho . . . to say the French flag will be hoisted in Corea, the English . . . at Chusan, and the United States in Formosa, and there remain till satisfaction for the past and a right understanding for the future are granted; but, being granted, these possessions shall *instantly* be restored, negotiation would no longer be obstructed and the most advantageous and desirable results to all concerned secured."[5]

To be fair, Dr. Parker added that such a program should only be put into effect as a last resort. The shock occasioned in Washington by this dispatch may be imagined. America, having so lately freed herself from the British colonial yoke, was hardly in the mood to assume a colonial stance herself. With the dispatch came other reports from England on the bellicose state of British relations with the Chinese in Canton. It was reported that the American consul in Hong Kong, James Keenan, had personally carried the United States flag over the wall of Canton when the British briefly occupied the city. Secretary of State Marcy replied at once that Parker was on no account to get mixed up with British quarrels, that he was to remove Keenan from office forthwith. Keenan, showing the spirit which doubtless carried him over the Canton wall, simply repudiated Parker's authority to remove him, and remained at his post.

At this time the Pierce administration in Washington was in its last weeks and little inclined to commit itself on any new China policy. Marcy's letter to Parker says the president hoped that the defense of American citizens would be assured "without [America's] being included in the British quarrel or producing any serious disturbance in our amicable relations with China." With Peter Parker in charge, relations were hardly amicable, but the secretary of state could not know that. Parker was left without specific guidance on the subject of the revision of the treaty. And, so it is said by a few authori-

5. This and later extracts from Parker's dispatches are quoted by Dennett.

ties, he had already heard a rumor of his impending loss of office. He seems to have determined then to press ahead with the plan to annex Formosa for his native land. The remainder of his time in China saw him frantically occupied in the attempt.

In February, 1857, he wrote to Washington: "The subject of Formosa is becoming one of great interest to a number of our enterprizing fellow-citizens, and deserves more consideration from the great commercial nations of the West than it has yet received; and it is much to be hoped that the Government of the United States may not *shrink* from the *action* which the interests of humanity, civilization, navigation and commerce impose upon it in relation to Tai-Wan [Formosa]. . . ."

After this Dr. Parker's dispatches followed one another thick and fast as annexation fever took hold of him. He summoned Commodore Armstrong from Hong Kong and succeeded in carrying that officer most of the way with him on the subject. They justified the projected annexation of the island "by the principles of international law, since grievances against the Chinese government justified reprisals; and because Formosa was an island most desirable to the United States." But it was admitted that occupation at the moment was impracticable and that such an action "in any other country than China . . . would be regarded as a virtual dissolution of avowed amicable relations."

Parker had by now discarded the idea that the occupation should be temporary. In a letter marked "Confidential," written in March, 1857, he said: "In the event of the establishment of a line of steamers between California, Japan and China, this source of supply of coal will be most advantageous. That the islands [Formosa and its small neighbors] may not long remain a portion of the [Chinese] empire is possible; and in the event of its being severed from the empire politically, as it is geographically, that the United States should possess it is obvious, particularly as respects the great principle of the balance of power."

Thereafter, as Tyler Dennett remarks, "the commissioner surrendered to his imagination completely." His next communication included the following flight of fancy, advanced in all seriousness: "Great Britain has her St. Helena in the Atlantic, her Gibraltar and Malta in the Mediterranean, her Aden in the Red Sea, Mauritius, Ceylon, Penang and Singapore in the Indian Ocean, and Hong Kong in the China Sea. If the United States is so disposed and can arrange for the possession of Formosa, England certainly cannot object." "Much reading of international law since the eye-doctor became the diplomat," says Dennett succinctly, "had made Dr. Parker a little mad." Thus he swept on, in his own mind at least, with schemes to occupy Formosa. His dispatches went by every available boat across the Pacific to a Washington which had by now made up its mind to disassociate itself from the mad doctor. When the Pierce administration was succeeded by that of Buchanan, Dr. Parker was dismissed and William B. Reed was appointed in his place.

Reed's instructions, quoted by Dennett, were explicit on the subject of American-Chinese relations: "This country, you will constantly bear in mind, is not at war with the Government of China, nor does it seek to enter into that empire for any other purpose than those of lawful commerce, and for the protection of the lives and property of its citizens. The whole nature and policy of our government must necessarily confine our action within these limits, and deprive us of all motives either for territorial aggrandizement or the acquisition of political power in that distant region. . . . You will not fail to let it be known to the Chinese authorities that we are no party to the existing hostilities, and have no intention to intervene in their political concerns, or to gain a foothold in their country. We go there to engage in trade, but under suitable guarantees for its protection. The extension of our commercial intercourse must be the work of individual enterprise, and to this element of our national character we may safely leave it."

Dr. Peter Parker now fades from the scene. Perhaps with a lingering hope that the turn of events might allow him to recoup his political losses, he made his home in Washington and spent the remainder of his life dabbling in such activities as were afforded him by the American Evangelical Alliance and the Smithsonian Institution.

The policy of the American government spelled out in Commissioner Reed's instructions was both admirable and clear. Freedom to trade and, although not explicitly stated, freedom to propagate the word of God were its main pillars. Almost every Western government whose citizens ever went to China would have subscribed to this very statement: and indeed there are instances of some which actually did, in words entirely similar. Most Western governments, including that of the United States, continued to say just this throughout the nineteenth century. But the statement was, and continued to be, naïve. It overlooked the problems which arose in achieving that freedom of trade and evangelism. It overlooked entirely the demonstrable fact that neither the Chinese government nor the Chinese people desired the Western Christian faith, and that the Chinese government did not desire Western trade for the simple reason that it impoverished China.

Already, in the treaties of Nanking and Wanghia, we have seen how the British got their way by going to war with China, and how America was quick to see and to take advantage of the British war without herself firing a shot. This pattern, by which European forces acted as spearhead to Western trade and missionary activity, and were followed by opportunist American seizure of the advantage so gained, was established in the time of Peter Parker. (The pattern, incidentally, was reversed in the case of the opening of Japan, where America led the way.) It continued with slight variations throughout the nineteenth century—until that day more than half a century later when, as a result of the Allied defeat of the Boxer Rebellion, the power of the Chinese government was finally shattered in all but name by soldiers of Europe, America, and Japan. The American historian John K. Fairbank, referring to United States policy toward China, sums it up in these words: "The contradictory idealistic and realistic elements in our China policy can be understood only if we remember that until the early 1920's our interests in China were junior to those of Britain, under whose leadership they had grown up. This allowed us the luxury of constantly denouncing British imperialism while steadily participating in its benefits."

But what of the American missionary effort in China? Making a scholarly start in the persons of Bridgman, Williams, and others, it cannot be divorced from Western missionary effort in general. Its aims and methods were all but identical with those of the European missionaries. The sole difference was that Americans poured into China as much, perhaps more, money than was contributed by all the other nations together. Some idea of the gargantuan scale of the American effort emerges from two simple figures. By 1930 there were more than three thousand American missionaries in China, and the annual total remittance from the United States to China missions was almost eight million dollars.

This commitment in men and money reflected the deep emotional involvement of ordinary Americans in the attempt to convert hundreds of millions of "benighted" Chinese to the Western God. In the end the effort failed as miserably as did that of all other missions. The success of Dr. Peter Parker in the strictly medical field, and the successes of later missionaries in medical and educational (especially scientific) fields in China, must stand as the only constructive and lasting result of all that money and human effort and ardent American hopes in China. The same can be said of the European effort.

There is, fortunately, another side to the coin of the missionary effort in China: though the missionaries may have contributed little to Chinese life, their efforts in the field of Chinese studies laid the foundations of our present-day knowledge of Chinese language and literature. Among them there were always a few dedicated scholars in the field of Chinese studies. We have already seen Matteo Ricci beginning the long and often painful process, and he was followed by many more—Americans, Englishmen, and other Europeans—of whom we shall mention but a single representative example. James Legge (1815–97) came to the Far East from his native Scotland in 1839 as a Protestant missionary and lived in Hong Kong for thirty years as head of the Anglo-Chinese College, where he undertook the monumental task of translating and preparing his well-known commentaries on the Chinese Classics, published in many volumes. In his later years he occupied

JAMES LEGGE, D.D., AND CHINESE STUDENTS. It was missionary-scholars such as Dr. Legge who laid the foundations of knowledge of the Chinese language. (Museum and Art Gallery, City Hall, Hong Kong)

the newly created Chair of Chinese Language at Oxford's Corpus Christi College. The Western world owes much to him and the other fine scholars who worked quietly in China through many long years.

As for the missionary effort itself, broadly and simply the process of its involvement in China can bè stated in the following terms: First there was coercion of the Chinese government, by Western diplomacy and Western force, to permit trading on Western terms; this was followed by agreements which admitted missionaries to varying degrees of freedom among the Chinese people. By successive agreements, all of which were extracted from the Chinese government by use of, or threat of, force, trading and mission activity spread until all but the whole of China was covered. Under the agreements the Chinese government was compelled to attempt to ensure the freedom and safety of the missionaries, and in so doing was equally compelled to suppress antimissionary and antiforeign sentiments and acts among the Chinese people.

The government, through acting against its own people and in favor of the Westerners, progressively lost its support among its own people. Apart altogether from the gradual economic bleeding of China by Western trade, it was this process, whereby missionaries and their (to the Chinese) hateful and alien ways of thought and belief were given free rein and power and judicial immunity in China, which was in large part the undoing of the Manchu dynasty.

It is a sad story, unpalatable to Westerners, incomprehensible to missionaries, bewildering to millions of Americans and Europeans who contributed their pennies in the belief that they were doing a Christian good in China. Little though they knew it, they were contributing a significant blow to the rain of Western blows that in the end prostrated China.

It is all too easy to be wise after the event, and Dr. Peter Parker can perhaps hardly be blamed for his bizarre excursions into diplomacy. To have singled him out from among others may seem invidious; but in following his story in South China we have prepared the canvas on which the next episode involving Westerners and Chinese was to take place, far away to the north in Peking.

PART FIVE

CHINA IN CHAINS

Vandals, Reformers, Heroes, and a Poet

CHAPTER SIXTEEN

LORD ELGIN AND WANG'S MOTHER

IN THE year 1860 the late summer in Peking was a fine one. The worst of the humid heat and the drenching showers that made nonsense of the drainage system were dying out. Another few weeks and it would be autumn. It was still warm under the pearly blue of the Peking sky. The breeze, when there was one, had nothing yet of the chill that would soon make cheeks tingle with its warning of the approaching Asian winter. Despite disturbances in the south (the Taiping Rebellion) and the humiliation of the capture of Canton two years previously by the foreigners, life in the capital went on much as it had done for most of the two hundred years since the Manchu armies had wrested it from the Ming.

In most ways Peking was still a medieval city. With the coming of night the big gates in the walls standing foursquare round the capital were closed. The watch was mounted on the gate towers under their triple roofs. The suburbs clustering outside each gate housed travelers who arrived after dusk, entertained them in their caravanserais and their eating-houses, pleasured them in their brothels. At dawn the gates swung open again. The traffic surged through—springless Peking carts with their jolted passengers, strings of camels loaded with coal from the Western Hills, with merchandise from all over Asia, pedestrians with their bundles, hawkers, beggars, loiterers—all entering the city for the commerce of the day. And, outgoing, came streams of people and vehicles bound for the suburbs and farther—the carriers of the Peking dead, the night-soil carts, the courier carrying an imperial decree wrapped in yellow silk, the riders with dispatches to the emperor lying ill at this time in the Summer Palace five miles to the northwest.

Inside the city about this time of the morning the bird-airers were leaving the courtyards of the houses they served, carrying covered cages, walking in the thick dust of the Peking streets to some place where the sun's first rays would soon slant over the gray roofs through the haze of cooking-smoke. In that ankle-deep and powdery dust (or mud, if it rained) here and there the sedan chair or curtained palanquin of some important person was borne by, moving expeditiously between the single-story houses of the *hutung* or narrow lanes, now passing beneath the florid ceremonial arches at intersections of major roads, its coolies trotting noiselessly, their shoulders taking the spring of the carrying-poles. Processions of the long, big-wheeled North China carts now began to pass, their gangs of hauling men chanting a heaving-song if the load was heavy. Now the beggars were rousing themselves from sleep in corners, scratching in their rags, gathering their rubbish, and shuffling off somewhere else.

As the sun topped the eastern city wall, reached over the corners of houses and courtyards, and pierced the trees, the bird-airers sitting with their cronies on some wall or other took off the covers of the cages, and the birds sang to the new day. Not far away one or two Peking-opera singers would be standing in the corner of two walls, practicing their voices in the thin dawn air—"lifting the voice," it was called. And in such quiet places there would be groups of young and old men performing the strange ritual of *t'ai chi ch'uan,* an ancient set of exercises in which their bodies moved in slow motion from one curious balletic posture to another. Some of them still wore the white summer gown, others had put on the long gray or blue gown, black cloth shoes, and dark cap of the colder months, against the possible chill of early morning.

Now the day was beginning, the volume of city sound slowly growing—the clang of a temple bell, the cries of wheeling birds, human voices, and the buzz of unidentifiable Peking sounds. Peddlers in the *hutung*—those blank-walled lanes in residential areas, pierced only by a doorway to each house, leading past a spirit screen to the first courtyard within—began to call their innumerable wares, striking little bells or hollow-voiced sticks of bamboo, clicking the metal instruments of their trades. The knife grinders, the trios of blind musicians with their two- and three-stringed fiddles and eyes as blank as the *hutung* walls, the match vendors, the sellers of thread and pins, the mattress makers, the repairers of pots, the metal workers, the man who replaced broken roof-tiles, the purveyors of potted plants—all now were out on their rounds.

It was the seventh lunar month of the Peking year (August by the Western calendar). At this time there was always a fair in one of the temples in the Outer City. The itinerant markets that set up for a day or two in the yards of various temples in succession in the city were changing some of their wares in anticipation of autumn. The lovers of crickets were looking forward to the arrival in Peking toward the latter part of the month of the "little golden bell"—the cricket whose chirp was delightful to the ear while one lay abed of a night, its voice thought to be so clear, hopeful, and somehow noble. But before that the young boys would have begun to collect lotus leaves and blooms from canal and pond, securing a candle in each; and as the evening of the fifteenth day arrived they would walk in little processions along the streets with their lanterns, singing:

Lotus leaf candles! Lotus leaf candles!
Today you are lit, tomorrow thrown away!

For the fifteenth was a Buddhist day, the lotus a Buddhist flower, and the paper boat burned in temple courtyards that day was the vehicle of the Buddhist faith.

More than a million people lived in Peking. As the morning warmed with the mounting sun, streets became busier, shops opened up—shops in which the produce of a whole vast and curious empire was gathered for sale. The teahouses had been open since dawn, frequented by the bird-airers on their way back, the birds hung up over their heads as the men gossiped with friends. The kitchens of famous restaurants here and there were already busy with the preparation of food for later meals. And at the eating stalls in many a street succulent smells rose to tempt the humbler passerby. A peddler would soon be on his rounds, bringing cooks to their doors with his cry: "Old chicken heads!"—for it was the season of water chestnuts, which resemble chicken heads in shape—"Old chicken heads, fresh from the canal!"

In the midst of this crowded and turbulent city of Peking there was embedded a second, the Imperial City, surrounded by its wall topped with yellow tiles. Inside the vast square of this enclosure was the great imperial park with its string of three lakes and their bridges, an artificial hill crowned by a white Tibetan *dagoba,* delicate lakeside pavilions and covered riparian walks, all the elegant simplicities of a dynastic *rus in urbe:* and much else. And within this Imperial City was yet another, the Forbidden City, enclosed by its tall red walls, studded with noble gateways, with corner towers of bird-bone fineness—concealing the sacred person and appurtenances of the emperor of China. Peking people knew little of what lay behind the rhomboid of the Gate of Heavenly Peace, had only the smallest idea of the vast courtyards, of the splendor of the pavilions, the apartments, the white translucent marble on which they stood, or of the life that went on there. With the ministers and with the hordes of palace eunuchs who were the servants of the Great Within, the people had almost no contact; with the emperor—none. But in the fine late summer of 1860 it was generally known that the emperor was not in Peking but at the Summer Palace outside the city, with, among others, the concubine Yi; and that he was in some apprehension for his own safety.

At this time in August a certain doctor of letters—Wang, let us call him[1] —an official of the Hanlin Academy and therefore privy to what went on in imperial circles, discovered that his old mother was ill. She had concealed

1. "Wang's" story appears in Bland and Backhouse, where his name is unfortunately not recorded. The quotations have been slightly adapted and lunar-calendar dates changed to Western ones.

this from him, but he noticed a prescription in the house and thus found out. "Five or six days after my mother fell sick, rumors began to circulate that the barbarians [English and French forces] had already taken Taku [a fortified place at the mouth of the Pei River]. There was a widespread feeling of uneasiness. . . . His Majesty was seriously ill but the concubine Yi . . . assured him the barbarians would never enter the city." Wang sent for his mother's doctor, in whom he had little confidence. "During the next few days people began to leave Peking . . . for the report was spread that our troops had been defeated at Taku . . . and were in the hands of the barbarians."

The date by the Western calendar was August 21, 1860. French and British troops were advancing to Tientsin, sixty miles nearer to Peking. They reached Tientsin between August 25 and September 5, without serious opposition. The crisis of the long, squabbling relationship between China and the West since the Treaty of Nanking had come at last. It was almost precisely eighteen years since the treaty was signed, and during that time neither side had observed either the letter or the spirit of the agreement. The English felt they had not obtained enough, and the Chinese resented that the English had obtained anything at all. Both sides felt the provisions were restrictive. In 1856 a minor difference over a Chinese trading vessel registered by the British at the new colony of Hong Kong was made the excuse for a quarrel at Canton. And this had culminated in the sack of Canton by British forces in 1858. Further treaties were exacted as a result—collectively called the Treaty of Tientsin, signed between the Chinese and the British, French, Americans, and Russians. The following year, 1859, British and French delegates on their way to Peking to exchange ratification papers were obstructed by the Chinese at the mouth of the Pei River. This led to a joint British and French expedition, under the eventual command of the diplomat Lord Elgin, to exact the ratification. It was this force that Wang heard had taken the Taku forts as his mother lay ill of dysentery.[2]

Lord Elgin, a Scots peer, descendant of the Scots hero Robert the Bruce and son of the man who rescued for posterity the matchless sculptures of the Parthenon frieze (commonly called the Elgin Marbles), was a sober, typically Victorian statesman. At thirty-one he had been governor of Jamaica, and then governor-general of Canada; he had declined a place in Palmerston's cabinet in 1854, and had gone to China in 1857, quelling a mutiny in India on the way, and had negotiated a commercial treaty with Japan, before returning to England. When he was again sent East to enforce ratification of the Treaty of Tientsin, he was already familiar with every

2. Easily the most concise and clear account of the years between the Treaty of Nanking and the culmination of the events we are now describing is that of Morse's *International Relations*. The works by Banno and by Teng and Fairbank give something of the picture from the Chinese side, thus balancing a confused and otherwise Western-angled story.

detail of Anglo-Chinese relations of the period. Behind his rosy face, which has been described as that of a bewhiskered cherub, lay a cool, calculating intelligence and a conviction that England was doing the right thing in China—and, indeed, in the East at large.

"On the 23rd September," Wang continues, "I noticed a change for the worse in my mother's condition, and straightway applied for ten days' leave of absence from my official duties. I kept her ignorant of the political situation and urged her to abstain from worry. . . . But every day the news was worse, and people began to leave the city in thousands." The allies were slowly approaching Peking. Already the imperial envoys had accepted the terms proposed to them by the allied force and already, according to the British, there had been a "breach of faith" on the Chinese side. Already the only battle of the brief war had been fought outside Peking, ending in the rout of the Chinese on September 18.

On September 24, Wang decided to take other medical advice. His mother was shortly afterward seized with a shortness of breath which frightened him, and he reverted to her own doctor, urging him to use "drugs less strong and more suited to a patient of my mother's advanced years. . . . My mother then bade me prepare her coffin . . . certain that her death was near. Fortunately I had bought the wood eight years before . . . and had stored it in a coffin shop in Peking." Now he had it fetched to the house, and carpenters set to work to make the elaborate casket required by tradition. Soon it was finished. "Never could I have expected that at such a time of haste and general disorder, so perfect a piece of work could have been produced." And next morning they began to put on the first of the many coats of lacquer. Wang then sent for the tailor to begin making the "ceremonial going-away dress" in which his mother would be interred.

She seemed a little better that day—perhaps the cooler weather helped her. But "rumors were now rife that the barbarians . . . were going to bombard Peking" soon. "Everyone was escaping who could leave the city. On that day [September 18], our troops captured the barbarian leader Pa Hsia-li. . . . " Harry Parkes, interpreter, not leader, was taken by the Chinese when he brought a letter acknowledging the earlier Chinese acceptance of the allies' conditions. Flung on the ground before a Manchu prince, he could only expostulate that he was merely the bearer of Lord Elgin's letter and not a negotiator—an attitude he maintained throughout his imprisonment. "The whole city was in an uproar," Wang says of this day. "And it became known that His Majesty was preparing to leave on a tour [!] northward."

On October 7 the allies demanded the release of Parkes and those captured with him, including the correspondent of the London *Times*. At the same time they stipulated that they must have the Anting Gate (in the north wall of Peking) surrendered to them as an earnest of good faith. The follow-

HARRY PARKES experienced much at the hands of the Chinese. Here, in 1856, as the departing British consul at Canton, he was ceremoniously bid goodbye by the "Hong" merchants, those Chinese designated by their government as the only ones with whom foreign trade could be conducted. Within three years he was held prisoner by the Manchu at Peking. (From the *Illustrated London News,* 1857)

ing day Parkes and some others were returned to the British and French, but in the meantime several of the prisoners, including the *Times* man, had been murdered by the Chinese. The next coat of lacquer was being applied to Wang's mother's coffin. Her doctor changed his prescriptions, but "the dysentery continued unabated."

Lord Elgin had already decreed that "the bad faith of the Chinese releases us from any obligations to restrict our advance." The day before the release of Parkes, the French and a few British had occupied the Summer Palace, from which the emperor had fled some days before toward Jehol to the north with his court and harem, including the concubine Yi. The official history of the dynasty remarks that he departed on an autumn tour of inspection.

Wang's mother remained much the same. He got another ten days' leave from his duties at the Hanlin Academy. Some days later she called him to her bedside. "I cannot possibly recover," she said. "See that all is prepared for the burial." Wang notes: "I felt as if a knife had been thrust into my vitals." Next morning "the confusion in Peking was hourly increasing, and huge crowds were hurrying from the city. Most of the gates were closed for fear of the barbarians," except one of those in the Outer City. "Our troops engaged the barbarians outside the Western gate. The van was composed of untrained Mongol cavalry who had never been in action," and they fled almost at once.

"Meanwhile my mother's condition was becoming critical. . . . Every official of any standing had either left the capital or was about to leave, and all the merchants who could afford it were sending their families away. The cost of transport was prohibitive for many. . . . In any case there could be no question of removing my mother, and there was nothing for it but to sit still and face the situation. . . . " In a few days she had "lost the power of swallowing, and at 11 p.m. she passed away, abandoning her most undutiful son."

With pathos that comes through the meticulous detail, Wang describes what he did and felt. "We arrayed her, then, in her robes. First a handmaiden put on the inner garments, a chemise of white silk, then a jacket of gray silk, and . . . a wadded robe of blue satin . . . then the robe and mantle of State, with the badge of her official rank, the jade girdle and necklace of amber. After the gold hair ornaments . . . the Phoenix hat was set upon her head; red mattresses were laid . . . and we placed her in a comfortable position . . . her head reclining on the 'cockcrow' pillow of red satin. Not a friend came near us . . . every door in the neighborhood was closed. Next morning I lined the coffin with red satin . . . padded it with straw to prevent it shaking, and at 3 p.m. I invited my mother to ascend to her 'long home.'

"The city was in a terrible tumult. A friend came in to advise me to bury my mother temporarily in a temple outside the city. It would not be safe . . . to inter her in the courtyard of this house, for the barbarian is suspicious by nature, and will . . . search every house in Peking as soon as the

A TYPICAL CHINESE COFFIN carried by four men. (From Favier)

city is taken. It was impossible for me to consider calmly what might happen if . . . they were to desecrate my mother's coffin. I remembered what has been told of their doings in Canton. . . ."

Wang sent his family out of Peking. "The stream of traffic . . . caused perpetual blocks in that gateway [the only one still open]. All the peddlers, hawkers, and barbers were fleeing . . . but still the large business houses remained open." Later he took the coffin to a temple outside the city. And on his return "there were but few people abroad, and these clustering together . . . and speaking in low voices. Suddenly, a little after midday an immense blaze was seen to the northwest and speedily it was reported that the barbarians had seized . . . the Summer Palace." Apparently there was some burning on the day the French began their looting there. Wang marvels that the half-million-strong Chinese army dare not oppose the barbarians' advance, even though the enemy have only a thousand cavalry.

On October 13, the Anting Gate was seized without trouble. The allies virtually held Peking. Victory was within their grasp. On the eighteenth, Wang says: "Vast columns of smoke were seen rising to the northwest. . . . The barbarians had entered the Summer Palace and after plundering the three main halls . . . had set fire to the buildings. Their excuse for this abominable behaviour is that their troops got out of hand. . . . After this they issued notices, placarded everywhere, in very bad Chinese. . . ." And he hears that the Sacred Chariot, as he calls the emperor's cortege, has reached Jehol.

Lord Elgin gave quite a different explanation of his order to burn the Summer Palace. He wanted to make a gesture which would remain hurtful to the Chinese, as a symbol of what he felt to be their intransigence, and at the same time as a symbol of British, or Western, power to exact the proper retribution. Charles Gordon, soon to lead the Manchu armies to victory against the Taiping rebels as "Chinese Gordon,"[3] had just arrived in Peking—too late, he says with regret, for the only battle of the campaign. But he saw the blaze. "After pillaging it [we] burned the whole place, destroying in a Vandal-like manner the most valuable property. . . . We got upwards of $48 apiece prize-money. . . . Imagine D— giving 16 shillings for a string of pearls which he sold the next day for $500. . . . The people are civil, but I think the grandees hate us, as they must after what we did to the palace. You can scarcely imagine the beauty and magnificence of the buildings we burnt. It made one's heart sore to burn them. . . . It was wretchedly demoralizing work for an army. Everybody was wild for plun-

3. He eventually went down in history as Gordon of Khartoum, the sentimental hero of the Victorian age. His generalship of the Manchu armies in quelling the Taiping Rebellion was in accordance with the British (and Western) decision that to keep the Manchu dynasty on the throne would best serve Western interests in China.

der."[4] He goes on to tell how the French smashed everything they could not take away. "It was a scene of utter destruction which passes any description." (But Gordon himself got one of the imperial thrones, which he eventually gave to his regimental headquarters in Chatham, England, where it still stands.)[5]

Another witness, more typical of the Western point of view, was the Reverend R. J. L. McGhee, whose book is called *How We Got to Peking*. "A temple . . . was in flames, and communicating destruction to the noble trees . . . around it which had shed their grateful shade over it for many a generation: its gilded beams and porcelain roof of many colors . . . all, all, a prey to the devouring element. You could not but feel that, although devoid of sympathy for its deity, there was a sacrilege in devoting to destruction structures which had been reared many, many hundred years ago. . . .

"Soon the wreath becomes a volume, a great black mass: out burst a hundred flames, the smoke obscures the sun; and temple, palace, buildings and all, hallowed by age, if age can hallow, and by beauty, if it can make

TWO PORTRAITS OF "CHINESE" GORDON. *Left:* in his mandarin robes, General Gordon is much less a romantic figure than he was made by the usual contemporary drawings (courtesy Royal Engineers' Museum, Chatham, England). *Right:* this cartoon from *Punch* of January, 1884, appeared when Gordon's success as the suppressor of the Taiping Rebellion was still his major claim to fame.

4. Quotations from Gordon's letters are taken from Hake.

5. Queen Victoria herself received, among other loot from the Summer Palace, a Pekinese dog, the first to be seen in the West. With a fairly typical insensitivity, she named it "Looty."

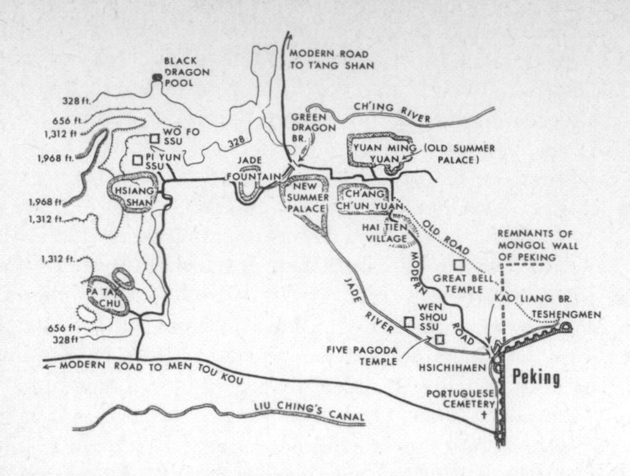

SKETCH MAP OF THE SUMMER PALACE AREA northwest of Peking. The contours of the Western Hills' lower slopes are shown and likewise the position of a few of the many temples built among their magically beautiful valleys. (Redrawn from Malone)

sacred, are swept to destruction. . . . A pang of sorrow seizes on you. . . . No eye will ever again gaze upon those buildings . . . records of by-gone skill and taste, of which the world contains not the like. You have seen them once and for ever . . . man cannot reproduce them."

But his pang of sorrow is soon tempered with another emotion. "Now back again in Peking, a good work has been done. Yes, a good work, I repeat it, though I write it with regret. . . . Stern and dire was the need that a blow should be struck which should be felt at the very heart's core of the Government of China, and it was done. It was a sacrifice of all that was most ancient and beautiful, but it was offered to the manes of the true, the honest, and the valiant, and it was not too costly. . . ." Even one of the murdered men's lives from the party the Chinese had captured, he says, was "worth it all."

No one at the time understood that this act of destruction was for the Chinese, then and for generations to come, a mere gratuitous insult. The Western powers were in any case about to humble them, not only by the series of treaties they were forced to concede, but by the occupation of Peking, the demonstration that China was utterly weak and could not even begin to defend her treasured way of life against the despised "barbarians." Neither Elgin nor anyone else seems to have understood this. All that can be said in the noble lord's favor is that he thwarted a French design to burn the Imperial Palace itself. At least the Chinese were left with this surpassing

RUINS OF THE SUMMER PALACE after it had been sacked by the French and British armies. The details show the baroque style of the part of the palace designed by the Jesuits. The photographs were taken in 1876 by Egerton Cleeve. (Courtesy Royal Photographic Society, London)

A Chinese Caricature of an English Sailor. When published in England, this was accompanied by an inscription, in poor Chinese, and an English translation: "This creature appears in the Tsing-teen-heen district . . . . Several troops of men surrounding it, it then changed into blood and water. Soldiers should shoot it with fire-arms, for bows and arrows are unable to injure it. When it appears, the people and troops should be informed that whoever is able to destroy or ward it off will be most amply rewarded. If the monster finds itself surrounded by soldiers, it turns and falls into the water. When it meets any one it forthwith eats him. It is truly a wonderful monster." (From the *Illustrated London News,* 1857)

example of the Ming architectural genius, even if they, and we, lost the unique example of Ming architecture, landscaping, and invented nature that was the Summer Palace.

Soon after, on October 13, Wang records: "The barbarians entered the city by the Anting Gate, occupying its tower and the adjoining wall. . . . With the exception of those . . . negotiating, not one official remained in the city. . . . Two days later I managed to leave by the western gate where I was nearly crushed to death in the enormous crowd." Placing a "nice wadded cover" over his mother's coffin in the temple, he returned to Peking. But not before he had seen a high government official hiding there with his chief concubine, both wearing poor people's clothes for disguise. "This is an example of the condition to which even the very highest had been reduced. The barbarians were now in full possession of the city. . . . Everyone in Peking—and there were still a good many people—was terrified." Manchu men were sending their women away to save them "from being outraged by the barbarian bandits. The condition of the people was deplorable in the extreme. . . ."

During all this time, as background to the warlike acts of the Western forces and the rearguard action of the Chinese, which was viewed by the Western commanders as treachery and intransigence (when in fact it was the only way the Chinese had in their power to stave off the West)—a long and complicated series of communications flew between the two sides. Eventually the Chinese capitulated on every point. There was nothing else they could do. An enormous indemnity in money was exacted for the murder of the prisoners. "The whole sixteen articles of the . . . demands," Wang says, outraged, "have . . . been accepted without modification. The only thing that our negotiators asked was the immediate withdrawal of the invading army, and to obtain this they are prepared to yield everything.

LORD ELGIN'S ENTRY INTO PEKING, October 24, 1860. The sketch, made from the roof of the Anting Gate, shows Peking looking very much as it did until a little over a decade ago. (From the *Illustrated London News,* 1861)

Therefore, the barbarians openly flout China for her lack of men. Woe is me; a pitiful tale, and one hard to tell!"

Woe, indeed, for China. Lord Elgin entered Peking with utmost pomp on October 24, passing through those ancient streets now for the first time (but not the last, had Peking known it) lined with British soldiers. The Convention of Peking was duly signed, the Treaty of Tientsin duly ratified. Next day Peking had its second opportunity of witnessing a conqueror's entry as the French, not to be outdone, entered with similar panache and signed, and accepted ratification, too. The subjection of China was all but complete.

The allied forces soon withdrew, partly because they were unable to live together in peace. The emperor never returned from Jehol. He died there, a dissolute wreck who was no loss to China, leaving a first-class dynastic dispute which was no help to his country either. From its toils there emerged the evil genius of the Manchu dynasty—Concubine Yi, who was later to be styled Empress Dowager.

Among the new territorial and commercial advantages now conceded to the British were: the opening up of the Yangtze River ports to trade (virtually allowing complete penetration of the Chinese hinterland), the cession of Kowloon Peninsula (the mainland adjacent to Hong Kong), the freedom of missionary and other travel and settlement and preaching in China, and the legalization of the opium trade.

But there was another concession. The Western powers were permitted to station in Peking resident ministers to represent their interests. And along with this unprecedented step went the formation of the first Chinese equivalent of a foreign affairs ministry, called the Tsungli Yamen. Its formation was the direct result of the long exchange which led to the capitulation at Peking; and with its birth China had been forced to respond in a Western manner, for the first time in her history, to a Western pressure. For her own protection it was a step she could well have taken several hundred years previously.

From the time of the Treaty of Nanking, and even more definitely after

PAYMENT OF THE CHINESE INDEMNITY. The first installment of the huge indemnity in silver extracted from China by the Treaty of Tientsin here arrives at Tientsin to be shipped out of the country. Was the artist, M. Claret, making an editorial comment when he gave such superior airs to the European officers depicted? (From the *Illustrated London News,* 1861)

British Opium Ship, 1860. The new treaty had legalized the opium trade, which was carried on in the latest-model ships such as this, for which the original caption proudly reads: "The new clipper steam-ship 'Ly-ee-moon,' built for the opium trade." (From the *Illustrated London News,* 1860)

the events in 1860 just described, the role of the individual in the conduct of Sino-Western affairs diminished sharply. Governments, through their ministers, took over. They dealt with the new, raw, often bewildered Tsungli Yamen, whose inexperienced officials did their best, often, to preserve what rights the Chinese still retained against the impositions and the pressures of the West.

This is not to conclude that the individual—Westerner or Chinese—did not matter. In fact the first British minister to Peking, Sir Frederick Bruce, wrote: "I should get on much better with this Government and make more progress toward a better order of things, were it not for the idea entertained by Admirals and Consuls that they may use force whenever they deem it expedient."[6] Individuals were frequently obstreperous, but finally it was governments and their attitudes which counted—just as was the case between other states. Relations with China, from the Western viewpoint, had been normalized as between any other two states. In the diplomatic field it was a better arrangement; in the cultural, it was disastrous.

The Chinese, however, did not regard relations as normal. The acceptance of the relationship had been thrust on them. Gradually they saw it had its points, but they never got over the fact that it was a shotgun mar-

6. Quoted by Oliphant.

riage, a disaster. For, in her own estimation, China could no longer afford to be the Middle Kingdom, center of the world. The West had demoted her to the status (which in some aspects, but not all, China recognized in private as true) of a backward, impoverished country. The trauma, in China, went very deep. It was gradually mortal. The height from which China fell, as far as her power among the nations of the world was concerned, was an eminence (by the mid-nineteenth century) of her own imagination and not of fact. But that did not soften the reality of the blow. The West failed utterly to understand this simple fact, this terrible hurt it had administered.

But there was much more to come—an agonized intermingling of dynastic and social convulsion and the determined struggle of the West to remain in the profitable though miserable land of China. One wonders if Wang lived on long enough to see that his "pitiful tale" was far from done.

CHAPTER SEVENTEEN

MRS. ARCHIBALD LITTLE

Unbind Thy Feet!

THE INTREPID Mrs. Archibald Little—Alicia was her Christian name—was not the first European woman to enter Peking, nor the first, either, to travel widely in China. But she was one of the first intelligent European women to do so and to favor posterity with books on the subject.

She had several obvious handicaps. First, she was a woman in a man's world, both East and West. Second, she knew remarkably little about the background to the Chinese life into which she intruded her revolutionary ideas. She was, for example, all but totally ignorant of Chinese history, imagining that Kublai Khan built the Peking she visited; and she knew almost nothing of Chinese language, art, or philosophy.

Fortunately for her, she also had a few advantages. She had honesty and directness matched by sympathy and a kindly nature which went a long way to compensate for her lack of knowledge. And, above all, she had immense enthusiasm—something so un-Chinese, so extraordinary from the point of view of the Chinese, that it was her greatest asset in her dealings with them. At those many public meetings which she addressed, she carried the Chinese with her in gales of laughter, without perhaps realizing that a high percentage of that mirth was simply embarrassment.

If her books make pleasant enough reading even today, it is because they can be read with a wry smile. But now and then, with a kind of innocent penetration to the heart of the matter, she puts down a comment which, with its horse sense, cuts immediately through the fog of Chinese politeness and the mist of European nineteenth-century romanticism.

"I see the scarred bodies of the men [gangs of coolies hauling the Peking carts along the rutted roads], the prematurely aged faces, their rough, rude manners. I have seen the sickly faces, the diseased heads and bad eyes of the children for whom the Sisters of St. Vincent care. People say the Chinese

MRS. ARCHIBALD LITTLE in a heroic pose. The original caption for the photograph read: "Excelsior!" (From Little's *Land of the Blue Gown*)

do not suffer, but laugh and are light-hearted. People said just the same thing of the negro slaves. They also laughed."[1]

And again, in her biography of Li Hung-chang: "With all the nations of the West contending who is to have its bones to pick, it is necessary that some nation or nations should in the first instance stand by China. But once let some great Western nation make it plain to the world that he who attacks China attacks her, and there will be no attack. And let China's feet but once be firmly set in the ways of progress, and there will be no going back.

"I conclude with the words of the man who, I believe to be the wisest statesman of the day. . . . Lord Salisbury in June, 1898, said, 'If I am asked what our policy on China is, my answer is very simple. It is to maintain the Chinese Empire, to prevent it from falling into ruins, to invite it into paths of reform, and to give it every assistance which we are able to give it, to perfect its defence or to increase its commercial prosperity. By so doing

1. Unless otherwise indicated, all quotations from Mrs. Little's writings are from *The Land of the Blue Gown.*

we shall be aiding its cause and our own.' . . . Where and when, may I ask, has the British Government acted on this policy laid down by the Prime Minister with the strongest following of any Minister of modern times?" For its times, that is quite a perspicacious statement.

The first chapter of her best book, *The Land of the Blue Gown,* all unconsciously tells a good deal about her character as a woman. It opens:

"On returning from Peking I still thought it the most wonderful place I had ever visited. On reaching Tientsin the first thing we saw was the then newly arrived Thevenet steam engine and rails, and . . . it seemed as if we had traversed centuries since, three days before, we rode stumblingly through the Peking gates. Steamers were shrilly whistling in Tientsin, men hammering, blue-jackets encouraging their donkeys . . . along the Bund in true English style, and the fair White Ensign floating from a real live man-of-war lying off the consulate door. . . . Three days before long strings of two-humped camels were the baggage wagons. . . .

"Wearied of London, and perhaps somewhat overladen with the cant of the day, aesthetic, hygienic, and social-economic, I can imagine nothing more tonic . . . than a sojourn in Peking. . . . Even quinine is bitter in the taking." She expatiates on the beauties of the city, and then on "the after-effects of mingling with a people so democratic as not even to distinguish between Bostonians and others, and yet without one touch of Radicalism, and always ready to make way for Acknowledged Merit in the person of a mandarin with eight bearers!"

China struck her as wonderful in its neglect. "And how congenial such neglect is to the human heart is . . . sufficiently shown by the way in which it grows upon the Europeans in China. . . . I never but once heard a lady say she had been to a picnic on Sunday . . . till I came to China. Here it seems the rule rather than the exception. 'It is the men's one day for getting away,' they urge. But this could be urged with more force in smoky, foggy Liverpool or London. Where in all the world will you find the European churches so little frequented as in China? I am reminded of a Commissioner of Customs' [Sir Robert Hart's] remark: "The Chinese have done more to heathenise the English than the English with all their missions to Christianise them.'

"Looking at that huge caravanserai Peking, I wondered what the subtle influence was that even had conquered the conquering Manchus, for at first sight everything seemed so overpoweringly repulsive, so beyond all exaggeration disgusting, that one would have thought [it] would rather serve as a horrible example. Does the common saying: 'The Chinese care for nothing but money, talk of nothing but money,' explain it all? So far I could not make out it was anything else the Europeans wanted to get out of the Chinese. . . . Even the very missionaries sent to teach that 'the love of

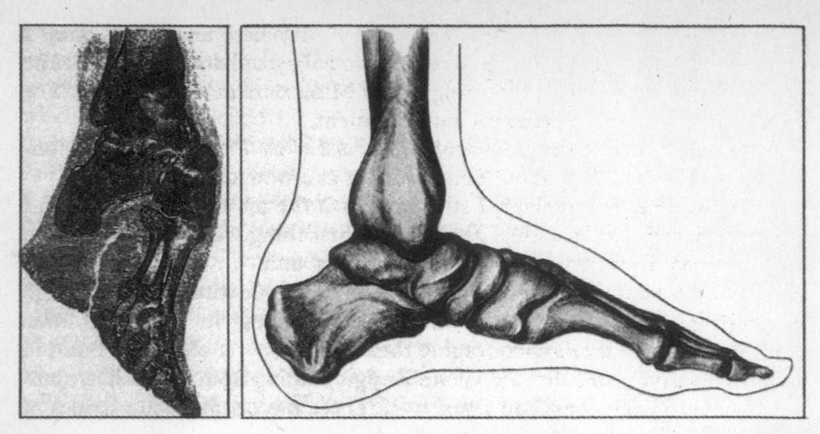

X-RAY OF A BOUND FOOT taken in 1900. The actual length of the foot from toe to heel is exactly four inches. In the drawing of the normal foot at the right, the upturned U-shaped line superimposed on the X-ray would be almost a straight horizontal line. (From Chinese Maritime Customs)

money is the root of all evil' seemed in many ways to have caught the infection. . . . Each nation gets accustomed to its own shortcomings, has wide-open eyes for [those of] its neighbours."

Mrs. Little belongs to a school of popular writers on China whose influence in Europe and America was probably much greater than sinologues would like to acknowledge. There were hundreds of such writers, beginning from the seventies of last century, when the events of 1860 in Peking had opened the country to all and sundry from the West. Most were impelled to write their lengthy, portentous books by what for them was the intense curiousness of the China they found, by the avid Western interest of the times in strange places and peoples and, in the case of the women writers, by the fact that almost none of them had anything better to do in the long Chinese days while their merchant and diplomat husbands were attending to the affairs of men. There is, of course, a separate group of writers from that time—the missionaries of both sexes—who, if we may follow the opinion of the learned sinologue R. F. Johnston, demonstrate how the opening of China to the Protestants brought to China hosts of people of little formal education. Mrs. Little's formal education (especially in Chinese subjects) was certainly small enough, but at least she spares us the cant. Her real forte was that she found a cause in China for which there was much Chinese sympathy. And it was a cause unconnected with Western commercial or territorial aggrandizement.

She came to China from her native England in the late 1880's. With her husband she spent the best part of the following twenty years in the country.

Mr. Little appears to have made a sufficient fortune in business in Shanghai, and, unlike most merchants, having done so, he decided to stay on. He turned into a kind of merchant-explorer with a spirit of adventure fully shared by Alicia. Together they navigated the first steamboat up the Yangtze River to Chungking. Together they lived on a farm some distance out of that city. Together they explored and enjoyed both the delights and the exasperations of China in the last two decades of the nineteenth century.

By 1895, Alicia was an Old China Hand, as the saying went. She had seen much, noted it, drawn her often superficial, but sometimes sensible, conclusions about the Chinese and their way of life. Her concern was, naturally, the women. And in her view one of the graver abuses to which Chinese women were subjected was footbinding. In her zeal to alter this barbarous custom she appears not to have noticed the lamentable position of women within the Chinese family system; but perhaps this was not surprising since in Victorian England the position of women was also rigidly set and confining, although not so cruel as that of their Oriental contemporaries. The other fact of which Mrs. Little apparently remained blissfully unconscious was that footbinding had a directly sexual significance. Luckily for her, nineteenth-century Chinese did not suffer from European taboos on the subject of sex in public discussion, or at least not to the same extent.

The origins of the custom are mysterious, but it seems to have begun at least a thousand years ago.[2] Girl children at a tender age—often at five or less—had their feet tightly bound with bandages so that the four little toes were compressed under the sole of the foot. Later the foot was bound in a different way so that the heel was brought closer to the ball of the foot, causing overall shortening. The feet were so deformed as to be mere irregular appendages of the legs, the skin shriveled, the bones of ankle and arch crushed into a lumpish shape. The muscles of the forelegs atrophied. The "golden lilies" (the Chinese term) were then squeezed into richly embroidered shoes which would have fitted a normal girl of three or four, from which they peeped seductively under the hem of an equally richly embroidered gown.

The results of this process were torture for all little girls, and often for grown women. But the tortured women had achieved the sexually arousing three-inch shape which sent the average Chinese male into transports of lust, and their chances of making a good marriage were greatly enhanced. The custom was a direct consequence of the Chinese family structure in

2. The best guide to the subject of footbinding is perhaps that by Matignon, a doctor working in a Catholic mission clinic in Peking at the time Mrs. Little was making her antifootbinding tour. Dr. Matignon gives firsthand clinical accounts of the process of binding and its results—both physical and psychological. His conclusions, with adequate documentary evidence in support, on the sexual significance of bound feet are highly interesting. See also the recently published work by Levy.

which a surfeit of girls was financially disastrous. The girls who lived through the process (and it is estimated by more than one Chinese that one in ten died, while others found their feet had rotted off) could only totter, supported at least by a stick, or usually by two maids—while young Chinese bloods observed their pathetic and painful movements with the mental equivalent of what is now called the wolf whistle.

Lord Macartney, commenting on the fetish with his usual dry humor, remarked that he could hardly complain too much about "squeezing . . . women's petitoes into the shoes of an infant," since "it is not a great many years ago that in England thread-paper waists . . . and tight lacing were in high fashion, and the ladies' shapes were so tapered down from the bosom to the hips that there was some danger of breaking off in the middle at any exertion." And a glance at a photograph of Mrs. Little herself would appear to confirm that the great antifootbinder was much constricted at the waist in deference to a later recurrence of the same fashion.

There had been several attempts to suppress the Chinese custom, which was never practiced by Manchu women—a notable one by the emperor K'ang-hsi—but all of them failed. In 1895 a band of ten Western ladies formed the Natural Foot Society and elected Mrs. Little their president. They at once drew up a memorial on the subject in the proper Chinese manner, which was sent by the American minister in Peking to the Tsungli Yamen (the Chinese foreign ministry). On the memorial were the names of almost all the foreign women in China. But the ministry felt squeamish (for reasons not explained) about submitting it to the Dowager Empress. The society was ferociously active, sending letters to viceroys and provincial governors, holding meetings in many big centers during which Mrs. Little and other ladies addressed huge audiences of women, and sometimes of men too. This in itself was almost a revolution—for never before in the history of China had a woman (more especially a foreign woman) stood up on a platform and talked to large numbers of Chinese, and very certainly not on such a delicate and embarrassing subject, related to sex and the most intimate life of the home, as bound feet. Mrs. Little was more courageous than she knew, because she was delving much deeper into the sexual life of Chinese men than she understood. She recognized that tiny feet were to them attractive, but in her book she seems to have classed the attraction with minor ones such as the fineness of the eyes or the excellence of the hair arrangement. This was, of course, long before Freud.

With her husband's connections, she had no trouble in arranging a free passage on the ships of the China Merchants' Company up and down the coasts of the country and along the Yangtze as well. Off she set, loaded with tracts and with little pieces of paper for people to sign, pledging their adherence to the Natural Foot. Her campaign was not only laudable, but also a great success. Her progress (and she is too artless to conceal a failure) was an uninterrupted series of vast public lectures to enthusiastic audiences.

"Those who remember their sensations as children when first taken to plunge into the cold sea, can realize . . . the feeling with which I contemplated starting off on a tour . . . among complete strangers to oppose footbinding, one of China's oldest, most deep-rooted, domestic customs."

Her first stop was Hankow, one of the cluster of three cities now called Wuhan, on the Yangtze. Here she hired the largest hall. "The audience began to come in, official after official, some with retinues, some without, some . . . with that tremendous swagger that makes one feel as if the man who thus walked could think of no subject in heaven or earth worthy of his interest. . . . Then I had to stand up . . . realizing to the fullest extent exactly how strange . . . these Chinese officials must consider a woman addressing them at all"—especially on that indelicate subject, women's feet. "I did not wonder my Chinese interpreter's courage gave way, and he became to all intents and purposes voiceless, for to him these officials were far more awe-inspiring personages than even to me." The audience took away over two thousand leaflets against footbinding and were asking for more. This was partly due to the fact that the viceroy of Wuchang, across the river, had made some encouraging remarks in print about the campaign, which were placarded in the hall.

Next she held a meeting, in Hanyang, of women who were supposed already to have been convinced on the subject of not binding. On asking how many had unbound their feet, she was astonished to find the whole audience rose. No question, the women of China were ready to unbind. They were only waiting for the chance.

Soon she was in Canton, where she had an interview with the viceroy—one of the great men of nineteenth-century China—Li Hung-chang. It is probable that in this she made history, for it seems unlikely that a viceroy had ever given an interview to a woman, a foreigner, before. With the viceroy's adopted son and an interpreter, she entered the great man's presence.

"A most imposing figure, six foot four in height, clad in an ermine-lined gown down to his feet, and a beautiful sable cape, with diamonds actually in the front of his cap as well as on his fingers. This at least my missionary friend told me with an American's quickness of sight for diamonds. . . ." In spite of the fact that she disapproved of him, "in but a few minutes he had so entertained and charmed as to have quite disarmed me. Footbinding . . . was of course the subject he tried to avoid." Li Hung-chang tried talking about her husband, but he had underestimated the determination of Mrs. Little—to whom her campaign was more important than all the husbands in China. Finally they got round to the topic.

"'No, I do not like to see little children suffer over having their feet bound,' grumbled out the genial viceroy. 'But then I never do see them,' he hastened to add.

"'I can tell you,' his adopted son said, 'of one who never has been and never will be, my own little girl.'" Li Hung-chang apparently thought it discreet not to seem to hear this as he proceeded: "'And you want me to unbind the feet of the women of China? Now that is really beyond my powers.'"

But from him she extracted a piece of the great man's calligraphy on the subject, which he wrote on her fan. It was one of the most useful weapons she had for the future.

"'You know,' Li said in parting, 'if you unbind the women's feet you'll make them so strong, and the men so strong too, that they will overturn the dynasty.'"

Li Hung-chang was quite wrong, and knew it. It was finally internal rebellion and the depredations of the European powers which overthrew the Manchu. But Mrs. Little certainly did not realize he was teasing her.

At Hong Kong, her next port of call, the City Hall was packed and her speech was greeted with much laughter, as was generally the case. Probably rightly, Mrs. Little remarks: "Chinese . . . think signs of sorrow unbecoming, and being affected must show it in some way." And later, the governor's lady threw open the doors of Government House for the first time to Chinese women for another meeting. This was history, indeed. Later still, Mrs. Little lectured to the Chinese youth of Queen's College in the colony. Over five hundred boys and young men sat at her feet. With what feelings they listened as she talked on about this indelicate subject we can only conjecture. Through their minds, doubtless, being young men like any others in the world, floated memories of those many pornographic Chinese drawings and paintings which demonstrated the use of the courtesan's bound feet to stimulate even the most jaded of her customers; and other matters of a similar nature which our modesty, like theirs, must suppress. No wonder they "applauded with such long-continued waves of applause and laughed such echoing, rolling peals of laughter, that it was almost impossible to get on with what I was saying." In the end, after seeing X-ray pictures of bound and unbound feet,[3] and also Li Hung-chang's calligraphy on her fan, the youths "simply stormed the platform, carrying away the railing. . . . It must surely be a long time," Mrs. Little suggests, "before Queen's College forgets that afternoon." Translating the scene into

3. It should be remembered that no Chinese man ever saw the bound feet of any woman in the unbandaged state, except those of his wife. Hence the boys' excitement at the pictures displayed. Even in the pornographic pictures common in China at this and former times, the feet are almost always shown rather dimly through some gauze-like material covering them. It would have been hard to hit on a subject more titillating for schoolboys.

FOOTBINDING. The nineteenth-century Chinese drawing at the left shows a woman binding her feet, after which she would have put on shoes like those at the right. From the point of view of the foot-fetishist (a term that applied to most Chinese males) these beautifully embroidered, tiny articles of female footwear were almost as exciting as a sight of the lady's actual feet. (From Matignon)

equivalent terms for Europe in the late nineteenth century, what Mrs. Little had been doing at the boys' college was to give something like a detailed lecture on the evils of sexual life in Parisian brothels. After their initial surprise, the boys' enthusiasm was not unnatural.

But with women, and with men of more sober years, her success was phenomenal. The remainder of an exhausting tour through the southern provinces of China was nothing short of a triumph. Certainly it was *her* triumph; but it demonstrated too the basic fact that Chinese women were just waiting for the chance to change the system of unholy torture which a man's society obliged them to undergo in order to please. The smallest thought on the terrors, the long toothache pains, of young girls as their feet were slowly crushed in the bandages makes it obvious that Mrs. Little was lighting the fuse to a very explosive substance. The bomb that went off was slow in detonating, damped down by tradition, by fear of financial or emotional privation, by innate conservatism on the part of the men. But the explosion, once begun, could not be stopped.

Soon a Chinese society was formed, called, literally, the Not Bind Foot Society, and the women of China gathered together and unbound. It was not quite so fast or so simple as that. But eventually all viceroys and governors issued proclamations against the practice. In 1902 the Empress Dowager, probably because at that moment she wanted to ingratiate herself with the ladies of the foreign diplomats in Peking, issued the last of the several recent Manchu edicts which had tried to discourage the practice. For various involved reasons the custom lingered on in several segments of Chinese society in parts of the country, and it is still possible to see a few women here and there in China and in Hong Kong who are probably not much over fifty and whose feet conform to the excruciating perfection of the old "three-inch golden lilies."

But we must salute Mrs. Little. Her efforts and the society of which she was president certainly led the way. How much influence they eventually had on altering this horrible custom and liberating Chinese women has not yet been definitely assessed; but it must have been considerable. Certainly no other European woman or group of women in China had any such fundamental effect on any other aspect of Chinese life. At a time when their menfolk in Peking were about to deal the final blow to the Manchu dynasty and inflict the final humiliation on Chinese people at large, chasing the court out of Peking for the second time, the efforts of Mrs. Little and a few others come as a welcome light in a darkening world.

CHAPTER EIGHTEEN

DEFENDERS OF THE FAITH, 1900

WE ARE coming toward the end of our story.

The very thing the Chinese always feared would happen had happened, almost exactly as their better minds had predicted: to concede any privileges to Westerners in China would bring hordes of them, and these hordes would be followed by Western armies. From the time of Pirès, from the time of Ricci, from the time of Weddell and later the East India Company at Canton—the same Chinese fear had been voiced. While they were in a position to do so and could deal with the West on their own terms, the Chinese either excluded or circumscribed the Westerners who came knocking. But the moment the Chinese gave way a little, the West poured in. And then there was no way to stem the flood. The West brought its technically superior armies and navies to back up its merchants and its missionaries. The West brought its armed might to protect what it then called its rights—rights created by those treaties it had extorted, by force or by threat, from a country able to offer only passive, evasive, "breach of faith" resistance.

As to the missionaries, there is ample evidence of the depredations of some of them, Catholic and Protestant alike. Doubtless the evidence should be balanced by many instances of charity and humility—but, alas, in the Chinese memory what stands out is the less worthy side. If we choose a single example from among the stories relating to Catholics, it is not because Protestants were above such conduct, but simply because the Catholics had been longer in China, and to that extent must bear a larger part of the blame. It is well also to keep in mind that the Chinese, who were not attuned to Christian theological arguments, regarded both groups as a single entity and judged them simply as representatives of the Western faith.

A responsible Chinese of that time[1] wrote: "Szechuan was the main bastion of Roman Catholic Christian endeavour into Central Asia. From Chengtu French priests went on long treks into Tibet and Mongolia. Their

1. A Western-trained engineer, the father of Han Suyin, who quotes his remarks in *The Crippled Tree*.

courage and ability to endure hardships might have been praiseworthy had it not been only too clear that the religious garb covered most unreligious actions. Catholic priests and bishops bought up whole villages in times of flood and famine, demanded and obtained on threat of military action the best land in cities for their churches, after evicting the inhabitants and paying no compensation. . . . Bishops were invested with the pomp and power of governor-generals. They used sedan chairs with eight carriers, a drummer going in front, and everyone in the street where they passed had to stop work, stand up, and unroll their headbands in obeisance . . . on pain of being beaten with the 'heavy bamboo.'

"No one dared to go to law against a priest, or one of the converts; the priests were immune, like diplomats, and the converts claimed the same immunity; protected by the Church, the converts could do anything they pleased, and escape the due process of ordinary law, for they always asked protection from the Church against what they called 'pagan customs.' "

Once begun, the process of Western commercial and military domination and intimidation inevitably increased. By the turn of the nineteenth century the decisive moment came, partly through internal conditions in China, partly through the ineptitude of the Manchu dynasty, and, precipitately, through the actions of the Western powers in Peking. The decision turned not on the Manchu administration, good or bad, but on the military and commercial power of the West, which was greatly superior to any in China. By 1900, China was at the mercy of the West. Only the inability of the Western powers to agree among themselves prevented China from becoming a colonial patchwork like Africa. In the manner of a classical tragedy, the end was ineluctable. The West pressed its advantage to the ultimate. Not content to break, they ground under their boots whatever was left of China. The last scene is generally referred to as the Boxer Siege of the Legations. But it was rather more than that.

After the events already described, the Westerners' scramble for territorial concessions and commercial rights grew more and more uninhibited. The Taiping Rebellion had raged over the southern and central provinces of the country, and was not suppressed until 1865, leaving its toll of disruption and costing perhaps twenty million Chinese lives. As if that catastrophe were not enough for any country to suffer in a half-century, others followed fast. The French war of 1882–84 against the Chinese tributary state of Tonkin and the destruction of the Chinese fleet off Foochow had resulted in the establishment of French sovereignty over the whole of what was then called Indo-China. Then the Sino-Japanese war of 1894–95 brought a humiliating defeat for China, and the loss of Formosa. Along her coasts the Manchu government was forced to concede land and trading rights and sovereignty to Western governments and their agents and traders, whose activities (as the Chinese could plainly see) were directed to extracting as much money

THE EMPRESS DOWAGER photographed in 1903 by a Manchu nobleman whose hobby was photography. She is supported by ladies in waiting, who wear the Manchu headdress. The scene is a snow-covered palace garden, probably in Peking.

from China as possible. The edicts signed by the young emperor in 1898, aimed rather naïvely at sweeping reforms in administration and other matters, so alarmed the Empress Dowager and her party, as well as the Westerners, that they were swiftly annulled after ninety days, whereupon the

ANTIMISSIONARY FEELINGS were prevalent enough in China before 1900 to give rise to rough-and-ready satires on them in the popular puppet shows organized by the Boxers. Here the pig with its trotters on the edge of the miniature stage represents the missionary. (A drawing by P. Frenzeny in the *Illustrated London News,* 1900)

Empress Dowager took the conduct of state matters into her own hands. From that moment on, it was with this in some ways able but reactionary and passionately conservative woman that the Chinese and the West alike had to deal.

The Boxers, a name that was on everyone's lips in late 1899, were a group that seems to have originated in a much older secret society whose name derived from their "Chinese boxing" type of semimilitary exercises. The complete picture of how and exactly when they arose has not yet been elucidated. Certainly in the tangle of fact and superstition surrounding their rise, real and potent factors were the decadence and corruption of the Manchu administration in the provinces and the deep resentment caused by the giving away of rights to foreigners. The Boxers' aim was stated to be "to drive the foreigners into the sea" and "to protect the country."[2] At first, protecting the country seems to have included driving out the Manchu dynasty. But later, when Boxer power increased, the Manchu administration saw in the Boxers its own salvation, and joined forces with them. The Boxers

A BOXER. The stalwart, idealized figure in this published drawing by a Westerner named H. Koekkock is a surprising comment in the context of 1900, but of all Western representations of Boxers it is one of the best executed. (From the *Illustrated London News*, 1900)

2. Purcell's study of the Boxer Rebellion is probably the most enlightened review for the general reader.

then generally dropped their anti-Manchu stand. Whatever the acts of the Boxers and the Manchu authorities then and through the convulsion that followed in the summer of 1900, the provocation for these acts was very real. A large proportion of the Chinese populace, in the coastal regions at least, accepted the justice of the cause and supported the Boxers in spirit if not always in deeds.

To Europeans the most insupportable Boxer provocations were the murders of Christian missionaries and Chinese converts in late 1899. Boxer fury against missionaries was only just stronger than their hate of "secondary Westerners"—the Chinese Christians. They hated the converts because those converts were often afforded a measure of privilege and protection against their nonconvert fellows. Many were "rice Christians"—accepting the religion on the surface for the advantages it brought in food and special privileges with which the Western missionaries provided them. Naturally, converts were thought to be on the side of the Westerners.[3]

Western protests about the murders met with evasive governmental replies. Edicts on the subject of the Boxers were seldom intended to be carried out literally. In Western eyes the Manchu government had failed in its treaty obligation to protect the foreigners in China. But both missionaries in the country and ministers in the legations at Peking were incredibly slow in comprehending the magnitude of the movement afoot, far less its passionate hate of the foreigner. Europeans in China simply did not know how painful a wound in the Chinese flesh they had inflicted in the nineteenth century, or how every move they made pulled at the wound's raw edges.

Not until the last day of 1899, when a British missionary was murdered, did Sir Claude MacDonald, the British minister in Peking, send a dispatch (by sea!) to his prime minister in London, noting that the Boxers seemed to be antiforeign and recording his own and other ministers' protests to the Tsungli Yamen. On May 17 of the following year, Sir Claude informed the prime minister (by telegraph this time) that the French minister had told him sixty-one Roman Catholic converts had been massacred only ninety miles from the capital. Except to warn the Manchu government and report home, there was not much the diplomats could do about it immediately.

Some ministers had ideas. E. H. Conger, the American minister, had stated a little earlier that he thought a Western naval demonstration off the mouth of the Pei River would impress the court with the West's displeasure and with Western power, a demonstration which would force the government to suppress Boxer activities. If this could not be agreed on (and in the quarrelsome state of the Western powers in Peking and China then, such agreement was unlikely), he wanted extra guards brought up from the coast

3. Cf. George B. Smith: "In lawsuits between [Roman Catholic converts] and non-Christian people, the latter had or thought they had, no chance . . . there was general complaint of the constant interference of the priests in litigation."

to strengthen the defense of the legations at Peking. The guards would come from the naval vessels of several powers now off the Pei River. Nothing, he thought, could be better calculated to chagrin the Chinese court than for its own people to see it was unable to protect foreigners and keep order in its capital city. The Chinese, he said, would do *anything* rather than suffer such an affront to their dignity. His judgment of the Chinese and the situation was quite superficial—as was that of all the other ministers in Peking. Their demands that Boxers be made criminal offenders by Chinese law were tantamount to asking the government to annul all its edicts on the subject. In Manchu eyes the Boxers were a much lesser evil than the foreigners. The Manchu did not comply.

Monseigneur Favier, the Roman Catholic bishop of Peking, on May 19, 1900, informed the diplomatic body of the "known Boxer intention to enter Peking at a given signal." He asked Pichon, the French minister, that forty or fifty marines be sent at once to protect his cathedral, Pei T'ang. The diplomats, whose main task in Peking seems to have been to elbow each other away from national advantages into positions of disadvantage, at last woke up to the fact that they were *all* equally threatened. By May 28 they were requesting Chinese permission to bring up guards. Contrary to the opinion of Minister Conger, the Tsungli Yamen agreed after only a token show of reluctance. The first detachment (much bigger than the agreed size) arrived from Tientsin on June 1—75 Russians, 75 British, 75 French, 50 Americans, 40 Italians, and 25 Japanese.

How grave was the situation at that moment? The bishop's report had alarmed the ministers. But what actually made them call up the guards was not the murder of missionaries or converts (no Westerner had been murdered for five months previously) but a Boxer attack on the Western-owned railway. They were perhaps alarmed more for the safety of Western commercial interests and for their own personal safety than for anything else. It is arguable that the passage of many armed Western soldiers through the country toward Peking, and the flurry of military movement on the coast entailed by their dispatch, were more potent factors in sparking off what soon happened than anything else. The Taku forts on the Pei River were taken by the Western powers. War was then inevitable.

The ministers telegraphed their home governments asking that their respective navies take steps to defend the legations at Peking. A force of about two thousand troops left Tientsin on June 13 but was forced by the Boxers to fall back to that city two days later. The popular antipathy the expedition encountered was widespread. Everywhere there were riots. The telegraph line to Peking was cut. In the Western Hills outside Peking the British summer legation was burned. The chancellor of the Japanese embassy was murdered at Ch'ienmen, a few hundred yards from the embassy. During the afternoon of June 13 a large force of Boxers in their red headbands and sashes swept into Peking. Before nightfall many churches were ablaze, many Chinese converts massacred. The whole city was in a turmoil of revolt and fear and indecision.

Not least the court and its ministers. A deputation from the Tsungli Yamen called at the British legation to urge that the relief force from Tientsin be stopped. On June 17 the Chinese sent notes to all Western powers in Peking to the effect that since hostilities had commenced at Taku they could no longer guarantee the safety of ministers and their families. The foreign community was asked to leave *en bloc* in twenty-four hours and proceed under Chinese escort to the coast. Councils in the legations were divided, but eventually it was decided to resist this demand. No reply was received to the Westerners' note stating their refusal to leave. The German minister, Baron von Ketteler, left his legation to pay a call on the Tsungli Yamen a mile or so distant. He was shot before he reached it. Chinese troops opened fire on the legations exactly twenty-four hours after the demand that they should be evacuated had been received. Most of the legation quarters were then abandoned, and their inmates crowded for ease of defense into the large British legation and adjoining areas. Besides some 473 civilians—diplomats, their families and staffs, and others—and a garrison of 400 men, there were also about 2,700 Chinese Christians and 400 servants crammed into the small space. June 20 came, and with it began the celebrated Siege of the Legations.

Of all the incidents in Chinese history in which Europeans were involved, the Boxer siege of the Western legations in the summer of 1900 has spawned the most voluminous printed comment. Literally hundreds of books, in English, German, French, Japanese, Russian, Italian, have bludgeoned the world outside China into accepting their view of the events. Most of those books were published in the first few years of the century while their authors were still smoking hot with rage and with assorted antipathies toward China and the Chinese. The drift of most of them is that a handful of Europeans and Americans were unjustly attacked, and defended themselves with the utmost heroism against a fanatical horde of Chinese Boxers, thereby saving the honor of the West for all the East to see and applaud.

While parts of this view may be so, what does not appear in more than two or three of those impassioned volumes is the fact that the defense of the legations was not on the whole so heroic as the books make out, that the Chinese never seriously tried to take the legations by force, and, further, that during the whole episode from the murder of missionaries, through the not very heavy bombardment of the legations, and afterward, until more normal times came to Peking and China at large, remarkably few Westerners met their death at the hands of the Chinese. But during this same period many thousands of Chinese men, women, and children were murdered, raped, ravaged, robbed, deprived of home and livelihood by European

REPELLING A BOXER ATTACK ON THE LEGATIONS. Temporary barricades such as this were strengthened later in the siege. The officer commanding the Russian troops is Baron von Rahden. This is one of hundreds of heroic representations of the legation siege published at the time in the West. (From the *Illustrated London News,* 1900; a painting based on a photograph)

soldiery, who turned Peking into a long nightmare—in order, presumably, to uphold that same Western honor and, it should be added, on an official level, to preserve the last ignoble remnant of the Manchu dynasty in China.

Furthermore, those writers of siege books, relegating everything but the legations' trials to limbo, afford hardly a mention to the other siege at Peking which was going on with real fury at the same time. The siege of Pei T'ang—the North Hall, as the Catholic cathedral was called—displays something much more like the heroism attributed to the legations' defenders.

To see the cathedral siege in the round we must first of all look briefly at events in the legations. The first attack came on June 20. The legations were defended by 20 regular officers and 398 armed men of various national forces, augmented by about 125 civilian volunteers from among the assorted missionaries, diplomats, and others who had taken refuge there. They were

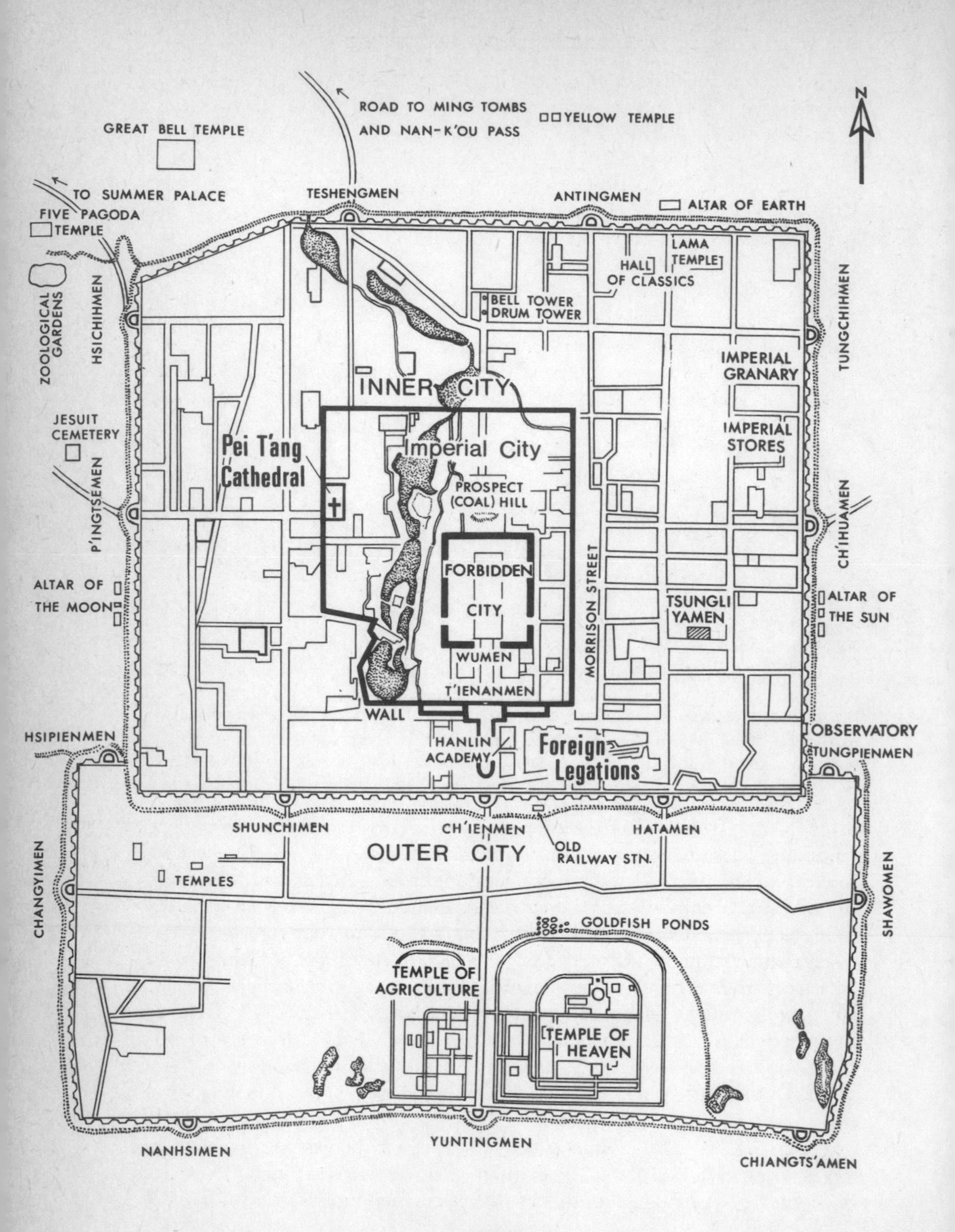
N
ROAD TO MING TOMBS
AND NAN-K'OU PASS
YELLOW TEMPLE
GREAT BELL TEMPLE
TO SUMMER PALACE
TESHENGMEN
ANTINGMEN
ALTAR OF EARTH
FIVE PAGODA
TEMPLE
LAMA
TEMPLE
HALL
OF CLASSICS
ZOOLOGICAL
GARDENS
HSICHIHMEN
BELL TOWER
DRUM TOWER
TUNGCHIHMEN
IMPERIAL
GRANARY
INNER CITY
JESUIT
CEMETERY
Imperial City
IMPERIAL
STORES
Pei T'ang
Cathedral
P'INGTSEMEN
PROSPECT
(COAL) HILL
CH'IHUAMEN
FORBIDDEN
CITY
ALTAR OF
THE MOON
MORRISON STREET
TSUNGLI
YAMEN
ALTAR OF
THE SUN
WUMEN
T'IENANMEN
WALL
HSIPIENMEN
HANLIN
ACADEMY
Foreign
Legations
OBSERVATORY
TUNGPIENMEN
SHUNCHIMEN
CH'IENMEN
HATAMEN
OUTER CITY
OLD
RAILWAY STN.
CHANGYIMEN
TEMPLES
SHAWOMEN
GOLDFISH PONDS
TEMPLE OF
AGRICULTURE
TEMPLE OF
HEAVEN
NANHSIMEN
YUNTINGMEN
CHIANGTS'AMEN

commanded by the British minister, Sir Claude MacDonald. Of the fifty-five days of siege, the legations were under fire sporadically for approximately thirty-nine days—there being three truces occasioned by complex factors in the Chinese conduct of the war, and by the all but continuous diplomatic exchanges carried on between defenders and attackers. Without question, had the legations been subjected continuously to the full weight of Boxer and imperial-army attack, they would have fallen. There is no reason to minimize the ordeal of the almost four thousand people enclosed in that smallish space. But the defense was conducted in the most disorganized manner, due partly to the fact that ministers and others suffered from bitter rivalries and national jealousies; and partly to the fact that, given those rivalries, it was seldom possible for the commander to hope that everyone would obey orders. The atmosphere pervading the higher ranks of legation defenders has been likened to that existing between gold prospectors in a Klondike saloon.

Before the siege was relieved on August 14, 66 Westerners were killed (not a few by their own stupidity and lack of discipline) and 150 were wounded. How many Chinese within the legations were killed or wounded seems to be a matter no one thought to record. No Westerner died of starvation, though many Chinese converts did in the compounds adjacent. One Scandinavian went mad, numbers of people appear to have hidden themselves away and were not seen until the relief. But on the whole the Westerners conducted themselves with ordinary dignity, except when they fell into national disputes. They were never for more than a few days cut off from communication with the outside world. There was a running exchange of notes with the Chinese government all the time, messages were sent clandestinely to Tientsin and the allied forces there, and replies were received. The ministers were at one stage even allowed to telegraph their home governments. For some time there was a thriving fruit and vegetable market conducted over the barriers, and although meat was desperately short and donkeys were eaten, there is no record of much worse than that (at least for the Westerners). The legations had some medical supplies, though not enough. They even managed to buy rifles across the barriers from some enterprising Chinese. They do not seem to have run short of ammunition.

With these few details in mind, let us make our way from the angle of the south and west walls of Peking, where the legations lay, diagonally across the city, across the corner of the Imperial City where the Dowager Empress fumed in the Great Within, to the site of Pei T'ang—where the bishop,

◀ SKELETON MAP OF PEKING, 1900. Imperial Peking in its last years still spread within its walls round the core of the Imperial City. Scattered about the town were the palaces of princes, and, apart from the enclave of the foreign legations, very little had changed since the Ming dynasty. (Redrawn from Bredon)

The Catholic Cathedral of Peking Today. The Chinese characters on the gate read "Peking City 39th Middle School." (Photo by Marc Riboud, 1965)

Bishop Favier in a photograph taken soon after the siege of the Catholic cathedral. (From Little's *Round About My Peking Garden)*

Paul Henry as a cadet, shortly before leaving France for China. (From Henry)

Monseigneur Favier, presided over his cathedral, its large compound of buildings, and several thousand Christian souls.

While the foreign community in Peking was taking refuge in the legations, the Chinese Christians of the city (those who did not go to the legations) were scuttling within the walls of the church, two miles away. Bishop Favier had achieved a guard for them, as he had long ago requested—forty-three French and Italian marines commanded by a French lieutenant named Paul Henry. About five hundred Christian Chinese males were recruited from the converts as militia, unarmed except for staves. There were a dozen or so priests (Chinese and Europeans), and some three thousand Chinese, including large numbers of children.

Here, it is apparent from all the sources within the cathedral precincts at the time, order prevailed. There is no hint of squabbling from first to last. Over all, the venerable, slightly roguish, imperturbable figure of the bishop ruled with care and intelligence. With his commander, Paul Henry, he conducted the affairs of this large flock of people during the terrible weeks to come. There was no shadow of a truce and the attack was considerably heavier than that on the legations.

Paul Henry came from Brittany in France. By the time he reached China on a French warship he was twenty-three, an *enseigne de vaisseau*—a junior lieutenant in the French navy. His character was a mixture in equal parts of fierce but orthodox Catholic piety and fierce and orthodox patriotism. Family, God, and country meant a lot to him. His loyalty to them was the motivation of his life. His whole life is easily traceable, upright, filled with pedestrian sentiments which are evidently sincere. He is quite unusually composed, sober, and temperate for a young man in his position. He looks on his own life as a small but definite dedication, and conducts himself with energy, intelligence, and a lack of fanaticism which is in sharp contrast to the attitude displayed in the confessions of most defenders of the legations.

Paul Henry's letters and diaries, together with writings of other participants, are to be found in two obscure books never translated from the original French.[4] They seem to have made little impression when they were published, perhaps drowned by the dam-burst of print that poured from the emotional reservoir of the legation siege.

Yet, if a document of heroism in all its slightly pathological dedication, its carelessness of personal safety, its naïve devotion to a cause, be needed to lend dignity and some sanity to one of the ugliest shows of Western arrogance in China, Paul Henry's words, and his conduct, with the diaries of a few comrades and priests in Pei T'ang, may perhaps provide it. It was for his faith he (and they) fought, and also for his country's honor in defending that faith. Not at all, ever, for his own life.

In Pei T'ang on June 22, as they came under siege, Bishop Favier wrote in his diary: "We are completely cut off and can no more communicate with the outside world. . . . For arms we have 40 rifles of the marines, seven or eight of all kinds in the hands of the Chinese, some miserable sabers, and 500 lances or rather long sticks trimmed with metal. That is all. The perimeter to be defended is exactly 1,360 meters."

Inside the perimeter stood the cathedral—successor to the church built long ago on ground given to the Jesuits by K'ang-hsi. And behind it lay the quarters of the priests, a school, a printing press, storerooms, a small hospital, a library of Chinese and European books, and gardens and courtyards. To the north, across a road, lay the convent of the Sisters of St. Vincent de Paul, around which were a number of buildings serving the same variety of purposes.

By the twenty-second, the day the siege began, the whole enclosure inside its existing Chinese wall had been fortified under the instructions of Paul Henry and Bishop Favier. The most vulnerable point, the great gate leading from a wide street to the cathedral, was buttressed with earth inside; the road was cross-trenched and outer barriers built. Other openings in the walls had been cemented shut by one of the marines who was a mason by trade. And in case the outer defenses were penetrated, an inner line of barricades made of basketry and earth had been thrown up to surround the cathedral as a last refuge. Various redoubts, from which the defenders could fire in comparative safety, had been made at strategic points—even a subterranean post. Thus the whole enclosure was put into good defense order. Inside, as the Boxers began their attacks, a very large number of persons began to organize themselves for the ordeal they knew was upon them. The forty-odd

4. By Bazin and Henry. Unless otherwise stated, the quotations in this chapter are from the latter, in my own translation.

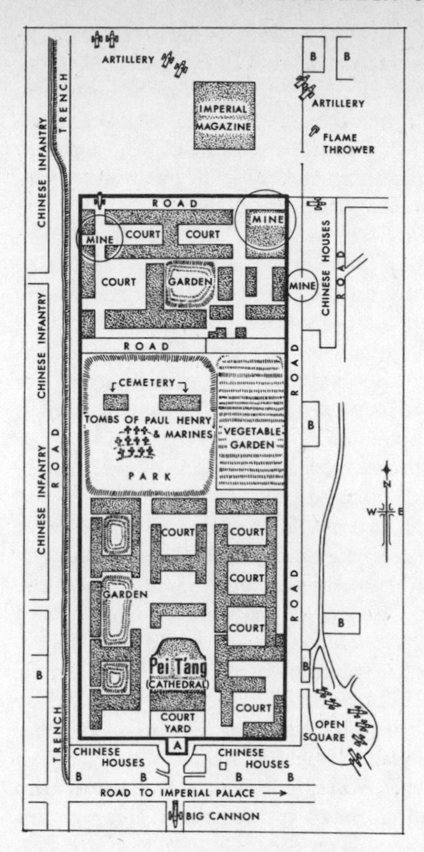

PLAN OF THE CATHEDRAL AND SURROUNDINGS. (Redrawn from Henry)

marines had two officers—Paul Henry and an Italian. The two bishops, Favier and his coadjutor Jarlin, had over a dozen European and the same number of Chinese priests, and twenty-two sisters of St. Vincent de Paul. There were 111 Chinese pupils from the school, 900 Chinese men, 1,800 Chinese women, 450 young Chinese girls from the sisters' school, and 51 babies.

Bishop Favier had arranged a special service for the twenty-second at six-thirty in the morning. "The priest on his knees at the altar was reading the first words of it when a loud cannon shot shattered the stained-glass window in the church, where everyone was gathered, killing a poor woman. Quite excusable panic broke out among the congregation: everyone crowded into the chapels and the sacristy to the west, for we had been attacked from the east side. The cannon shots went on, one every minute, and we evacuated the church at speed. Fourteen pieces of Krupp artillery threw shrapnel shells of the latest type at us without interruption. Several . . . pairs of windows took wing in splinters; the facade of our cathedral was badly damaged; the bell turrets were smashed to pieces, but the marble cross still stood. Toward three-thirty in the afternoon the attack was so violent that we thought our last hour had come. Toward five o'clock an ordinary Chinese cannon, trained some three hundred meters from our main gates, sent us a solid shot, which removed one leaf from the door. . . ."

This cannon was loaded and fired by Chinese who, under rifle fire from the marines, retreated into shelter after each shot. Paul Henry decided he must take the weapon. But first he wanted to test the morale of his men. He walked out alone through the broken main gate and stayed outside a full ten minutes. As he returned he said quietly to his men: "You see, it's as simple as that." Several diaries record the words, and as several journals of the priests recount, not one of his marines ever hesitated to throw himself into the fight, however intense the opposition.

Paul Henry then organized the sortie. With eight marines disposed in fours on either hand, he left the cathedral wall again, sword in hand. Bishop Jarlin followed with twenty Chinese Christians. As they neared the cannon, its gunners fled under fire from the marines on the Pei T'ang walls; but furious rifle fire from the Chinese houses surrounding the place met the party as it approached the abandoned weapon. Three times the Chinese and marines attempted to take it, and at last, spurred on by Jarlin and Henry, they managed to drag the cannon up the slope toward the cathedral, their rear covered by the marines and the lieutenant. The Boxers, infuriated by this audacity, let loose a "*fusillade endiablée*"—a withering fire. But, according to several diaries, it seemed as if an invisible shield protected the defenders. Only one Chinese was killed and three wounded. And the marines who provided covering fire for the hauling party miraculously escaped unscathed.

It was about four in the afternoon. Paul Henry drew breath and congratulated his company, regretting only that he could not follow up the exploit. "If I had a hundred marines like these thirty [Frenchmen], I would march on the Imperial Palace."

The cannon was a curious piece. It carried the date 1601. Favier and Jarlin, translating the Chinese characters on it, explained to the marines that it must have been made by Jesuits for the Chinese. Now it had returned to the spiritual descendants of its makers, to serve them. Fortunately there was a former gunner of the Chinese army among the defenders. Putting him in charge of the cannon, Paul Henry himself learned how to give the correct orders in Chinese for its firing.

"The night," he notes in his diary of the twenty-second, "passed relatively quietly, the enemy just keeping us on the lookout by salvos once each hour or so, without doing us very much damage. Altogether during the day we received at least 531 cannon shot, without counting the big guns of the rampart and the incessant musket fire. We ourselves fired sparingly; but we killed a good number of Boxers and soldiers. . . . Our losses: one woman and three Chinese killed, three wounded and another lightly wounded."

He gives only a passing mention to the conflagration in the huddle of Chinese houses south of the cathedral wall as they were set alight by the Boxers in an attempt to burn out the defenders. Over by the legations, the next day, they did much the same thing—only there the attackers set fire to the Hanlin Academy, where the flames consumed all but five hundred or so of the ten thousand manuscript books of the *Yung-lo Ta-t'ien,* the huge encyclopedia ordered by the great emperor, which had proved too big to print. How much the world lost in this fire at the Forest of Pencils Academy, we can only guess.

The following day was Saturday, June 23. Back at the cathedral, there were two hours of fierce attack in the morning. Then Chinese cannon took up the siege from another quarter. One Pei T'ang Chinese was killed. There were only 275 rounds of ammunition left for each man with a gun.

On the twenty-fourth, one of the priests noted in his journal: "The Manchu soldiery . . . keep trying to hit the people in Pei T'ang. . . . They are evidently afraid of our marines. . . . The brave commandant Henry is in good spirits; he says we can hold out for a long time. . . . I wonder what they are thinking about Pei T'ang in the legations?"

Apart from some of the French at the legations no one was thinking anything at all about Pei T'ang. They were focused on their own squabbles, their own predicament, their diplomatic exchanges with the Chinese—none of which seems to have mentioned any relief for the cathedral. And the following day, June 25, saw the first of the legation truces and twenty-four hours of comparative peace there.

In Pei T'ang that day the defenders noted the quiet over by the legations. "Since June 10," a priest wrote, "we have not really known what is going on around us. A wall of bronze separates us from the rest of the universe, which seems no longer to remember us." But they were relieved when that night they heard the bombardment begin again at the legations; for then they knew the Westerners there were not lost.

In Pei T'ang the shells and the bullets fell on them for most of each day. A strict watch had to be kept, twenty-four hours out of every twenty-four, on the movements of the attackers. Each move of theirs entailed some improvised countermove in the cathedral precincts. On the twenty-sixth the Boxers again tried to burn them out. And on the twenty-seventh they again attacked the vulnerable front gate. Salvos from the marines left many of the attackers on the ground. Paul Henry sent a party of his Chinese protected by marines to collect the discarded rifles. His account of the sparing use of ammunition in that day's heavy attacks is appended to the day's entry in his diary:

Bullets expended in the attack	55
Others	23
Total from 6 A.M. to 6 P.M.	78
Remaining bullets: 8,191	

On the twenty-eighth, toward the evening of a hot day, the Boxers tried a new tactic. At the west gate of the precincts they stationed a fire pump which threw not water but flaming kerosene onto the roofing of the gateway. Soon they augmented this with another more powerful jet nearby. And then came a rain of strange (to the French and Italians but not to the Chinese) projectiles—the traditional flaming arrows, their tails soaked in burning kerosene. The danger was acute. Everyone held their breath as they waited to see the fate of the gateway standing between them and the Boxers. At nine in the evening Paul Henry climbed onto its roofing to replace another marine. There he stood directing the rifle fire toward the place where the archers hid behind a wall. "Clad in white and surrounded by flames of burning kerosene, he moved in a luminous halo where he was the point of aim of the enemy rifles. But he was sure that the hour when his life was to be sacrificed had not yet sounded. Under the unremitting fusillade, in the center of a hail of bullets, he turned to his assailants and called out loud enough to be heard by his own marines: 'You can fire, you band of skunks! You won't be getting my skin today.' " Nineteenth-century stuff it may be, but it had the desired effect on the defenders.

It took six hours of fighting before those two pumps were put out of action, six hours of fire-fighting with bands of Chinese carrying water from the well, six hours of hot but highly organized work in smoke and flames and desperation as arrows fell all over the compound. But they won. And suddenly it seemed that a general panic seized the attackers without the walls. Perhaps they thought the besieged were imperishable, impregnable, unburnable (as popular belief supposed the Boxers to be): for they began to run from the cathedral walls, yelling in Chinese: "Let's get away! Never, never come near these devils inside again!"

But there was no time for congratulations. Paul Henry at once made a sortie and captured a case of bullets and another of gunpowder, some clothes, but few rifles. Then in the lull, the marines and Chinese began destroying the Boxer hideouts and setting fire to the remainder of the kerosene they found. Two pumps in working order were dragged back—a godsend for the fire-fighting they might soon have to do again.

The following day was quiet. "You would have thought," says the bishop in his journal for that day, "that they wanted to let us have our little time of joy. . . . At ten at night a great storm broke, and the thunderbolts seemed to be falling on the palace" adjacent to the cathedral. Over the heads of defenders and Boxers alike there broke one of those violent Peking summer cloudbursts. In a moment everything was drenched, the ground flooded, the air degrees cooler. There was a little time to regroup, to repair the defenses, to invent others, to gather the spirit together for the next phase of the assault which would surely come.

Paul Henry at such moments took himself off, after he had seen to whatever needed his attention, to his room to play the harmonium that Bishop Favier had put there. He liked to accompany himself as he sang. One of the bravest of his comrades says: "It was the only pleasure he took. . . . The venerable bishop also affirms that Paul was always cheerful. He had the power to master himself completely. . . . He sometimes sang a canticle or a religious hymn, sometimes a fine song like the 'Binniou Paimpolaise' by Botrel, or 'The Marine's Girl-friend' or tunes he had picked up" with the troops. "Or perhaps a Breton song."

On July 1 the bishop was writing: "About eight o'clock we heard numerous cannon shots to the south. Could it be the reinforcements? One hopes against all hope. For the first time we begin to eat donkey. The mules and horses will come next—we have eighteen. Smallpox has broken out among the children. Seven or eight die each day." And on July 2, Bishop Favier wrote: "The attacks are less lively the last day or so, but the food is much worse; no more vegetables, no more pickled greens for our poor Christians. . . . Temperature up to 100 F., humid air, already ten days without any news. How long it is!" One of the priests lamented the season of Peking rains which, he felt, would delay a relief force.

Unknown to the defenders, the *Peking Gazette* of that day published an imperial edict: ". . . The Throne exhorts the Boxers now to render loyal and patriotic service, by taking their part against the enemies of the country. . . . The missionaries must once and for all be sent away to their own countries. . . . "

Soon in Pei T'ang they were burying fifteen children a day in the garden. One of the Christian converts surreptitiously left for the outer world—whether to satisfy his craving for opium or to inform the Boxers of what went on inside, no one could guess.

So it went on, day after day. Food became scarcer despite rationing; tobacco was nonexistent. One of the sisters later recounted (to Mrs. Archibald Little) how she tore up all her letters to use as cigarette papers for the men—the "tobacco" being the dried leaves of a pear tree. The toll of dead—the starved and diseased infants, wounded men and women and children for whom there was now no medicine—mounted as the days passed. The sisters, with their troops of several hundred orphans and foundlings, "all starved and suffered through the hot summer months. . . . The sisters used to make the children follow them in a long train to this side or that [of the cathedral area], wherever the firing seemed least heavy. . . ."[5]

Then the Boxers began to tunnel under the walls. There they set off mines, in the chaos of whose explosions many of the defenders were killed and even more died from being buried in the rubble. Fire bombs thudded down on compound and roof. The kerosene pumps came into action again and again. The heavens seemed to pour shells and bullets, and to put a foot out of a door was to invite to death. There was no truce, hardly more than a few hours' lull; no peace, only snatched hours of troubled rest for the defenders.

From July 18 to 28 the legations enjoyed one of their truces. By accident a cart laden with fruit and fresh vegetables sent by the Dowager Empress to cheer the Boxers and troops attacking the legations was delivered to the legations themselves! The defenders inside the walls were in any case buying fruit and vegetables from a thriving market set up outside the barriers. They had their alarms and moments of danger, but they were hardly cut off.

In the shattered enclave of the cathedral the inmates were as effectively sealed as prisoners in a condemned cell. They listened as best they could, now and then, for sounds from the direction of the legations, alarmed when they heard no firing there, cheered when it resumed—for then they knew the legations were holding out. Mined, scorched by kerosene missiles, under heavy fire, half-starved, surrounded by militant hordes of Boxers whose chant, *Sha! Sha!*—Kill! Kill!—pulsed rhythmically over their walls, the defenders of Pei T'ang awaited their fate without much hope of relief.

But—perhaps it was a blessing—they were never idle. Chinese mines were countered by Christian mines. Fires were put out, the defense was continually reorganized. Spirits were cheered by the priests and the sisters. No food found its way to them, though they made sorties in search of some. The marine force dwindled by sudden death, or by lingering, gangrenous stages. The converts prayed and starved. They "sickened and died, of wounds, of hunger, their lips and faces swollen, trying to keep themselves alive upon the leaves of trees, rose-leaves so bitter, and other leaves yet more unwholesome. I myself," says Mrs. Little when she visited the wreck of Pei T'ang after the siege, "saw the trees stripped of their bark, all gnawed off." The diaries of the defenders bear her out.

It is impossible to read the laconic diaries of Henry, Favier, and others in Pei T'ang and not realize that there, in the rubble of their uncertain fortress, a long, grisly hell was theirs. Like animals fallen into the hunter's trap, surrounded by sharp stakes, they twisted and turned for long weeks. But all

5. From Mrs. Little's *Round About My Peking Garden.*

the time they were clawing back with a spirit quite unanimal—with a prolonged, fine, human courage. Their spirit was not vitiated by dissension. They had one purpose.

On July 18 Paul Henry wrote his will. On the twenty-third several thousand Boxers made an unusually violent attack at two places simultaneously. With superhuman effort they were repulsed, leaving about 150 dead and as many wounded. Five days later the diary of Paul Henry reaches its last entry. "Toward eleven, bombardment from northeast and east. . . . To the northeast a strong fusillade. Intermittent firing. Salvos in the northeast. Cannon went on all evening. Neither killed nor wounded."

Next day, July 29, one of the sisters wrote: "The Boxers and soldiers advanced right to the foot of the encircling wall, yelling *'Sha! Sha!'*—Death! Death! Today, Sunday, we had Mass to the sound of cannon. . . . We could all be massacred from one moment to the next. One has to be ready."

In the night the Boxers set up a huge new cannon near the northeast gate. Paul Henry determined to capture it. He gathered a dozen French and Italian marines and addressed them: "I have chosen you, friends, because I know you are bold, courageous, and capable. I am going with you to take the gun. . . . We'll rush it without firing a shot. Once there, we'll fire a salvo while our Chinese are yelling *'Sha! Sha!'* If we have to withdraw, we'll do it slowly, firing and bringing back our arms and our wounded."

Just at that moment the Boxers launched a strong attack. As usual, Henry mounted to a higher level to assess what was happening. Two of his best men were with him. Almost at once one of the marines was struck a glancing blow on the shoulder by a bullet, which ricocheted and entered the throat of Henry at one side, ripping through to the other and severing his carotid artery. He managed to tell his comrades to get down. Then another shot struck him in the side. He jumped to safety and fell into the arms of one of his men. "Seeing blood on his tunic, they opened his collar. Then a jet of blood spurted from the wound. . . . He made a sign that he wanted to speak . . . but no sound came. . . . His head fell inert."

Paul Henry, the gallant commander, was dead. But the siege was far from over. Bishop Favier and the Italian marine officer took over. The military situation remained the same. They held their own, even inflicting heavy casualties on the Boxers, and suffered their own with a dignity which came perhaps not only of conviction but also from the knowledge that there was no alternative but surrender and an even more hideous (if swifter) death.

There was almost no food. "The horrors of the famine make themselves harshly felt," says the bishop. "Bodies lengthen, color vanishes, an air of death reigns among us. Walking corpses still move about, but with effort. . . . There is nothing to lose now: leaves of trees of every sort, remains of every kind, all are gathered up with care, with rapacity one might say. . . . Our Christians are so enfeebled that they lie out on the verandas, thin, pale, almost dead."

For more than two weeks after Paul Henry died, the siege went on with unabated horror and fury.

In the evening of August 13 the defenders heard for the first time the Boxers calling to each other: "The European devils are coming!"

On August 14 they heard the distant machine guns of what they felt must be the relief force. And that day, though they had no means of knowing it, the legations were relieved. The same day the French in the legations prepared a relief force for Pei T'ang, but its orders were suddenly canceled before it could leave to cross the city.

Over there on the west side of the palaces, at Pei T'ang, they observed the Chinese flags on the Peking wall near the legations coming down. They heard the spate of firing as the distant siege was relieved. But they could not really be sure what had happened. They thought perhaps the legations might finally have been conquered.

On the next day, the fifteenth, they could hear the Boxers and imperial troops over the walls quarreling and accusing one another. "It's you," said the imperial troops, "you Boxers who're the cause of all our troubles." After a time the soldiers and Boxers began firing at each other. On this day, though they did not know it in Pei T'ang, the court fled from the palace.

On Thursday, August 16, one of the marines as usual mounted on a roof as soon as it was light. "I . . . saw Japanese soldiers crouched on the roofs near the gate. . . ."

The defenders ran up the French flag. They sounded the "Marseillaise" on a bugle. They waited for a sign from the outer world. The only reply was a cannon shot from the Boxers. Their fears increased.

"A few instants after that, Japanese soldiers distinctly appeared. Then an officer of the same nation came to speak on the barricade. The bishop embraced him with joy; some of the happy inmates even managed to touch his hand. We were saved. . . ."

A quarter of an hour later, French soldiers arrived. The long agony of Pei T'ang was over. The fifty-six-day siege was over at last.

Apart from the two volumes in French by René Bazin and Léon Henry, which deal with the life of Paul Henry and, of course, with his defense of the cathedral, there seems to be not a single volume on the siege itself. Various writers, such as the journalist Savage-Landor, who arrived with the relief forces in Peking, give it a paragraph or, at most, a chapter or two. Savage-Landor is so inaccurate that he puts dates into the mouth of Bishop Favier, in reporting an interview with him, which do not tally with those in the bishop's own diary. Morse devotes a balanced paragraph to the Pei T'ang siege. Mrs. Little understood the heroism of the defenders and their priva-

First Aerial Photograph of Peking, taken from a balloon after the Boxer siege of 1900. Here the whole enclave of the Catholic cathedral is visible. The balloon was put up by the British Royal Engineers under the command of Captain Hume. (From Henry)

tions. The American president of the Chinese Imperial University of Peking, W. A. P. Martin, an inmate of the legations during the siege, is brief but more honest than most. "The defence of the cathedral forms the most brilliant page in the history of the siege." He could have gone further and called it the sole brilliant page.

On August 15 the court had fled Peking. In the small hours of the morning the Dowager Empress and her prisoner the emperor donned the clothes of poor Peking citizens and left unobtrusively in a Peking cart. The emperor's favorite concubine, who had dared to suggest that the emperor should stay in Peking, was thrown down a well on the orders of the Dowager Empress. Eventually the old woman and the emperor reached Sian, the ancient capital of Ch'ang-an, near the middle Yellow River, and set up a court there.

If few wrote on the cathedral siege, even fewer cared to record the events of the two or three weeks after it was over, when "Peking was humbled in the dust," as the historian Morse wrote a few years afterward. Almost alone among Western partisans, he stands back a pace. Peking, he writes, "suffered as Asiatic cities have always suffered when taken by assault by Asiatic armies—but now the invading forces were the armies of the Western powers. Strong men fled: those not so strong fled if they could find the means of

"PRESENTS FOR QUEEN VICTORIA: marking the trophies destined for Her Majesty during the dismantling of the Lama Temple at Peking." This sketch and caption appeared in the *Illustrated London News,* December 29, 1900.

transport, but that was scanty; many of the Manchu nobility and countless thousands of women put an end to their lives; many men were killed in a wild orgy of slaughter; the survivors who remained cowered like whipped hounds in their kennels. The city had been plundered, first by the Boxers, then by the Chinese soldiers, but they had only skimmed the surface. The foreign troops now set to work to sack it riotously, but systematically. At first they spread over the city . . . wherever their fancy led them; then each contingent plundered the district assigned to it for policing. . . ."

Savage-Landor, who denies reports of atrocities committed by the Western soldiery, nonetheless chronicles their other acts: "Looting commenced at once all over Pekin, although the capital of China did not suffer quite so much in this respect as Tientsin city. Attempts were made to keep soldiers and civilians in order, and threats of 'shooting looters at sight' were issued, but never carried out *except in the case of the Chinese* [my italics].

"Long lines of mules were to be seen all day long, carrying away loads of silver, grain, and silk from various Government and private stores to the quarters of the Allies. Missionaries, male and female, were frequently observed collecting curios in uninhabited houses, and soldiers helped themselves freely to anything they liked."

There is one extended account by a European of the orgiastic and murderous few weeks in Peking before order was restored by the Western commanders. Putnam Weale, the *nom de plume* of the Englishman B. L. Simpson, an employee of the Chinese Customs Service, which had for long been run by the English, wrote a fictionalized account under the title *Indiscreet Letters from Peking.* The book appeared in America in 1907 and enjoyed a *succès de scandale.* Simpson is suspect in Western eyes because he had his knife into a number of important Westerners in Peking. He himself took a very active part in looting, and in fictional guise admits it. A warrant for his arrest was at one time issued by the British minister, but it was never carried out. Yet before that, he had been commended by the same British minister immediately after the relief of the legation siege. Whatever his personality defects, it is obvious that his story of those weeks of unrestrained license in Peking is not all invented. Incidents may be colored here and there, but they fit very accurately with what we know from other fragmentary sources. It makes depressing reading—the story of Westerners who turned themselves into a scourge as horrible as the Boxers had ever been. For the few Western lives lost they exacted the lives, the livelihood, and personal integrity of thousands of innocent Chinese in Peking and Tientsin and all the way between the two cities.

Sir Robert Hart, chief of the Chinese Customs, was less discursive than his erstwhile employee Simpson, and somewhat more Olympian. Four days after the relief he wrote: "Confusion has reigned since the 14th . . . and the Chinese think the end of the world has come." (Quite naturally—for many it had.) And later, on August 30, he wrote: "Such confusion and disorder I have never lived in."[6] McCarthy, another writer of the time, was more precise: "In the Russian district the most unspeakable excesses are said to have been committed. The same is said of the district entrusted to the French troops. . . . It is admitted that order was first established in the district entrusted to the Japanese, and soon afterwards in the British district . . . and in the American quarter."[7] The nationalities who, by general con-

6. Quoted in Morse's *International Relations.*

7. From Michael F. G. McCarthy's *The Coming Power, 1898–1905,* as quoted in Morse's *International Relations.*

sent came out of the disorder with less hideous records than most were the Japanese and Americans.

Thus, then, did the defenders of the Western faiths, the commercial and the religious, withstand the sieges of Peking. Whatever may be history's eventual judgment, the opinion of Chinese is both relevant and understandable. To them, whose country, whose capital, was the stamping ground of the Western armies of relief, it seemed the West was unjust, and that its faith was hypocrisy. The Christianity it preached spoke of peace, of justice, and brotherly love, yet it condoned the commercial rape of their country and the actual rape of Peking and other areas.

They saw the Boxers, for all their ferocity, as a patriotic movement dedicated to driving the foreigner into the sea. Convulsively, but convincingly, the Chinese had demonstrated their abhorrence of the Western presence in China. But, under the effete Manchu dynasty, with their total lack of modern technology and military resources, they could manage no more than this bloody demonstration, in which they themselves got the worse of the argument.

"For these offences," says Morse, meaning the attacks of the Boxers on Westerners, "retribution was inflicted on the guilty officials, and on the troops and Boxers who had been the active agents in massacre and destruction. . . . The empire was humiliated as it had never been from any of its previous wars, for it had lost, not only prestige, but reputation as well. . . . [It]was now called upon to enter on a diplomatic battle, as a result of which the nation was subjected to still deeper abasement."

The Western allies exacted in money from the Chinese the equivalent of

"'Foreign Devils' in a street in Peking: a salute from guttersnipes." Such was the caption of this illustration in the London *Graphic* of May 18, 1901. The arrogance imputed to the soldiers requires no further comment.

W. A. P. MARTIN as he arrived in New York in October, 1900, in the "siege costume" he wore through the defense of the legations. (From Martin)

about 340 million U. S. dollars—the money to be raised from customs and salt taxes.[8] A firm grip was therefore now obtained on the Chinese economy—a grip which allowed the Western powers effectively to control the destinies of China, to force the Chinese government to suppress any anti-foreign policy or uprising. The legations were to have strong defensive armed forces, and turned themselves into fortresses. The forts on the Pei River were razed, leaving Peking defenseless from the sea. China was in all but complete Western control.

Martin, the president of the University of Peking, whom we noted praising the Pei T'ang defenders, left Peking for home as soon as he could after the legation siege was relieved. "On reaching New York in the actual costume which I wore during the siege," he wrote, in a book hastily dictated to cash in on the avid market, "I called a boy to carry my packages. . . . As I was carrying a gun, the lad remarked:

"'You must have been hunting somewhere?'

"'Yes,' said I, 'in Asia, beyond the sea.'

"'What kind of game?' he enquired.

"'Tigers,' I replied—I ought to have said hyenas."

8. A portion of the indemnity was belatedly returned by the allies, first by America in 1908 and 1924.

THE EMPRESS DOWAGER GIVING AUDIENCE to the ladies of the diplomatic corps at Peking. The emperor sits at the right of the dais, receiving the curtsy of Lady MacDonald, wife of the British ambassador. Note the wife of the Japanese ambassador in the row of waiting ladies. The original caption to this sketch by "Captain Wylde, H.M.S. 'Undaunted' " adds a society-column note to the occasion: "The Dowager Empress wore yellow silk, with blue and gold embroidery. Lady MacDonald . . . wore an astrachan cape with large collar, black hat with feathers, light dress, and white gloves. The other ladies were all in furs, the day being very cold. The ladies of the Court were dressed in pink, light green, and light blue, with dark embroidery. They wore flowers and hairpins." But no one bothered to mention the emperor or his dress. (From the *Illustrated London News,* ca. 1902)

He was referring to the Boxers. But it would perhaps not be an unwarrantable transposition to apply his epithet hyenas to the Westerners he had just left in the sad city of Peking.

The epilogue to the unsavory tale of Western acts and Chinese rearguard action in Peking in that first year of the twentieth century reveals both sides in equally unflattering colors. When the treaty was signed and the last troops had left, the Empress Dowager returned from exile. The last stage of her journey was made in a train that arrived at the newly built terminal within the old Ming walls. The imperial procession stopped just outside Ch'ienmen, the front gate of Peking that bore the scars of fire from the events of the siege, and the Empress Dowager alighted in order to sacrifice at a shrine. She stood for a moment or two looking up at the parapet where hundreds of Westerners stood, cameras at the ready, and waved her handkerchief to them. Again, as she was on the threshold of the shrine, she stopped and made several little bows toward them. A year or so later the empress gave the first of a series of tea parties for the wives of the diplomats accredited to the court of China. Western triumph over the ancient traditions, over the tight, xenophobic world of court and mandarinate, was complete. And Chinese humiliation, Chinese impotence in the heart of their own land, was also complete. For it was now the barbarians and not the mandarins who ruled.

THE EMPRESS DOWAGER WAVING. This memorable photograph was taken from the south wall of the legation quarter upon the occasion of the Empress Dowager's return from her Sian exile. A eunuch at the left carries the jade imperial scepter. The carrying poles of the imperial palanquin are seen at the lower right. (Radio Times Hulton Picture Library, London)

CHAPTER NINETEEN

THE WRITERS FROM THE WEST

WITH THE events of 1900 and their outcome, the pattern of Sino-Western relations took its final form. Individual persons were now of still less importance than their governments. In common with other individuals, the writers who went to China after that time had no direct influence on the pattern of their age. But they have a relevance to our story since, in the writings of this tiny minority of Western travelers to China, it seems possible to discern the first change in the color of Western opinion about the Chinese. The writers we will follow a little way were in China in the decades after the dynastic system had vanished, a time when a bewildering and grim struggle was being waged for dominance in the country. These were the years of a hapless republic, of feuding warlords and Japanese aggression and civil war. While the gaze of the writers turned mostly on that postdynastic scene, yet the past—as ever in old civilizations, both East and West—intrudes and qualifies their views on China. Through the eyes of that handful of Western writers, from whom we have chosen a small sampling, there emerges a new approach to China and to the Chinese. It is still only the very beginning of an approach and grows uneasily. But it is entirely different in quality from that of the writers' contemporaries both in China and in the West. We have not discovered anything like the words that follow, in any former travelers—not even in Ricci:

"When I went to China, I went to teach; but every day that I stayed I thought less and less of what I had to teach them and more of what I had to learn from them. Among Europeans who lived a long time in China, I found this attitude not uncommon, but among those whose stay is short, or who go only to make money, it is sadly rare. It is rare because the Chinese do not excel in the things we really value—military prowess and industrial enterprise. But those who value wisdom or beauty, or even the simple enjoyment of life, will find more of these things in China than in the distracted and turbulent West, and will be happy to live where such things are valued. I wish I could hope that China, in return for our scientific knowledge, may give us something of her large tolerance and contemplative peace of mind."

This quotation comes from a long-forgotten book by Bertrand Russell, *The Problem of China,* which dates from 1922. The author was then a man in his prime and teaching in a Peking university.

He continues: "New arrivals [in China] are struck by obvious evils; the beggars, the terrible poverty, the prevalence of disease, the anarchy and corruption of politics. Every energetic Westerner feels at first a strong desire to reform these evils, and of course they ought to be reformed. But the Chinese, even those who are the victims of preventable misfortunes, show a vast passive indifference to the excitement of foreigners; they wait for it to go off, like the effervescence of soda-water. And gradually strange hesitations creep into the mind of the bewildered traveller; after a period of indignation, he begins to doubt all the maxims he has hitherto accepted without question. Is it really wise to be always guarding against misfortune? Is it prudent to lose all enjoyment of the present through thinking of the disasters that may come? . . . Should our lives be spent in building a mansion that we shall never have the leisure to inhabit?

"The Chinese answer these questions in the negative, and therefore have to put up with poverty, disease and anarchy. But, to compensate for these evils, they have retained, as industrial nations have not, the capacity for civilized enjoyment, for leisure and laughter. . . . The Chinese, of all classes, are more laughter-loving than any other race with which I am acquainted; they find amusement in everything, and a dispute can always be softened by a joke. . . .

"The Chinese, from the highest to the lowest, have an imperturbable quiet dignity, which is usually not destroyed by a European education. They are not self-assertive, either individually or nationally; their pride is too profound for self-assertion. They admit China's military weakness . . . but they do not consider efficiency in homicide the most important quality in a man or a nation. I think that, at bottom, they think that China is the greatest nation in the world, and has the finest civilization. A Westerner cannot be expected to accept this view, because it is based on traditions utterly different from his own. But gradually one comes to feel that it is, at any rate, not an absurd view; that it is, in fact, the logical outcome of a self-consistent standard of values. The typical Westerner wishes to be the cause of as many changes as possible in his environment; the typical Chinaman wishes to enjoy as much and as delicately as possible. This difference is at the bottom of most of the contrast between China and the English-speaking world. . . .

"The old indigenous culture of China has become rather dead. . . . Confucius does not satisfy the spiritual needs of a modern man. . . . The Chinese who have had a European or American education realize that a new element is needed to vitalize native traditions, and they look to our civilization to supply it. But they do not wish to construct a civilization just like ours;[1] and it is precisely in this that the best hope lies. If they are not goaded into militarism, they may produce a genuinely new civilization, better than any that we in the West have been able to create."

For its time—indeed for any time—such a balanced statement is rare. A long way back we came upon another European who wrote succinctly that the Chinese were *"di nostra qualità."* Accepting the great differences between Chinese life and that of the West, this is precisely what Bertrand Russell is saying. Yet in the centuries that lie between the Corsalis of 1515 and the Russell of 1922, and in the decades since the latter's remarks, it is hard to find an exponent of this theme.

We have seen something of the mutually exclusive viewpoints of Chinese and Westerners in the quotations scattered through the foregoing many pages, showing how rigidly these views were often held. So it is surprising when we come to the twentieth century that in a few Westerners there is an almost revolutionary change of attitude. The deep-rooted view that Eastern people are in some sense inferior has yielded a little to more intelligent and objective approaches.

Let us follow Bertrand Russell a little further as he completes his picture of the Chinese in 1922: "So far, I have spoken chiefly of the good sides of the Chinese character, but . . . China, like every other nation, has its bad sides also. It is disagreeable to me to speak of these. . . . But for the sake of China, as well as for the sake of truth, it would be a mistake to conceal what is less admirable. I will only ask the reader to remember that, on the balance, I think the Chinese one of the best nations I have ever come across, and am prepared to draw up a graver indictment against every one of the Great Powers. Shortly before I left China, an eminent Chinese writer pressed me to say what I considered the chief defects of the Chinese. . . . I mentioned three: avarice, cowardice, and callousness. . . ."

On avarice he continues: "I think this defect is probably due to the fact that, for many ages, an honest living has been hard to get; in which case it will be lessened as economic conditions improve. I doubt if it is any worse in China now than it was in Europe in the eighteenth century. I have not heard of any Chinese general more corrupt than Marlborough. . . . It is . . . quite likely that changed industrial conditions will make the Chinese as honest as we are, which is not saying much." How prophetic these words were, any returning European businessman will recount today. And as to cowardice, probably Bertrand Russell has come to the conclusion by now that this too is in some measure a product of conditions. One does not think of cowardice in the Chinese when (for example) one thinks of the Korean war. It is a question of what one is fighting for, and whether one believes in

1. The reader may wish to pause a moment on this phrase. Odd though it may seem, history apparently shows that almost no non-Western nation wishes "to construct a civilization just like ours"—whether "ours" be European or American.

A Portfolio of
Henri Cartier-Bresson's Photographs

The earliest photographs of China date from the decade following Lord Elgin's order to sack the Summer Palace. From that time in the 1860's the cameras of many Westerners were clicking busily, if largely at random, in China. After the siege of the legations in 1900 there were journalists with cameras recording the scene. But most were interested in the picturesque, the curious, and the "quaint." It was not until the 1920's and a little later that work of real photographic merit was done, and then it was more documentary than creative. Photography was still tied very largely to the concept of the easel picture and used as an adjunct to the printed word.

Henri Cartier-Bresson was the first man in China to use the camera as a means not only of recording the contemporary scene but—in the manner of the creative writer—as a means of interpreting that scene, as a means of social comment.

In Peking, where he arrived almost on the eve of its fall to the Communists, "the daily routine remained fixed in time, a peaceful island within war-torn China. . . . My teahouse was silent: its clients meditating rather than talking. A sign on the wall read: 'Track down the unstable elements,' a euphemism for Communists. . . . The calm detachment, still so great a Chinese virtue, was being maintained on the eve of the downfall of their city." But outside the charmed circle of the teahouses people were reading the newspapers on placards, trying to decide what was true and what false; mothers were searching the ranks of conscripted soldiers for their sons and husbands; and while a few of the surviving eunuchs from the Empress Dowager's long-defunct court still smiled their turtle-like, pathetic smiles, dogs and thousands of the populace combed the garbage dumps for scraps of food.

From Peking, Cartier-Bresson went to Shanghai and saw the final agony of a huge starving city in which money lost its value overnight and the gutters were littered with the dead. He was little interested in the picturesque: it was hardly the time to enjoy the beauties of nature and architecture that Chinese genius had welded into one. It was a time of human dismay, of human powerlessness against corruption and cynicism.

In his work in those few months Cartier-Bresson was the first Westerner to use the photograph as a means of incisive comment on the Chinese *condition humaine*. Many of the resulting photographs are by now landmarks in the history of photography, of the interpretation of life with a camera. With the work of a few writers of the period, they are also the most poignant and sympathetic Western comment on the last struggles of the old China as it succumbed to the new.

The photographs that follow were chosen especially for this book by the photographer himself. Almost all of them are here published for the first time.

READING THE NEWS. Government-controlled newspapers kept saying the KMT armies were winning victories, even after the Communists had encircled Peking.

A BEGGAR outside a Peking antique shop selling glazed pottery tomb figures of the T'ang dynasty.

CONSCRIPTS EXERCISING on a cold morning as their relatives watch forlornly.

TWO AGED EUNUCHS, survivors of the four hundred who used to serve the Empress Dowager, talking to an acquaintance.

A MOTHER PLEADS with a soldier outside one of Peking's gates to find her son among the troops.

it; and whether men trust their leaders not to sell out to the enemy—a factor ever present in Chinese strife in the past.

"I have been speaking of the Chinese as they are in ordinary life, when they appear as men of active and sceptical intelligence. . . . There is . . . another side to them: they are capable of wild excitement, often of a collective kind. . . . It is this element in their character which makes them incalculable, and makes it impossible even to guess at their future. One can imagine a section of them becoming fanatically Bolshevist, or anti-Japanese, or Christian, or devoted to some leader who would eventually declare himself Emperor."

Of all that torrent of material which poured from European presses in the years between the Boxer siege of 1900 and the Communist take-over in 1949, it is hard to match Bertrand Russell for an objective and prophetic view of China.

It is tempting to quote other passages from the books of the period, because they are for the most part either hilariously funny or pathetically ill-informed—or both. But probably no special purpose would be served. There were, however, a few—among that very large number of Europeans and Americans who went to China during the first fifty years of this century—who illuminate corners, large and small, of the Chinese scene. There were many who came to love China—and as we have seen, almost no Westerner disliked Peking.

It would be reasonable to expect that Shaw, with his Fabian socialism, would have found food for thought, and even polemic, in China. But he produced nothing of the kind. Blanche Patch, in her biography of him, tells how, when he was flying over the Great Wall, "they unexpectedly found themselves hovering over a battle which raged below." Shaw went dumb until the end of the flight when, "Thank God we're safely landed!" he said.

Hesketh Pearson recounts a later conversation with Shaw:

"Tell me, were you deeply impressed by anything you saw on your travels," I asked him.

"No, one place is very much like another."

"By anybody?"

"No, they're all human beings."

"Did you see the Taj Mahal at Agra?"

"No. The rest of the party tried to see all India in a week. I stayed in Bombay, where I found that my religion is called Jainism."

"Did you see the Great Wall in China?"

"I flew over it in an aeroplane."

"Interesting?"

"As interesting as a wall can be."

Doctor Johnson and William Alexander must have turned in their graves.

Somerset Maugham is another disappointment. It is obvious from his novels set in the East that he views Orientals as inferior people, using them as a kind of animated backdrop to the activities of his Occidental principals. But would it have been too much to hope that that intense, almost febrile empathy Maugham displays in portraying the expatriate Europeans might have found in some Chinese a worthy object for its exercises? Apparently it would be too much to hope—for there is nothing but superficial observation on the Chinese in the only book he wrote on his time in China, *On a Chinese Screen*.

But the Chinese were readily understood by at least two other Europeans who were writing in China somewhat later, as the Japanese intensified their campaign of attrition in the country during the years before World War II. By Harold Acton, for one, who wrote in his book *Memoirs of an Aesthete:*

"I walked through the dumb streets deserted and ominous; even the hawkers were silent. I was reminded of Ping Hsin's poem:

> Silence
> Is filled with the triumphal song of the victor.

"We were encircled by the victor.

"Pei Ta [the university where the writer taught English poetry] was migrating to Ch'ang-sha in Hunan, a long trek from which many would not return, and various students dropped in to say goodbye. They were nervously calm, their faces set in an expression of fixed resolution. Though they had to leave most of their chattels behind, their precious rubbings of old calligraphy and collections of postage stamps, their eyes glittered like black steel with an excitement which was otherwise suppressed. They were bubbling and effervescing, inwardly relieved that matters had reached a climax. China's humiliation had been their humiliation: now they were going to cut it like a cancer out of their system. Even the most delicate of them, who had thought me eccentric for walking when I could use a rickshaw, were now prepared to trudge for months on end. Their smiles were radiant. . . . It was I who found it difficult to smile. I could not share their optimism. They were thinking in terms of geography and manpower, of China's four hundred and fifty million against Japan's seventy million, and they were remembering China's historical assimilation of her conquerors. They had forgotten the machine. . . .

"Sung Cheh-yuan, the quasi-independent general in command of the Peking area, was putting up a fight against overwhelming odds. His ill-equipped soliders were bombed and machine-gunned. . . . It was the mechanized twentieth century against the age of chivalry. The wounded were deluging the hospitals; some eighty of them were deposited in a vacant temple behind my house. I went to see them every day. . . . They asked me eagerly about the war. Would England help China against the dwarf-bandits? I did not tell them that our 'commercial stake' was not large enough for our government to bother. . . .

"They looked like children; one veteran was only thirteen years old. . . . They had no modern amenities and their bandages looked none too clean, yet the atmosphere was not gloomy. . . . None wore more than a loincloth, and their hairless ivory bodies were strangely serene and ascetic, like young Buddhas in meditation.

"Those who were able to read took turns to read to their comrades, and as I stood in the doorway I was hushed by the rapt attention of the listeners, their oblique eyes turned towards the reader, so dignified and conscious of his role as interpreter of bygone acts of heroism, his voice rising and falling in a higher timbre than ours, like a rhythmic chant. . . . He was describing Yueh Fei, the pattern of a self-sacrificing patriot, who lived in the twelfth century and strove to keep the Golden Tartars out. His horse had collapsed in the heat of battle and he was urging the beast to rise. The horse makes a supreme effort, blood gushing from its nostrils. . . . And I felt that these soldiers with the clear vision of childhood were seeing as well as hearing. They were living in another age, when warfare was relatively human. . . . And you could tell from their eyes they were living in it proudly. . .

"They had all come from villages five hundred to a thousand miles away. . . . Their spirits were extraordinary; they would rejoin the army [now defeated in the north] wherever it was and chase the dwarf-bandits into the sea. . . . The thirteen-year-old veteran with the bandaged arm could neither read nor write but he could play chess. I have never seen such an indomitable little fellow. He was tense with patriotism, almost grim. . . . An orphan, he had never known what it was to be really young. He sat brooding over the chessboard with puckered brows, a solitary strategist."

How few Europeans who put pen to paper thus felt at one with the Chinese, or regarded them as ordinarily human. Most, like Maugham, looked upon them with a kind of detached clinical eye—as if China were a vast suffering casualty ward in a hospital-cum-mental-clinic called the East.

For those glimpses of China in the inter-war years we have turned to Harold Acton; but there is also John Blofeld, whose youth and scholarly outlook, whose love of youth and of the scholarly outlook in the East, permit him to see something of the reality there. As in the pages of Russell, there is in his no breath of superiority. The absence of that Western barrier dissolved the inherent Chinese barrier.

Blofeld's picture[2] is of a different Peking—of the net of the demimonde carelessly thrown over the city. His is the world of the singing-master of the

2. His *City of Lingering Splendour,* although published in 1961, describes the Peking of the late 1930's.

Peking opera and his epicene boy-pupil, the world of that strange theater of mime and male heroines; of opium dens, of the "flower-girls"—the prostitutes of the brothels in the area of theater and pleasure outside Ch'ien-men, the city's south gate. There are picnics with those girls and with scholars in the glory of the autumnal Western Hills, there is a miraculous scene at the white marble platform of the Altar of Heaven as it lay under snow; there are discussions with a Taoist adept; descriptions of the city in the night, and at dawn. At twenty-two Blofeld experienced his Peking with a kind of rapture, and with a total acceptance that it contained only people *di nostra qualità*. It was a Peking, a China, very far from the resolute Western-ness of that gilded (and heavily guarded) cage of the legations. "Most inmates of the Legation Quarter led lives so detached from the currents of life swirling around them that they were always the last to know what was happening, if they ever came to know it at all. . . . In 1948, only a month or two before the Red Army's bombardment of Peking's airfield ushered in the first stages of a siege that was to end in total victory for the Communists, there were foreign consular officers so ignorant of the opinions of every Chinese in the city that they still accepted Nationalist assurances that there was nothing to fear. Apparently they trusted Government-inspired news in Nationalist newspapers as fully as the Singapore authorities trusted to their seaward-facing guns!"

The intelligent Western public in Europe and America did not, unfortunately, read the learned books of historians, nor did it take up, except to glance at the pictures, the volumes on the history of Chinese art. Opinion about the Chinese was formed by newspapers, magazines, and the more popular books. Questions of politics, of art and history, of the development and achievements of Chinese civilization as a whole were, and to a lesser extent still are, generally regarded as too abstruse for pleasurable reading. Of those many books whose writers' intentions were to please, to inform, to persuade, there is perhaps only one written by an acknowledged literary figure which contains a penetrating picture of the Chinese and their way of life, and also ranks as literature in the strict sense. It is called *Escape with Me!* and was published in 1939. Osbert Sitwell, who wrote it, gives his own reasons for going:

"I journeyed to China . . . very largely to escape from Europe, but more especially in order to see China, and the wonderful beauty of the system of life it incorporated, before this should perish; I did not go there to observe the form taken by the Social Struggle (though one could not help seeing the increasing grip in those days of the communist creed upon all the younger and more intelligent students of the universities; a result of despair, of the hopeless situation in which China patently found herself), nor out of pure love of wandering, nor, alas! in response to a request from my publishers to write a strong left-wing book about that country."

He continues: "But . . . while confessedly I am no sinologue, I grew, inevitably, to comprehend certain facts about China and Chinese life which are almost impossible to understand without visiting the country. . . .

"I think no single day passed [during my stay in Peking] except I wandered at least once in some part of the Palace, sometimes only for twenty minutes, sometimes for hours, and I have seen it in winter and summer, under rain and sun and snow. And the Forbidden City is the heart of that metropolis I came to know and love, in similarly watching its aspect change through the seasons from winter to full summer. On arrival there, all save the Forbidden City seemed a bare, Breughel-like world of brown lanes, squat and narrow, of ribbed brown roofs, and of tall, naked trees posing their neat but web-like intricacies above them against a deep-blue sky (except when a dust-storm whirled down from the Gobi Desert, carrying its load a thousand feet in the air, overcasting the sun with a thin yellow cloud, and insinuating a fine layer of sand upon every chair and table, and even between the pages of my notebooks), and of figures in padded blue robes, or patched blue canvas, and crowned, many of them, with triptychal fur hats that framed faces in a new way: when I left, it was a sighing, young summer forest, the gardens were full of blossoms and on the stone paving stood plants, moulded to the fashion of the trees on a Chinese wallpaper, and large earthenware bowls of goggling goldfish, engaged in their eternal skirt-dance of flowing fins and veils, while figures in the thinnest silk gowns fanned themselves beneath the tender, quivering shadow of young leaves. I stayed there long enough . . . to appreciate . . . the truth of what Dr. Derk Bodde says. . . ." And he quotes: "'. . . the sentiment . . . of the essential oneness and harmony of man and universe. It is a sentiment which permeates much of the greatest Chinese art and poetry, for in the Chinese, as perhaps in no other people, has been developed a keen consciousness and awareness of the movement and rhythm of nature. . . . It is an awareness which has made them deliberately subordinate their own activities to that of the force of nature. . . .'"

In this way, in the preface to his book Osbert Sitwell lights with perfect sureness on one of the salient facts in Chinese existence. At the same time he sets down in a sentence or two what everyone who has ever lived in Peking will at once agree is a remarkable short evocation of the city in its changing moods. We lamented, reading Marco Polo, that his was not the poet's eye. It is sad that there were no Western poets in Peking in all the ages until so late in the piece, until just as the ancient, dusty, sweet, and hieratic body of China was about to enter a new and different chapter of its history. But it was not quite too late when Sitwell came to Peking. The understanding, the acuity, and the enviable gift of translating the messages conveyed by those qualities into a poetry fitting to the subject, allow us at this last moment in the old Peking to see it—so strange, yet almost as if it were a place we have always known:

"You must picture to yourself [in the Imperial Palace] immense deserted court after immense deserted court, spacious enough for the most splendid ceremonials. Each large building, or group of buildings, is flanked by smaller arcaded cloisters, of great extent —cloisters that seem to call for occupation by the Imperial Guard, for the sight of their uniforms and the sound of their cries. In front, in the middle, a high, slanting-roofed hall, tiled with orange or yellow (the tiles change from one colour to the other with the play of the light), stands pitched like a tent on triple-decked terraces of white marble, that seem, in the same way that the arcades evoked the Imperial Guard lounging in the light of dead suns, by their beauty of modulation and colour and surface, to demand groups of Mandarins in Court dress, kowtowing like painted tortoises before the Emperor himself. . . .

"Across the first and greatest of the courts, runs down its marble channel in a wide and most delicate curve, the celebrated canal, Golden Water River, with three fine marble bridges crossing it. To me, for one, observing this prodigious enclosure, it always seemed that . . . in the line of the canal and the poetry of its waters, flickering reflected sunlight under perfect spans, in the workmanship and design of the balustrades, in the vista of lofty, superb halls in front with their scarlet and green eaves high in the air, and still above, in the vista of roofs fitting, one into another, with supreme art, the Palace reached its culmination. The theme is one of ascent, flowing and progressive from flight of steps to flight of steps, from terrace to terrace, from lower deck to upper deck, from one building to another. The similarity and slight difference between court and court, roof and roof, outline and outline, invests it with the depth and immensity of a night sky when, gazing at it, you try to identify a single star. Yet each court constitutes a world of its own, remote in feeling as the moon; a world wherein nothing happens except the ceaseless, clock-like sweep of cold, golden light from east to west, across it, across the marble floors, across the broad and shallow flights of steps, across the canal, lying like a scimitar below, across the cracking, sagging terraces, their surface dry and powdery with age.

"As we stay there, gazing about us, and noticing how one court leads with line and colour, into the next, we begin to grasp a little the difference between the Western European and Chinese conceptions of architecture: for each offers to the other, as do the two methods of life of the races who evolved them, a complete alternative."

It is a pity to stop there in the quotation, for the following two pages contain the most accurate, subtle, and unmatched analysis of the genius of Chinese architecture ever penned. "Court opens out of court, hall out of hall, united by a marked symphonic similarity. It is an art of suave and subtle modulation, intensely civilized and well-bred, as against one of opposition. The risk—as a rule triumphantly avoided—which through its very nature Chinese architecture incurs, is monotony; that to which European is liable, is extravagance." Chinese architecture "strives less. It does not, as we do, build for eternity. (In the same way, the Chinese does not aspire to the same degree as Western man does to *personal* immortality, to the striving of an enlarged ego to survive *as an ego* and dominate through eternity, but relies, instead, wholly on a modified, articulated survival, from father to son, for countless generations.)"

One can only read on, pinned by the magic accuracy with which Osbert Sitwell deduces, from the manner of their architecture, the manner of their minds and the feeling of their hearts. Only a poet can do this. And Sitwell is the only Western poet of Peking.

"Just as one has only to exteriorize such processes [of thought], to look at a goldfish, a flowering tree or a Pekingese dog, to be convinced, once and for all, of the difference between them and their equivalents in Europe, so with the ways of thinking which, over a long period, produced these objects and creatures. . . . Thus, for example I do not believe that either Emperor or Republic, or any General, was ever popular in Peking." And he makes the distinction between "Chinese emperors" and "emperors of China" who "belonged to one more of the several foreign dynasties . . . which had imposed themselves by force (and the Chinese, though tolerant, despise both foreigners and force) for a term of two or three centuries upon a peaceful people; while on the other hand the Republic was a new idea (and the majority of Chinese are by nature violently opposed to new ideas). As for the Generals in the Civil War, one was as good as another. And so, in Peking, all kinds of flags, in addition to Imperial and Republican, were always, though ingeniously hidden, kept in readiness in nearly every house, in case they should be wanted: for, to the Chinese, the way of the Vicar of Bray is the Path not only of Wisdom, but of Virtue, and I am only surprised that there is not a temple to that ecclesiastic among those countless shrines to strange gods that are to be found everywhere in Peking. Because you should, they hold, as a matter of principle, accept and adapt yourself to events and not hurl yourself suicidally against them. . . . Thus the citizens of Peking are willing to hang out any flags required by expediency, and to render unto Caesar even the things which are not Caesar's, if he demands them. . . . Whichever—or whoever—entered the city in triumph would be received with triumph; but it would signify nothing. . . . For the rest, people will fall in with almost any plan for their own domination, or regeneration, because they know that in the end it will fail. They have seen so much in the long life of their country. Upon one thing only are they united and resolute: they will not allow outside events to break in upon their lives. Even so rigid a faith as Communism, if for the sake of convenience it had temporarily to be accepted, would find itself powerless to alter the national character; on the contrary, the national character would very soon modify Communism to suit itself, or even assimilate it, as it has always assimilated foreign conquerors."

In the light of current events, when Communist ideology has been so thoroughly indigenized by the Chinese as to appear to them pure and unadulterated only in their own country, these remarks are interesting. Part, at least, of the reasons for the rift between Russia and China—the Chinese necessity to have a monopoly of the only right way of civilized life—is foreshadowed by Osbert Sitwell.

It is an evocation, this book—one as haunting as the city itself—an evocation of Peking and of China as the last seconds of that long hour, the Chinese past, vanished, and the uninscribed space of the present appeared.

Refreshing as the sentiments and opinions of these writers are, such men with open minds were rare and, in all important senses, little regarded. Nothing comparable to the intellectual revolution of the Enlightenment in eighteenth-century Europe followed on the publication of their books. The significance of these twentieth-century writers is really that the opinions they expressed were far in advance of those generally held at that time before the revolution in China. The popular, and also the official, Western view of China and the Chinese continued to be that of a degraded people wallowing in the hopeless morass of their broken civilization. And deriving from this view was another: that the only way with China was to control it by the threat or the use of Western force. Mercantile profits were certainly dwindling as the break-up of the old order in China proceeded, but merchants and their governments still hoped for better times. Better times were, broadly, construed as some form of internal stability in China, whether produced by a monolithic warlord government backed by Western arms, or by some other more direct Western control.

As China's bloody internal struggle, and its other struggle against the forces of the external world that dominated the economy, continued, the enlightened ideas of a handful of Western thinkers made virtually no impact in their own countries, dedicated as these were to the suppression of all that was progressive and antiforeign in China. Yet, like the tale of Jesuit and other occasional comprehension of the greatness of the Chinese, of their civilization and its potential, the record left by those few Westerners who attempted to run a course against the winds of Western incomprehension, is worth setting down. For it is a human testimony still disbelieved in the hearts and in the capitals of the non-Chinese world. When the time comes for dispassionate thinking on China, it will be recalled that those few men who knew and loved the country and its peoples raised their small, reasonable protest against the pervading inhumanity of Western judgment.

CHAPTER TWENTY

THE PAST, THE PRESENT, AND THE FUTURE

ALL STORIES must come to an end. In reality, of course, no story ever ends. But always there comes a moment when the point is clear. The moment came in China with the death of the Empress Dowager in 1908 and the fall of the Manchu dynasty three years later. For the Chinese, even for the Westerners, for everyone capable of looking back and looking forward in history, it was a moment of intense drama. The body of the past, writhing in its death agony for a hundred years or so, now gave a final convulsive twitch and expired. Only the carapace, enduring as that of a turtle, finely marked by two thousand years of Chinese civilization, was left. No greater empire, none whose achievements in life's many avenues were more profound, had ever been continuously for so long a time on the face of the earth. That it had, in many departments, fossilized in the early Ming days, that its last century had been one of disruption and bankruptcy, that the incursions of the West had eaten its foundations, that its last rulers were alien Manchu people—none of this invalidated its essential strength. The Chinese way of life remained to the last deeply embedded in the hearts and the heads of all Chinese—or almost all. No other empire had ever contained an overwhelming majority of people who felt themselves united in belonging to a single race as did the great Han, as they called themselves.

When the end came in 1911, China changed its mind for the first time in two millenniums. The dynastic system, almost unaltered in its fundamentals from the time of Ch'in Shih-huang-ti, before the time of Christ, suddenly crumbled. And Dr. Sun Yat-sen's brash, ill-conceived, and, as it turned out, hopeless republic put on the mantle of the emperors. Now there was no more a Son of Heaven on the Dragon Throne at Peking, and the Chinese were suddenly no longer the Seed of the Dragon. They discovered they were citizens of a republic—to them an incomprehensible form of rule. Their bewilderment, naturally, was complete. In their hearts and often in their heads they resisted the change.

But there it was, this new pattern. The West—the hated and (to the Chinese) barbarous West—had somehow won the day. The ugly, hairy, big-nosed barbarians of the Western Ocean, smelling unpleasantly of beef, abrupt in their manners, devastating in their armaments, with their immunity from the law and their puerile religion—these strange people had finally won. The Chinese were dimly aware that it was not at all Western ideas, philosophies, or religion which had pushed their own way of life over the precipice. They began to understand that it was Western science, Western arms, which had conquered the old system in China. And having understood this, they found they had almost no means to draw level with the West in those departments of life.

The brief republic of Dr. Sun Yat-sen lasted a mere few years before it broke up into warlord activity. The West remained in China, infinitely stronger there than ever it had been. With the help of the West the warlords tore apart the remainder of what was once China. With Western money and Western arms they were more powerful than any of the warlords of old. With them—smiled on, frowned on, as they were, one after the other by Western powers who understood nothing of China and whose main preoccupation was to stay in the country and reap an ever increasing harvest from it by supporting the winning side—a new era of change began. Soon the Japanese arrived, waging a war of attrition and infiltration. And then, as the factions inside China grouped themselves, came civil war between the Nationalists and the Communists.

Of all the decades of internal strife in the Chinese past, there never was one such as this. Not even in the river of blood and destruction that was the Taiping revolt, with its twenty million dead, had there been anything comparable to this forty-year hell.

From this total chaos there emerged eventually a force capable of unifying and of ruling—the Communists, who took over in 1949. As soon as the Communist government assumed power, they expelled the Westerners from China. Once more, and for the first time since the end of the Ming dynasty in 1644, the Chinese took into their own hands the destinies of their country. Once more, as in Ming times, they drew a clear line round the borders of the Chinese lands and looked inward on themselves, and set about their own Chinese business. What confronted them was a task perhaps greater than that attempted by any nation in the history of the world. For, to all intents and purposes, China had stopped in her tracks several centuries before. It was as if the Europe of the sixteenth century had been suddenly confronted with an alien civilization already developed to twentieth-century standards, and had had to make the attempt to bring itself up to date. Whatever the failures and whatever the successes of the Communist regime in China since 1949, they have to be measured against that simple fact.

Unlike the great Ming, however, the new regime did not look back to a former glorious Chinese age as a model. Their creed was Communism—a peculiarly Western set of principles. Gradually repudiating the Russians (the only Westerners not at first excluded from the country), the Chinese as gradually, and with a determination comparable to that of the early Ming, began not only the process of turning the sixteenth century into the twentieth, but at the same time began to make of Communism an Oriental, a peculiarly Chinese thing. Just as in the past every idea the Chinese had taken from outside the frontiers (Indian Buddhism is a good example) had been indigenized and, so to speak, naturalized, so with Communism. By now, in a very real sense, Communism has been converted by the Chinese into a deeply Chinese style: and of course the Russians, the last Westerners in China, have had to go.

The Middle Kingdom, the greatest civilization on earth, the people in possession of the ultimate right thinking—we have seen those apparently timeless ideas making the texture and forming the basis of the outlook of the Chinese all the way down the centuries since they first came in contact with the Western travelers. Now it would seem that China is convinced that *her* Communism is the only right and true Communism. It is hardly surprising. With the great emperor Ch'ien-lung, and in much the same resounding phrases as he wrote to the English king George III, China of today seems to be saying to the rest of the world: "Swaying the wide world, I have one single aim: to maintain a perfect governance. . . . I set no value on strange or ingenious objects. . . . It behoves you . . . to respect my wishes . . . so that by perpetual submission to our Throne you may secure peace and prosperity for your country hereafter. . . ."

From that time in 1949 a new story began—the story of China able to deal (as Ch'ien-lung was not, and as the Chinese until 1949 were not) with the West from a position of some strength. Now it is not the travelers from the West who make the West-East story. It is governments with rival ideologies. The barbarians and mandarins have turned into democrats (of one sort or another) and Chinese communists.

So our story is ended. To round off the tale of that long dialogue between the barbarians and the mandarins, perhaps we may turn a little way back to 1727 when the emperor Ch'ien-lung's predecessor sat on the Dragon Throne. In an undistinguished reign, this emperor, Yung-cheng, wrote one fine day a comment on which, pondering the past, we may well conclude: "The people of the world are bigoted and unenlightened: invariably they regard what is like them as right, and what is different from them as wrong, resulting in mutual recrimination. . . . They do not realize that the types of humanity are not uniform and that their customs are also not one, that it is not only impossible to force people to become different but also impossible to force them to become alike."

Maybe that is the simple truth.

BIBLIOGRAPHY

THE LISTING that follows may strike the average reader as somewhat formidable. This arises from its twofold purpose: 1) to include a number of books of basic importance for the reader who wants further background information, and 2) to give full bibliographical information for publications that have been referred to, in shortened form, elsewhere in the book. Moreover, as the reader was gently warned in the preface, this bibliography represents the merest particle in a world of documentation and comment in many languages on our subject. Anyone with time to dip into a dozen or so of the books listed will discover, first, that they in turn list hundreds of additional sources, and second, that the present volume has done no more than etch the main lines of the complex picture of the East and West in confrontation. Books especially recommended for further background on the subject are marked with an asterisk (*).

Acton, Harold. *Memoirs of an Aesthete.* Methuen, London, 1948.

Backhouse, E., and J. O. P. Bland. *Annals and Memoirs of the Court of Peking.* Houghton Mifflin, New York, 1914; Heinemann, London, 1914.

Bacon, Roger. *The Opus Major.* Trans. by Robert B. Burke from *Opus Majus.* U. of Pennsylvania Press, 1928.

*Baddeley, John F. *Russia, Mongolia, China: Being some record of the relations between them from the beginning of the seventeenth century to the death of Tsar Alexei Mikhailovich, A.D. 1602–1676.* 2 vols. Macmillan, London, 1919.

*Banno, Masataka. *China and the West, 1858–1861.* Harvard U. Press, 1964.

Barrow, John. *Travels in China,* 2nd ed. Cadell & Davies, London, 1804.

Bartoli, Danielo. *Dell' historica della compagnia di Gesù;* Vol. I-IV, *La Cina.* Florence, 1832.

Bazin, René. *L'enseigne de vaisseau Paul Henry.* Tours, 1905.

*Beazley, C. R. *The Dawn of Modern Geography.* 3 vols. Murray, London, 1897, 1901, 1906.

Bell, John. *A Journey from St. Petersburg to Pekin, 1719–22.* Ed. by J. L. Stevenson. Edinburgh U. Press, 1965.

——. *Travels from St. Petersburg in Russia to Diverse Parts of Asia.* Foulis, Glasgow, 1763.

*Bernard, Henri. *Le père Mattieu Ricci et la société chinoise de son temps.* Mission de Sienhsien, Tientsin, 1937.

Bland, J. O. P., and E. Backhouse. *China Under the Empress Dowager.* Heinemann, London, 1910.

Blofeld, John. *City of Lingering Splendour: A Frank Account of Old Peking's Exotic Pleasures.* Hutchinson, London, 1961.

Boxer, C. R. *Fidalgos in the Far East, 1550–1770.* Mijhof, The Hague, 1948.

——. *Macau na epoca dea restauracao (Macau Three Hundred Years Ago).* Imprensa Nacional, Macao, 1942.

Bredon, Juliet. *Peking.* Kelly & Walsh, Shanghai, 1922.

Bretschneider, E. *Notes on Chinese Medieval Travellers to the West.* American Presbyterian Mission Press, Shanghai, 1875; Trubner, London, 1875.

Bronowski, J., and B. Mazlish. *The Western Intellectual Tradition.* Hutchinson, London, 1960.

Brumfitt, J. H. *Voltaire: Historian.* Oxford U. Press, 1958.

Cameron, Nigel, and Brian Brake. *Peking: A Tale of Three Cities.* Harper & Row, New York, 1965; Weatherhill, Tokyo, 1965.

Cartier-Bresson, Henri, and Barbara R. Miller. *China: Photographed by Henri Cartier-Bresson.* Bantam, New York, 1964.

*Cary-Elwes, Columba. *China and the Cross: Studies in Missionary History.* Longmans Green, London, 1957.

Chambers, Sir William. *Designs of Chinese Buildings, Furniture, Dresses, Machines, and Utensils, to which is annexed a Description of their Temples, Houses, Gardens, etc.* London, 1757.

Chang Hsin-pao. *Commissioner Lin and the Opium War.* Harvard U. Press, 1964.

Chang T'ien-tse. *Sino-Portuguese Trade from 1514–1644.* Brill, Leyden, 1934.

Ch'en Yuan. *Westerners and Central Asians in China Under the Mongols: Their Transformation into Chinese.* Trans. and annotated by L. Carrington Goodrich. Monumenta Serica, U. of California Press, Los Angeles, 1966.

Cheng, J. C. *Chinese Sources for the Taiping Rebellion, 1850–1864.* Hong Kong U. Press, 1963.

*Cohen, Paul A. *China and Christianity: The Missionary Movement and the Growth of Chinese Antiforeignism, 1860–1870.* Harvard U. Press, 1963.

Coleridge, H. J. *The Life and Letters of St. Francis Xavier,* 2nd ed. Burns & Oates, London, 1881.

Collis, Maurice. *Foreign Mud.* Faber, London, 1946.

——. *The Grand Peregrination.* Faber, London, 1949.

——. *The Great Within.* Faber, London, 1941.

Conquêtes des chinois. Paris, late 18th century.

*Cordier, Henri. *Histoire générale de la Chine.* Geuthner, Paris, 1920.

Cox, Edward G. *A Reference Guide to the Literature of Travel: Including Voyages, Geographical Description, Adventures, Shipwrecks, and Expeditions.* 2 vols. U. of Washington Press, 1935.

*Cranmer-Byng, J. L. (ed.). *An Embassy to China: Being the Journal Kept by Lord Macartney During His Embassy to the Emperor Ch'ien-lung, 1793–1794.* Longmans Green, London, 1962.

Cronin, Vincent. *The Wise Man from the West.* Hart-Davis, London, 1959.

Davis, Andrew M. *Certain Old Chinese Notes.* For the American Society of Arts and Sciences; Littlefield, Boston, ?1915.

Davis, John F. *Chinese Miscellanies.* London, 1865.

*Dawson, Christopher. *The Mongol Mission.* Sheed & Ward, London, 1935.

de Bry, Theodor. *Indiae Orientalis.* Frankfurt, 1607.

de Lange, Laurence. His journals are quoted in Bell's *Travels.*

*d'Elia, Pasquale M. (ed.). *Fonti Ricciane: Documenti originali concernenti Matteo Ricci e la storia delle prime relazioni tra l'Europa e la Cina (1579–1615).* 3 vols. Libreria dello Stato, Rome, 1942–49.

——. *Galileo in China.* Harvard U. Press, 1960.

de Mailla, Joseph Anne Marie Moyriac. *Histoire générale de la Chine.* 13 vols. Paris, 1777–85.

*Dennett, Tyler. *Americans in Eastern Asia: A Critical Study of United States Policy in the Far East in the Nineteenth Century.* Barnes & Noble, New York, 1922.

di Cora, John. *Livre de l'estat du grant caan.* Quoted in Yule's *The Book of Ser Marco Polo.*

d'Orleans, Joseph. See Ellesmere.

Downing, C. Toogood. *The Fan-Qui in China in 1836–7.* Colburn, London, 1838.

*Dowson, Raymond (ed.). *The Legacy of China.* Clarendon Press, Oxford, 1964.

Dunne, George H. *Generation of Giants.* U. of Notre Dame Press, 1962.

*Eberhard, Wolfram. *A History of China.* Trans. by E. W. Dickes. U. of California Press, 1950; Routledge & Kegan Paul, London, 1950.

Ecke, G., and P. Demiéville. *The Twin Pagodas of Zayton: A Study of Later Buddhist Sculpture in China.* Harvard U. Press, 1935.

Edkins, J. *Description of Peking.* Shanghai Mercury, Shanghai, 1898.

Elimosina, Friar. Quoted in Yule's *Cathay and the Way Thither.*

*Ellesmere, Earl of (ed. and trans.) *History of the Two Tartar Conquerors of China, including the two journeys into Tartary of Father Ferdinand Verbiest in the suite of the Emperor Kang-hsi: from the French of Père Joseph d'Orleans. To which is added Father Pereira's journey into Tartary in the suite of the same emperor, from the Dutch of Nicolass Witsen.* (Originally published as *Histoire des deux conquérans tartares, qui ont subjugé la Chine;* Paris, 1688.) Hakluyt Society, London, 1854.

Ellis, Henry. *Journal of the Proceedings of the Late Embassy to China.* Murray, London, 1817.

Fa-hsien. See Giles.

*Fairbank, John K. *The United States and China.* Harvard U. Press, 1962.

Fan Tseng-chung. "Dr. Johnson and Chinese Culture." Included in *Nine Dragon Screen.*

*Favier, Alphonse. *Pèking: Histoire et description.* Imprimerie des Lazaristes au Pé-T'ang, Peking, 1897.

Ferguson, Donald. *Letters from Portuguese Captives in Canton, Written in 1534 and 1536.* Education Society's Steam Press, Bombay, 1902.

*Fitzgerald, C. P. *China: A Short Cultural History.* Cresset Press, London, 1935 (rev. ed. 1961); Appleton-Century, New York, 1938.

* ——. *The Chinese View of Their Place in the World.* Oxford U. Press, 1964.

Fleming, Peter. *The Siege at Peking.* Hart-Davis, London, 1959.

Fox, Grace. *British Admirals and Chinese Pirates, 1832–1869.* Kegan Paul, London, 1940.

Fox, Ralph. *Genghis Khan.* Harcourt Brace, New York, 1936; Lane, London, 1936.

*Gallagher, Louis J. See Trigault.

Gernet, Jacques. *Daily Life in China on the Eve of the Mongol Invasion, 1250–1276.* Trans. by H. M.

Wright. Allen & Unwin, London, 1962.

Giles, H. A. (trans.). *The Travels of Fa-Hsien.* Routledge & Kegan Paul, London, 1956.

*Goodrich, L. Carrington. *A Short History of the Chinese People.* Harper, New York, 1943; Allen & Unwin, London, 1948.

Grantham, Alexandra E. *A Manchu Monarch.* Allen & Unwin, London, 1934.

*Grousset, René. *The Rise and Splendour of the Chinese Empire.* Trans. by A. Watson-Gandy and T. Gordon. Bles, London, 1952; U. of California Press, 1953.

*Gutzlaff, Karl. *Journal of Three Voyages Along the Coast of China in 1831, 1832, and 1833.* Westley & Davis, London, 1834.

Hake, A. E. *The Story of Chinese Gordon.* Remington, London, 1884.

*Han Suyin. *The Crippled Tree.* Jonathan Cape, London, 1965.

Hazard, Paul. *The European Mind, 1680–1715.* Trans. by J. Lewis May. (Originally published as *La crise de la conscience europêene.* Boivard, Paris, 1935.) Penguin, Harmondsworth, 1964.

Henry, Leon. *Le siège de Pé-t'ang dans Pékin en 1900: Le commandant Paul Henry et ses treinte marins.* Peking, 1908.

*Hertslet, G. E. P. *China Treaties.* 2 vols. H. M. Stationery Office, London, 1908.

Hibbert, Eloise T. *K'ang-hsi, Emperor of China.* Kegan Paul, London, 1940.

Hirth, F. *China and the Roman Orient.* Kelly & Walsh, Shanghai, 1939.

*Honour, Hugh. *Chinoiserie: The Vision of Cathay.* Murray, London, 1961.

Huc, M. *Christianity in China, Tartary, and Thibet.* Longmans, London, 1858; Kennedy, New York, 1897.

Hudson, G. F. *Europe and China: A Survey of Their Relations from the Earliest Times to 1800.* Arnold, London, 1931; Beacon Press, Boston, 1961.

Ibn Battuta. *Travels in Asia and Africa.* Trans. by H. A. R. Gibb. Routledge & Kegan Paul, London, 1929.

John of Marignolli. Quoted in Yule's *Cathay and the Way Thither.*

Johnson, Reginald F. *Confucianism in Modern China.* Gollancz, London, 1934; Appleton-Century, New York, 1935.

*Josson, H., and L. Willaert (eds.). *Correspondance de Ferdinand Verbiest de la Compagnie de Jésus.* Palais des Académies, Brussels, 1938.

Journey of the Emperour of China into East Tartary in the Year 1682, A. See Verbiest.

Komroff, Manuel. *Contemporaries of Marco Polo.* Boni & Liveright, New York, 1928; Jonathan Cape, London, 1928.

Krausse, Alexis S. *Russia in Asia, 1558–1899.* Holt, New York, 1899; Richard, London, 1899.

Lach, Donald F. *Asia in the Making of Europe.* 2 vols. U. of Chicago Press, 1965.

Lamb, Alastair. *Britain and Chinese Central Asia.* Routledge & Kegan Paul, London, 1960.

Langdon, William B. *A Descriptive Catalogue of the Chinese Collection, now exhibiting at St. George's Place, Hyde Park Corner, London, with condensed accounts of the genius, government, history, literature, agriculture, arts, trade, manners, customs and social life of the people of the Celestial Empire.* London, 1842.

*Latourette, Kenneth S. *The Chinese, Their History and Culture.* Macmillan, New York, 1960.

* ——. *A History of Christian Missions in China.* Macmillan, New York & London, 1929.

Lattimore, Owen. *Inner Asian Frontiers of China.* American Geographical Society, 1940; Beacon Press, Boston, 1962.

——. *Nomads and Commissars: Mongolia Revisited.* Oxford U. Press, New York, 1962.

le Comte, Louis. *Nouveau memoire sur l'état présent de la Chine.* Paris, 1697.

le Gobien, C. (trans.). *Edifying and Curious Letters of Some Missioners of the Society of Jesus, from Foreign Missions.* 1707. (See also Querbeuf.)

Levy, Howard S. *Chinese Footbinding: The History of a Curious Erotic Custom.* Rawls, New York, 1966; Weatherhill, Tokyo, 1966.

Li Yu. *Jo Pu Tan* (The Prayer Mat of the Flesh). Trans. by Richard Martin under the title *The Before Midnight Scholar* from the German version; Deutsch, London, 1965.

Little, Mrs. Archibald. *The Land of the Blue Gown.* Unwin, London, 1902.

——. *Li-Hung Chang: His Life and Times.* Cassell, London, 1903.

——. *Round About My Peking Garden.* Lippincott, Philadelphia, 1905; Unwin, London, 1905.

Liu Kwang-ching. *Americans and Chinese.* Harvard U. Press, 1963.

Macartney, Lord. See Cranmer-Byng.

McGhee, R. J. L. *How We Got to Peking: A Narrative of the Campaign in China of 1860.* Bentley, London, 1862.

Mahler, Jane G. *Westerners Among the Figurines of the T'ang Dynasty of China.* Istituto Italiano per il Medio ed Estremo Oriente, Rome, 1959.

Malone, Carrol B. *History of the Summer Palaces Under the Ch'ing Dynasty.* U. of Illinois Press, 1934.

Martin, W. A. P. *The Siege in Peking: China Against the World.* Revell, New York, 1900; Oliphant, Anderson & Ferrier, London, 1900.

Matignon, Dr. J.-J. *La Chine hermétique: Superstitions, crime et misère.* Geuthner, Paris, 1936 (first published 1898).

Maugham, Somerset. *On a Chinese Screen.* Doran, New York, 1922; Heinemann, London, 1922.

Mayer, Karl A. *Neapel und die Neapolitaner.* Oldenburg, 1840–42. Quoted in the Prandi edition of Ripa's *Memoirs.*

Medhurst, W. H. (trans.). *"The Shoo King"* [Shu Ching] *or Historical Classic.* Mission Press, Shanghai, 1846.

*Mendoza, Juan Gonzales de. *The History of the Great and Mighty Kingdom of China.* Trans. by R. Parke; ed. by Sir George T. Staunton, with an introduction by R. H. Major, Hakluyt Society, London, 1853.

Meng, S. M. *The Tsungli Yamen: Its Organization and Functions.* Harvard U. Press, 1962.

Meskill, John (ed.). *The Pattern of Chinese History: Cycles, Development, or Stagnation?* Heath, Boston, 1965.

*Morse, Hosea B. *The Chronicles of the East India Company: Trading to China, 1635–1834.* 3 vols. Harvard U. Press, 1925; Clarendon Press, Oxford, 1926–29.

* ——. *The International Relations of the Chinese Empire.* 3 vols. Longmans Green, London, 1910–18.

*Moule, A. C. *Christians in China Before 1550.* Macmillan, New York, 1930; Society for Promotion of

Christian Knowledge, London, 1930.
——. *The Rulers of China, 221 B.C.–A.D. 1949.* Praeger, New York, 1957; Routledge & Kegan Paul, London, 1957.
Mundy, Peter. *The Travels of Peter Mundy in Europe and Asia, 1608–1667.* Hakluyt Society, London, 1919.

*Needham, Joseph. *Science and Civilisation in China.* 4 vols. Cambridge U. Press, 1954–62.
Neill, Stephen. *A History of Christian Missions.* Penguin, Harmondsworth, 1964.
*Nieuhof, John. *An Embassy from the East Indian Company of the United Provinces,* etc. Trans. by John Ogilby. London, 1673.
Nine Dragon Screen. Selected reprints of papers read to the China Society. China Society, London, 1965.

Odoric of Pordenone. Quoted in Yule's *Cathay and the Way Thither.*
Oliphant, Laurence. *Narrative of the Earl of Elgin's Mission to China and Japan in the Years 1857-59.* Blackwood, London, 1880.
Olschki, Leonardo. *Guillaume Boucher: A French Artist at the Court of the Khans.* Johns Hopkins U. Press, 1946.
Opthalmic Hospital at Canton, The 16th Report of the, for the years 1850–1851. Canton, 1852.

Parker, Peter. See Stevens and Marwick.
Pastor, Dr. L. von, *History of the Popes.* Kegan Paul, London, 1891.
Patch, Blanche. *Thirty Years with G. B. S.* Gollancz, London, 1951.
Pearson, Hesketh. *Bernard Shaw: His Life and Personality.* Methuen, London, 1961.
Pegolotti, Francis Balducci. Quoted in Yule's *Cathay and the Way Thither.*
Pereira, Father. See Ellesmere.
Pfister, Louis. *Notices biographiques et bibliographiques sur les Jésuites de l'ancienne mission de Chine, 1552–1773.* 2 vols. Shanghai, 1932.
Pinkerton, John (ed. and trans.). *A General Collection of the Best and Most Interesting Voyages and Travels in All Parts of the World; many of which are now first translated into English, digested on a new plan.* 8 vols. Longman, Hurst, Rees, Orme & Brown, London, 1808–14.
Pirès, Tomé. *The Suma Oriental of Tomé Pirès.* Trans. by Amando Cortesão. Hakluyt Society, London, 1944.
*Polo, Marco. *Description of the World.* Of the many editions available mention may be made of: *The Travels of Marco Polo,* ed. by W. Marsden, rev. by T. Wright; Everyman Library, Dent, London, 1908. *Marco Polo (1254–1325): The Description of the World,* trans. and annotated by A. C. Moule and P. Pelliot; Routledge, London, 1938. And the works of Henry Yule, particularly *The Travels of Marco Polo.*
Pratt, Sir John T. *China and Britain.* Collins, London, n. d.
Prawdin, Michael. *The Mongol Empire.* Trans. by Eden and Cedar Paul. Allen & Unwin, 1940
Previté-Orton, C. W. *The Shorter Cambridge Medieval History.* 2 vols. Cambridge U. Press, 1962.
*Purcell, Victor. *The Boxer Uprising: A Background Study.* Cambridge U. Press, 1963.

*Querbeuf, Y. M. M. de (ed.). *Lettres édifiantes et curieuses, écrites des missions etrangères de la compagnie de Jésus.*

Vols. 16–20 concern the China mission. Paris, 1780–83. (See also Le Gobien.)

Ricci, Matteo. See Bernard, d'Elia's *Fonti Ricciane,* Trigault, and Venturi.

*Ripa, Matteo. *Memoirs During 13 Years Residence at the Court of Peking in the Service of the Emperor of China.* Trans. by Fortunato Prandi. Murray, London, 1846.

——. *Storia della fondazione della congregazione e del collegio de' Cinesi sotto il titolo della S. Famiglia de G. C.* 3 vols. Naples, 1832.

Roberts, F. M. *Western Travellers to China.* Kelly & Walsh, Shanghai, 1932.

*Rockhill, William W. *Diary of a Journey Through Mongolia and Tibet in 1891 and 1892.* Smithsonian, Washington, 1894.

* ——. See William of Rubruck.

Russell, Bertrand. *The Problem of China.* Allen & Unwin, London, 1922.

Saeki, P. Y. *The Nestorian Monument in China.* Society for Promotion of Christian Knowledge, London, 1916.

Savage-Landor, A. Henry. *China and the Allies.* 2 vols. Heinemann, London, 1901.

Schall, Adam (Joannis Adami). *Historica narratio, de initio et progressu missionis apud Chinenses, ac praesertim in regio pequinensi. Ex litteris.* 1665.

——. *Yuan-ching shuo* [The Telescope]. Peking, 1626.

Schutte, J. F. *Valignanos Missiongrundsätze für Japan.* Rome, 1951.

Scott, John Lee. *Narrative of a Recent Imprisonment in China.* Dalton, London, 1841.

Secret History of the Mongols, The. Trans. by Arthur Waley. Allen & Unwin, London, 1963.

Semedo, Alvarez de. *The History of the Great and Renowned Monarchy of China.* Trans. by "a person of Quality." Cook, London, 1655.

Servitá, Viani. *Istoria delle cose operate nella Cina de monsignor Gio. Ambroglio Mezzabarba.* Briasson, Paris, 1739.

Sitwell, Sir Osbert. *Escape with Me!* Macmillan, London, 1939.

Smyth, George. *The Crisis in China: A Collection of Articles from the North American Review.* New York, 1900.

Society for the Diffusion of Useful Knowledge in China, The Second Report of the. Office of the Chinese Repository, Canton, 1837.

Sogabe, Shizuo. *Shihei hattatsu shi* (The History of Paper Currency). Tokyo, 1951.

Staunton, Sir George Thomas. *Memoirs.* Privately printed, London, 1856.

Stein, Sir Aurel. *On Ancient Central Asian Tracks.* Ed. by Jeanette Mirsky. Pantheon, New York, 1964.

*Stevens, George B., and William F. Marwick. *The Life, Letters, and Journals of the Rev. and Hon. Peter Parker.* Boston, 1896.

*Teng Ssu-yu, and John K. Fairbank. *China's Response to the West: A Documentary Survey, 1839–1923.* Harvard U. Press, 1954; Atheneum, New York, 1963.

*Tong Te-kong. *United States Diplomacy in China, 1844–60.* U. of Washington Press, 1964.

*Trigault, Nicola. *China in the 16th Century: The Journals of Matthew Ricci, 1583–1610.* Trans. by Louis J. Gallagher. Random House, New York, 1953.

Trumbull, Henry C. *Old Time Student Volunteers: My Memories of Missionaries.* New York, 1902.

Tun Li-ch'en. *Annual Customs and Festivals in Peking as Recorded in the Yen-ching Sui-shih-chi*. Trans. by Derk Bodde. Vetch, Peiping, 1936; 2nd ed., Hong Kong U. Press, 1965.

Vander, Pierre. *Voyages curieux, faits en Tartarie, en Perse, et ailleurs*. Trans. by Sr. de Bergeron. 1729.

Van der Wyngaert, Anastase. *Jean de Mont Corvin: Premier évêque de Khanbaliq (Pé-King), 1247–1328*. Société Saint-Augustin, Désclée, de Brouwer, Lille, 1924.

*Varg, Paul. *Missionaries, Chinese, and Diplomats: The American Protestant Missionary Movement in China, 1890–1952*. Princeton U. Press, 1958.

*Venturi, P. T. *Opere storiche del P. Matteo Ricci*. 2 vols. Giorgetti, Macerata, 1911.

*Verbiest, Ferdinand. *A Journey of the Emperour of China into East Tartary in the Year 1682*. Excerpts from Verbiest's letters; trans. unknown. Collins, London, 1686. For other writings of Verbiest see also Ellesmere; Josson and Willaert.

Waley, Arthur. *The Opium War Through Chinese Eyes*. Allen & Unwin, London, 1958.

Weale, B. L. Putnam. *Indiscreet Letters from Peking*. Dodd Mead, New York, 1907.

Wieger, Léon. *Textes historiques*. 2 vols. Mission Press, Hsienhsien, 1929.

*William of Rubruck. *The Journey of William of Rubruck to the Eastern Parts of the World, 1253–55, as narrated by himself*. Ed. and trans. by William W. Rockhill. Hakluyt Society, London, 1900.

Williams, Samuel Wells. *The Middle Kingdom*. Allen, London, 1883.

Witsen, Nicolass. See Ellesmere.

Xavier, St. Francis. See H. J. Coleridge.

*Yule, Col. Sir Henry (trans.). *The Book of Ser Marco Polo*. 2 vols. Murray, London, 1873.

* ——. *Cathay and the Way Thither*. Ed. by Henri Cordier. 4 vols. Hakluyt Society, London, 1913–15.

* ——(trans.). *The Travels of Marco Polo*. Murray, London, 1903.

ACKNOWLEDGMENTS

IN A WORK covering such a long span of time and such a wide field of human endeavor, the author must necessarily be indebted to many people and institutions for help in a hundred ways. This book and this writer are no exceptions. It is pleasant for me to list here all those who have given me substantial help, and to thank them most heartily for it.

The following institutions have supplied microfilms, illustrations, and other material as indicated in the footnotes, captions to the illustrations, or in the text itself: Achivo Segreto Vaticano, Rome; Bavarian National Museum, Munich; Bibliothèque Nationale, Paris; Bodleian Library, Oxford; British Museum, (departments of Oriental Antiquities and Prints and Drawings, and the Library), London; Duke University Library, Durham, North Carolina; Louvre Museum, Paris; Museum of Far Eastern Antiquities, Stockholm; Musées Nationaux (Château de Versailles), France; National Maritime Museum, Greenwich, England; National Palace Museum, Taiwan; New York Public Library; Residenzmuseum, Munich; Riccardian Library, Florence; Royal Geographical Society, London; Royal Photographic Society, London.

It is perhaps invidious to make a select listing of persons who assisted me in many ways, but I hope that those whose names are not mentioned will accept my thanks, along with those listed here: Mrs. Mildred Archer, India Office Library, London, for help with William Alexander's drawings; Mr. Glen Bowersox, Asia Foundation, Kabul, for acquiring illustrations; Professor C. R. Boxer, London, for help on the Portuguese in South China; Mr. Ted Brake, London, for reading early drafts of the MS; Mr. Brian Brake, Hong Kong and MM. Henri Cartier-Bresson and Marc Riboud, both of Paris, for their photographs; Rev. Columba Cary-Elwes, St. Louis, Missouri, for help with illustration research; Lt.-Col. H. S. Francis, O. B. E., Curator of the Museum of the Institution of Royal Engineers, Chatham, England, for letting me examine and photograph relics of "Chinese" Gordon; Dr. Han Suyin, Hong Kong, for her encouragement; Miss E. Liversige, Stirlingshire Country Library, for help on John Bell; Mr. Arthur Miller-Stirling, London, for allowing me to examine and photograph relics of John Bell; Miss Mary Lee Morse, Rome, for help with research both there and, in Naples, at the remains of Ripa's Chinese College; Padre Edmondo la Malle, Chief Archivist of Istituto Storico della Compagnia di Gesù, Rome, for assistance with the Ricci MSS, etc.; Dr. Joseph Needham, Cambridge, for encouragement and stimulating criticism; Bishop Stephen Neill, Bonn, for information concerning Gutzlaff; Miss Vera Nielson, Royal Asiatic Society Library, London, for frequent help; Mr. H. A. Rydings, Librarian, and Mr. G. W.

Bonsall, Deputy Librarian, University of Hong Kong, for their help; Mr. George Savage, London, for advice on European figures in *blanc de chine* porcelain; the late Sir Osbert Sitwell, London, for sparing time to talk about China, and for permission to quote from his book *Escape with Me!;* Mr. J. L. Stevenson, Bonn, for research suggestions; Mr. Douglas Tunstell, New York, for research assistance; Miss Marjorie Wadsworth, London, for assistance with illustrations; Miss H. M. Wallis, Map Room, British Museum, for advice on portolan charts and for allowing me to spend an afternoon examining the Longobardi world globe.

Many of the books from which I have quoted have long been in the public domain—a fact that in no way lessens my gratitude to all the authors cited. For longer quotations of copyrighted material, the following publishers and organizations have kindly given their permission: Allen & Unwin, London; Barnes & Noble, New York; J. M. Dent & Sons, London; Hutchinsons, London; Macmillan, London; Random House, New York; Sheed & Ward, London; the China Society, London; the Society of Authors, London.

Finally, I want to thank three persons most particularly: Professor L. Carrington Goodrich, of Columbia University, for his careful reading of the MS and his constructive suggestions and criticisms; Mr. Meredith Weatherby, Tokyo, for his meticulous editing of the MS, which has resulted in material improvements; and Miss Freda Wadsworth, London, without whose intelligent and imaginative research and fine work on maps and plans this book would be much the poorer and its author still laboring to complete it. Needless to say, any faults the book has are mine, not theirs.

INDEX

NOTE: An "i" following a page reference indicates an illustration; an "n" indicates a footnote.

Abeel, David, American missionary, 334
Acton, Harold, on China, 409–10
Ahmed, Moslem finance minister to Kublai Khan, 83
Alans, appeal of, to Pope Benedict XII, 104–6, 117
Albuin, Gerard, bishop of Zayton, 101, 102
Albuquerque, Alfonso de, Portuguese viceroy of India, 129
Alexander, William, artist with Macartney embassy, 295–96, 305, 306 i; sketches by, 297 i, 305, 306–10 i; describes Peking, 299; regrets not seeing Great Wall, 302
Alexis, tsar of Russia, 225–26
Almeida, Jesuit priest, 166, 167, 169
Alopen, Nestorian traveler to China, 18–23
America: and opium trade, 316, 323, 333; missionaries from, 319, 329–33, 334–35, 338–39, 340–41; and Treaty of Wanghia, 326, 333, 334; Chinese name for, 333; power politics of, in China, 333–39; and trade with China, 334, 335, 338; and Treaty of Tientsin, 348; and Boxer Rebellion, 377, 395–96, 397
Amherst embassy to Peking, 320, 322
ancestor worship, 186–87, 267
Andrew of Perugia, bishop, 101, 102–4, 112
angel-type doll, 27 i
Anglo-Chinese war, *see* Opium War
Antermony House, 283, 284
antiforeignism as official Chinese policy, 146–48; under Ming, 123, 124, 131, 143–44, 201–2; under Manchu, 294, 312–15, 375–76; under Communists, 418–19; *see also* Chinese attitudes toward foreigners and the West
antimissionary feelings in China, 197–99, 201, 237–38, 281–82, 371–72, 374 i, 376
Anting Gate, 349, 352, 356
Arabs in China, 112
architecture, Chinese, 87–88, 356, 413–14
Areskine (Erskine), physician and councilor to Peter the Great, 275
Arghun Khan, Mongol ruler of Persia, 85, 86, 91, 94; signature of, 86 i
Armagh, bishop of (Richard Fitz Ralph), 118
Armenian, T'ang figurine of, 20 i
Arnold of Cologne, Franciscan priest, 96, 98
Ascelin, friar, 29
astronomy: Chinese, 159, 187–88, 206–7; European (Jesuit), in China, 187–88, 203, 206–7
auspices, see *feng-sui*

Bacon, Roger, 59
Baikoff, Theodore Isakovich, Russian envoy to Peking, 225–27, 230, 234
Baikoff embassy, *see* Baikoff, Theodore Isakovich
Baldwin III, emperor of Constantinople, 33
balloon (gift of Macartney embassy), 301
Bartholomew of Cremona, Franciscan priest, 32, 52, 58
Basil, English captive of Mongols, 53
Batu, Mongol general, 29, 31, 43–44
Beatrice of Bavaria, 119
Bell, John, Scottish traveler to China, 275–81, 282–84; background of, 275; estimate of his book, 275; in Peking, 276–81; describes audience with K'ang-hsi, 277–78; describes life in Peking, 278–80; describes Great Wall, 280–81; later life of, 283–84; tsar's gifts to, 283 i
Benedict XI, pope, 125
Benedict XII, pope, 104–6
Benedict the Pole, papal envoy to Mongols, 29, 31
Biddle, American commissioner at Canton, 334–35
Bintang, sultan of, 142–43, 144
Blofeld, John, on Peking, 410–11

Board of Rites: and Ricci, 180, 184; position of Paul Hsu on, 206, 208, 209; and Dutch mission, 228–29; and Schall, 237; and Spathary embassy, 252–57
Board of Works, Verbiest vice-president of, 242
Bodde, Derk, on Chinese sense of nature, 412
bogtak (Mongol female headdress), 115
Book of Descriptions (Pegolotti), 126, 126 i
Boucher, William, Parisian goldsmith in Karakorum, 51, 53, 55–56, 57, 58
bound feet, *see* footbinding
Bouvet, priest-mathematician in Peking, 281
Boxer puppet shows, 374 i
Boxer Rebellion, 375–99, 379 i, 394 i, 396 i; motivation of, 375–76; and Siege of Legations, 378–81; and siege of Pei T'ang, 381–92; results of, 396–97; Western writers on, 393–96
Boxers, 375 i, 375–76, 377; *see also* Boxer Rebellion
Bridgman, Elijah, American missionary, 334, 339
British opium ship, 359 i
Bruce, Frederick, British minister to Peking, 359
Buddha as term for emperor, 272
Buddhism, Buddhists: in T'ang dynasty, 21, 23, 24, 25–26, 27, 180; and Nestorianism, 23, 27; as faith of Mongols, 48, 57, 68, 75; under Western Liao, 97; and Ricci, 156, 187
Buglio, Portuguese Jesuit priest in Peking, 232, 233, 237, 240, 259
Bulgai, Mongol grand secretary, 49, 51, 52
Bureau of Astronomy, 206, 238; Verbiest president of, 242

caging of prisoners, 198–99, 201 i
calendar, Chinese: Ricci plans revision of, 187–88; revised by Jesuits, 206, 209, 224; Verbiest and Yang Kuang-hsien's version of, 238–40
Can Grande della Scala, Veronese "khan," 125–26
cannons, made for Chinese by Jesuits, 224, 242, 386
Canton: as seen by Odoric of Pordenone, 111; Portuguese in, 130, 134–39, 142, 143–45, 210; sketch of river at, 135 i; as seen by Pirès, 137; merchants of, 139, 210, 213, 214–15, 298–99; Westerners restricted to, 146, 294, 316; Dutch in, 201, 227, 229; maps of, 213 i, 321 i; British in, 214–15, 316, 324, 336, 348; and opium trade, 298–99, 316, 323–24; in Opium War, 324; Peter Parker in, 329–33, 335–36; Americans in, 333, 334–36; American missionaries in, 334–35; sacked by British, 345, 348; Mrs. Little in, 367
caricature of English sailor, 356 i
Cartier-Bresson, Henri, 403; photographs by, 403, 404–7 i
Castiglione, Jesuit priest-painter, 296
castle of Soldaia, 33 i
Catalan Map, 87
Cathay, 48, 53, 188; origin of name of, 53 n
Catholic Church, Catholicism, in China: under Mongols, 94, 98, 100, 104, 106; under Ming, 209; under Manchu, 224, 237–38, 248–49, 281–82, 371–72; *see also* Dominicans, Franciscans, Jesuits
Cattaneo, Jesuit priest, 169, 173, 189
Chambers, William, eighteenth-century writer on Chinese arts and crafts, 292–93
Ch'ang-an as T'ang-dynasty capital, 17, 21, 22, 23
Ch'ang Ch'un Yuan, country palace near Peking, described by Bell, 277
Chang Hsin-pao, Chinese-American historian, on Opium War, 323, 324
Chao-ch'ing, 154; governor of, 154, 155, 156–57; Ricci and Ruggieri in, 155–67; plan of 165 i
Charles I, king of England, 202–3; and Courteen Association, 210
Charles IV, emperor, as employer of Marignolli, 117
Cheng Ho, Ming admiral, 124, 147, 159
Ch'eng Tsung, *see* Timur
cherub-type doll, 27 i
Chia-ch'ing, Manchu emperor, 320, 322
Chiang Kai-shek, 221
Ch'ien-lung, Manchu emperor: character of, 296; receives Macartney embassy, 302–4, 311; his letter to George III, 312–14
Ch'ienmen, front gate of Peking, 377, 399, 411
China Inland Mission, 319
Chinese attitudes toward foreigners and the West: under T'ang, 20–21, 25–26; under Ming, 123–25, 131, 146–47, 216–17; under Manchu, 234–35, 261, 294, 312–15, 359–60, 396; under republic, 401–2, 414–15, 418; under Communism, 418–19; *see also* antiforeignism as official Chinese policy, Mongol attitudes toward foreigners and the West
Chinese character, Bertrand Russell on, 401–2, 408
Chinese Classics, 163, 164, 169
Chinese College in Naples, 263–64, 286–87, 286 i, 288, 295
Chinese dictionary (Ricci), 170 i

Chinese Exhibition, London (1842), 326, 327 i, 328
Chinese Gordon, *see* Gordon, Charles
Chinese Hospital, Canton, *see* Ophthalmic Hospital, Canton
Chinese in fourteenth-century Europe, 125 i
Chinese officials, ranks of, 155 n
Chinese opera, 279, 279 n, 411
Chinese Repository, missionary journal, 334
Chinese Republic, 417–18
Chinese rites, *see* rites and Rites Controversy
Chinese writing system, 53; Samuel Johnson on, 291
Ch'ing dynasty, *see* Manchu dynasty
chinoiserie, 235, 289, 292–94, 292–93 i
Ch'in Shih-huang-ti, emperor, 417
Christianity in China, status of, and Chinese attitudes toward: under T'ang, 27; under Mongols, 47–48, 75, 95–96, 106; under Ming, 162, 163, 186, 187, 197; under Manchu, 237–38, 248–49, 262, 267–68, 281–82, 340–41, 371–72; *see also* Catholic Church, Nestorianism, Protestant missionaries
Ch'uan-chou-fu, *see* Zayton
Ch'u T'ai-su, student under Ricci, 171–72; destroys his books and manuscripts at Ricci's behest, 172
Chu Yuan-chang, first Ming emperor, 123
City of Lingering Splendour (Blofeld), 410 n; quoted, 410–11
Classics, *see* Chinese Classics
clavichord (Ricci's gift to Wan-li), 180, 183, 184
Clement V, pope, 101
clocks as gifts from the West, 163, 180; Wan-li and, 181–82, 183
coal, 84
coffin, Chinese, 351 i
Coiac, Nestorian Christian, 42
Coleridge, Samuel Taylor, 70, 71
Columbus, Christopher, 66, 87, 132
Comans (South Russians), 38, 41
Communism, Communists, in China, 411, 414–15, 418–419
compass, Chinese, 88, 129, 130 i, 172, 280
concubines and concubine system, 75–76, 171–72, 244
Concubine Yi (later Empress Dowager), 347, 348, 350, 358
Confucianism: as orthodox moral code and philosophy, 17, 140, 170–71, 186, 244, 294; and Mongols, 95; in conflict with Christianity, 163, 186, 197, 201; Ricci and, 163, 170–71, 186; admired in eighteenth-century Europe, 235, 289–90, 291; *see also* rites and Rites Controversy
Confucius, 95, 170
Conger, E. H., American minister, 376–77
Convention of Peking, 357
Corsalis, Andrew, Florentine traveler, 132–33
cosmas (mare's milk), 38, 40
cosmas churn, 38 i
Courteen Association and its venture in China, 210–16
crusades, 28, 29
culam (wild asses), 46
curriculums, Chinese and Jesuit, comparison of, 164
Cushing, Caleb, American commissioner in China, 333
Cushing Treaty, *see* Treaty of Wanghia

da Gama, Vasco, 87, 127, 129
d'Andrade, Fernão Peres, Portuguese mariner, 130, 133, 138; in Canton, 134–35, 139, 143
d'Andrade, Simão, Portuguese mariner, 143, 144
da Vinci, Leonardo, 127
de Goyer, Peter, Dutch merchant, 227
de Keyzer, Jacob, Dutch merchant, 227, 231
De Lange, Russian representative at Manchu court, 276, 283
de Mailla, Jesuit in Peking, 276
Dennett, Tyler, American historian: on Treaty of Wanghia, 326, 334; on Peter Parker, 333, 337
Description of the World (Marco Polo), 88; *see also* Polo, Marco
Designs of Chinese Buildings, Furniture, Dresses, Machines, aud Utensils (Chambers), 291–92
De Tournon, papal legate to Peking, 266-67
de Ursis, Sabatino, Jesuit priest in Peking, 190
Diaz, Manuel, Jesuit priest in Peking, makes terrestrial globe, 204, 206
di Cora, John, Dominican priest, 96, 104
dictionary, Chinese, compiled by Ricci, 170 i
Dominicans in China, 154, 209
Downing, C. Toogood, on Ophthalmic Hospital in Canton, 329
Dunn, Nathan, American collector, 326
Dutch: Chinese relations with, 201–2, 227–34; Portuguese rivalry with, 201–2, 229, 232; embassy of, to Peking, 227–34, 235

earthquake of 1665 (Peking), 238
East India Company, 210, 211, 298; and opium trade, 298, 316, 317, 319, 323

eclipses, Jesuit prediction of, 206, 237
edict of toleration of Christianity (1692), 248–49
Elgin, Thomas Bruce, British diplomat: signs Treaty of Tientsin, 1 i; background and personality of, 348–49; mission of, in China, 348; orders burning of Summer Palace, 352; enters Peking, 357, 357 i
Elimosina, Franciscan friar, 90
Ellis, Henry, member of Amherst embassy, 320
embassies, Chinese treatment of, 79, 139–42
embassies to Peking from the West, table of, 236
emperors, Chinese, names of, 195 n
Empress Dowager (Tze-hsi): called Old Buddha, 272 n; as Concubine Yi, 347, 348, 350, 358; and footbinding, 369–70; usurps power of emperor, 373, 375; photographs of, 373 i, 398 i, 399 i; and Boxer Rebellion, 390, 393, 399; death of, 417
encampment, Mongol, see *ordu*
England, Chinese relations with: Courteen Association venture, 210–16; Macartney embassy, 294–97, 299–316; opium trade, 297–99, 323; Amherst embassy, 320, 322; Opium War, 322, 323–25; Treaty of Nanking, 324–25; British occupation and sack of Canton, 336, 345, 348; Treaty of Tientsin, 348; Elgin expedition, 348–58; Boxer Rebellion, 376, 377, 378
English sailor, caricature of, 356 i
English warship, 211 i
Enlightenment, attitudes of, toward China, 289–92
Escape with Me! (Sitwell), 411; quoted, 411–15
Ethiopians in Peking, 127 n
eunuchs and eunuch system, 176, 177–78, 184, 225, 406 i
European attitudes toward China, *see* Western attitudes toward China
events in Europe and China from 1610 to 1644, table of, 198–99
Ex illa die, papal bull on Chinese rites, 281–82
extraterritoriality, 325, 333

Fabri, Jesuit in Peking, 268
Fa-Hsien, Chinese Buddhist priest, 229 n; quoted, 228–29
Fairbank, John K., American historian, on American policy toward China, 339
Favier, Alphonse, bishop of Peking, 377, 382 i, 383, 384, 385, 389, 391
feng-sui (auspices), 156
Fitzgerald, C. P.: on Chinese treatment of Europeans, 146; on opium trade, 298
footbinding, 88, 115, 270, 364 i, 365–66, 369 i; Ripa on, 270; Mrs. Little and, 365–70; K'ang-hsi and, 366; Empress Dowager and, 369–70
Forbidden City (Great Within, Imperial Palace), Peking, 347; Ricci in, 181; described by Sitwell, 412, 413–14
foreign affairs ministry, *see* Tsungli Yamen
"foreign devils," 202, 396 i
"foreign mud," *see* opium and opium trade
foreign policy of China, *see* antiforeignism as official Chinese policy, Chinese attitudes toward foreigners and the West
foreign relations of China, *see under specific countries; see also* antiforeignism as official Chinese policy, Chinese attitudes toward foreigners and the West
foreigners, Chinese attitudes toward, *see* Chinese attitudes toward foreigners and the West
foreign trade of China: under T'ang, 20; under Mongols, 82–83; under Ming, 132–34, 138–39; under Manchu, 294, 297–98, 316, 323, 358
Forest of Pencils Academy, *see* Hanlin Academy
Formosa: Dutch in, 202; Parker's plan for American annexation of, 336, 337; ceded to Japanese, 372
fountain of Kublai Khan, 79
fountain of Mangu Khan, 55–57, 56 i
fountain seen by Odoriç, 115
France, Chinese relations with: Treaty of Tientsin, 348; Elgin expedition, 350, 352, 354, 357; Tonkin War, 372; Boxer Rebellion, 377, 383, 396–97
Franciscans, 90, 108; in Peking, 94–104, 106, 114, 116; in Zayton, 102–3, 112; in India, 108–9; in Hangchow, 113; in Yangçhow, 113; in Canton, 155; *see also* Andrew of Perugia, John of Marignolli, John of Montecorvino, Odoric of Pordenone, William of Rubruck
Frederick II, Holy Roman Emperor, 31

Galileo, 203, 207–8, 209, 216
galleons, Portuguese, 132 i, 136 i, 151, 152–53
Gate of Heavenly Peace, *see* T'ien An Men
Genghis Khan, 30, 31, 54, 78, 97
Genoa as maritime rival of Venice, 66
George, Nestorian king, 70, 97–98
George I, king of England, 285
George III, king of England, 288, 296–97, 320; his letter to Ch'ien-lung, 296–97; Ch'ien-lung's letter to, 312–13
Gerbillon, Jesuit priest in Peking, 248, 262, 268

gifts of Macartney embassy to Ch'ien-lung, 300–301, 301 i
globes, *see* terrestrial globes
Goa, 129, 151–52
Goes, Bento, priest in India and China, 188
goldfish, 278–79
Goodrich, L. Carrington, on Treaty of Wanghia, 333
Gordon, Charles (Chinese Gordon), 210, 352–53, 353 i
Gosset, member of William of Rubruck's mission, 32, 42, 44
governor of Chao-ch'ing, 154, 155, 156–57; his plaque for Ricci's mission, 157 i
Grand Canal, 175, 176; described by Ricci, 177
Great Wall, 70, 88, 94, 202, 206, 209, 221; described by Bell, 280–81; Alexander regrets not seeing, 302; Samuel Johnson on, 302 n; Shaw on, 408
Great Within, *see* Forbidden City
Gregory IX, pope, 31
Gregory X (Tedaldo Visconti), pope, 67–68
Gregory XIII, pope, 151
Gutzlaff, Karl F. A., Prussian missionary and opium trader, 317, 318 i, 319–20

Hakluyt's *Principal Navigations, Voyages, Traffics, and Discoveries of the English Nation,* 188
Hangchow: described by Marco Polo, 76, 84–85, 88; described by Odoric of Pordenone, 112–13
Hankow, Mrs. Little in, 367
Hanlin Academy (Han-lin-yuan), 24, 347; burned by Boxers, 387
Han-lin-yuan, *see* Hanlin Academy
Hanyang, Mrs. Little in, 367
Han Yu, T'ang scholar and anti-Buddhist, 180
Hart, Robert, British commissioner of Chinese customs, 363, 395
Helen, Ming empress dowager, 222–23, 222 i; her letter to Innocent X, 223 i
Henry, Paul, French lieutenant and hero of siege of Pei T'ang, 383–91, 383 i
Henry the Navigator, 87
Hermenian monk, *see* Nestorian monk from Hermenia
Hickey, Thomas, artist with Macartney embassy, 302, 305
Hideyoshi, Japanese military dictator, 176
History of Christian Missions in China (Latourette), 197
Holland, Chinese relations with, *see* Dutch
Hollar, Wenceslas, chinoiserie engraver, 235
Holyroods, John, first English author published in China, 191
Homo Dei, dragoman with William of Rubruck's mission, 32–33
Hong Kong: ceded to British, 325; as British colony, 348; Mrs. Little in, 368
Ho-shen, minister under Ch'ien-lung, 311–12, 314
How We Got to Peking (McGhee), 353–54
Hsi Hsia, kingdom conquered by Mongols, 48
Hsu, Paul (Hsu Kuang-ch'i), scholar and Jesuit convert, 189–80, 190 i, 194, 197; imports Portuguese cannon founders, 202; made vice-president of Board of Rites, 206; proposes translation of Western scientific books, 206–7; made president of Board of Rites, 208; invites Portuguese soldiers for protection of Peking, 209–10
Hsuan Tsung, T'ang emperor, 24
Hsu Kuang-ch'i, *see* Hsu, Paul
Huang Chi-shih, governor of Peking, 194
Huc, Catholic missionary in China, 197, 208
Hui-t'ung-kuan (Palace of the Four Barbarians), place of detention for foreign embassies, 184; Dutch embassy in, 227; Spathary embassy in, 252, 256; described by Bell, 276
Hung Mao (redheads), epithet for foreigners, 202
hunting parties: of Mongol khans, 73, 115; of K'ang-hsi, 245–47
hutung (residential lanes), 81, 345–46

Ibn Battuta, Arab traveler, 85, 93, 112
idols, Mongol, 38 i
Imperial City, 347
Imperial Palace, Peking, *see* Forbidden City
Impey, L., on site of Shangtu, 70
In, John, Ripa's favorite pupil, 284, 285
indemnity: under Treaty of Nanking, 325; under Treaty of Tientsin, 356, 358 i; after Boxer Rebellion, 396–97
India: John of Montecorvino in, 92–93; Odoric of Pordenone in, 108, 109, 111; martyrdom of Franciscans in, 108–9; Portuguese in, 129, 151, 152; Pirès in, 133; Chinese trade with, 139; Ricci in, 151–52; Courteen Association expedition in, 211; Ripa in, 266; and opium trade, 297, 298; as British colony, 312, 315; Elgin in, 348
Indo-China, 372
infanticide in China, 269–70, 269 i
Innocent IV, pope, 31
Innocent X, pope, Empress Dowager Helen's letter to, 223, 223 i
Innocent XI, pope, 242
Islam, *see* Moslems in China

Ismailoff embassy to Peking, 275–78, 280
isolationism, *see* antiforeignism as official Chinese policy, Chinese attitudes toward foreigners and the West
Italians in Boxer Rebellion, 377, 383

James of Padua, Franciscan martyr in India, 109
Japan, Japanese: piracy by, 131, 139, 143, 146, 155; Francis Xavier and, 150; Portuguese trade with, 152, 153; invasion of Korea under Hideyoshi, 172, 176; America and, 338; at war with China (1894–95), 372; in Boxer Rebellion, 377, 392, 395, 396; at war with China (1937–45), 400, 409, 418
Jarlin, bishop in Peking, 385, 386
Jehol palace, 273, 274 i; Ripa at, 273; Macartney at, 302–4, 311; emperor at, during Elgin expedition, 350, 352, 357
Jesuit cemetery near Peking, plan of, 248 i
Jesuits in China: missionary strategy of, 154, 161–62; and Chinese rites, 186–87, 244–45, 267, 281–82; as scientists, 188, 203–9, 224, 241–42, 243–44; persecution of, 197–99, 201–2, 237–38, 282; and new Manchu regime, 221–22, 224; intrigue of, against Dutch in Peking, 232–33; position of, under K'ang-hsi, 240, 241–45, 247–48; treachery of, against Manchu government, 256–58, 259–60, 262; Ripa on conduct of, 267–68; summary of their achievements, 282; *see also* Almeida, Buglio, Castiglione, Cattaneo, Gerbillon, Koffler, Longobardi, Magalhaens, Pedrini, Ricci, Ripa, Ruggieri, Schall, Semedo, Terrentius, Vagnoni, Valleat, Verbiest
John of Marignolli, Franciscan priest, 90, 106; in China, 117–18
John of Montecorvino, Franciscan priest, 90–106, 114, 119; in Persia, 91; is sent to China, 91; in India, 92–93; in Peking, 94–104; his churches in Peking, 96–97, 98–99, 100; buys forty Chinese boys, 96–97; and Nestorians, 97–98; becomes archbishop of Peking, 101; death of, 104
John of Pian de Carpini, Franciscan friar, 29, 31
Johnson, Samuel: on China, 290–92; on Great Wall, 302 n
Johnston, R. F., on Protestant missionaries in China, 364
Jordanus, Franciscan friar, 109

K'ang-hsi, Manchu emperor, 225, 238, 239 i, 281 i, 290, 314; character of, 240–41, 247; his suspicions of the Jesuits, 240; his friendship with Verbiest, 240–42, 245–49; on safari, 245–47; betrayed by Verbiest, 256–50, 262; receives Spathary, 259, 260; and Ripa, 268, 271–73; his private life described by Ripa, 272–73; receives Ismailoff embassy, 277–78; and Rites Controversy, 281–82; his edicts against Christianity, 281, 282; and footbinding, 366
Kao Tsung, T'ang emperor, 23–24
Karakorum, Mongol capital, 54–55; plan of, 54 i
Keenan, James, American consul in Hong Kong, 336
Ketteler, German minister in Peking, 378
Koffler, Andrew Xavier, German Jesuit, 222–24
Korea: first mentioned in Western literature, 53; invaded by Japanese in sixteenth century, 172, 176; proposed as French base by Parker, 336; Chinese at war in (1950–53), 402
Kowloon, ceded to British, 358
kowtow, 43 i, 140–42; William of Rubruck and, 43–44, 49, 50; Marco Polo and, 73; Dutch embassy and, 230, 231; Spathary and, 251, 259, 261; Ripa and, 268; Ismailoff and, 277–78; Macartney and, 300, 304; Amherst and, 320, 322
Kublai Khan, 65, 66–67, 72 i, 73–80, 83; receives elder Polos on their first journey to China, 65; makes request to pope, 66–67; his palace at Shangtu, 70–72; receives Polos in Shangtu, 73–74; described by Marco Polo, 74–75; his attitude toward Christianity, 75; his harem, 75–76; his palace in Peking, 76–79; condition of China under, 83; death of, 86, 94; pope's letter to, 93–94
K'ung Fu-tzu, *see* Confucius
Kurakin, Ivan Semenovich, governor of Siberia, 195
Kuyuk, Mongol khan, 31
Kwei, exiled Ming prince, 222
kwei (ceremonial ivory tablet), 115

Lamaism, 95
Land of the Blue Gown (Little), quoted, 363
Langdon, William B., on Western treatment of China, 326–28
Latourette, K. S.: on Portuguese in China, 145–46; on number of Chinese Christians, 197
Lattimore, Owen, on foreign students of Chinese, 74
Legation Quarter, Peking, 411; *see also* Siege of the Legations
Legge, James, Scottish missionary and scholar, 339–40, 340 i
Leibnitz, on China, 290
Li, Leo, *see* Li Chih-tsao
Liao dynasty, 53 n, 97

Li Chih-tsao (Leo Li), Jesuit convert and scholar, 19, 190–91, 194, 197
Li Hung-chang, Chinese viceroy, and Mrs. Little, 367–68
Lin Tse-hsu, commissioner, 317, 323–24, 328, 332; his letter to Queen Victoria on opium trade, 324, 332
Li Po, T'ang poet, 27
Lisbon, plan of, 128 i
literati, Chinese, and Ricci, 162, 163
Little, Alicia, *see* Little, Mrs. Archibald
Little, Mrs. Archibald (Alicia Little), 361–70, 362 i; her views on China, 361–64; describes Peking, 363–64; and footbinding, 365–70; on siege of Pei T'ang cathedral, 389–90
Li Tzu-ch'eng, leader of anti-Ming rebels, 216
Li Ying-shih, scholar, 172; destroys his books at Ricci's behest, 172
Longobardi, Nicolo, Jesuit priest in Peking, 197, 199; makes terrestrial globe, 204, 206
Lo Ping-wang, T'ang poet, 24
Louis IX, king of France, 28–30, 65–66; plans to convert Mongols, 29–30
Louis XIV, king of France, sends mathematicians to China, 281
Lucolongo, Peter of, *see* Peter of Lucolongo

Macao, 128 i, 200 i; established as Portuguese colony, 145, 164; Ricci in, 152, 153, 154; in sixteenth century 153; fortified by Portuguese, 201, 202; attacked by Dutch, 201, 232; Weddell at, 211, 214, 215
Macartney, earl of (Lord Macartney), 288, 295 i, 307 i, 308 i; his embassy to China, 294–304, 311–315; character and background of, 295, 296; sketches made during his embassy to China, 297 i, 305, 306–10 i; in Peking, 299–301, 312; gifts brought by, 301 i; in Jehol, 302–4, 311; his audiences with Ch'ien-lung, 302–4, 307 i, 311
Macartney embassy, *see* Macartney, earl of
MacDonald, Claude, British minister in Peking, 376, 381
MacDonald, Lady, 398 i
Magalhaens, Portuguese Jesuit, 232, 233, 237, 240, 259
Ma-ho-ma-sha, Chinese-Arab architect, 77
Major, R. M., on Portuguese in China, 134, 138
Malacca: as Ming tributary and trading port, 124, 133; captured and occupied by Portuguese, 129, 131, 138, 142, 143; as Portuguese trading port, 134, 135; Portuguese misconduct in, 144
Manchu: attacks of, against China, 202; invasion of China by, 209; capture of Peking by, 221; establishment of Manchu (Ch'ing) dynasty by, 221; eviction of Chinese from Peking by, 224; *see also* Manchu dynasty
Manchu (Ch'ing) dynasty: establishment of, 221; foreign policy of, 234, 261–62, 294, 312–14; corruption of, 298–99, 326, 372; Western domination of China under, 372–73; fall of, 417
"mandarin," origin and definition of, 135 n
Mandeville, John, historical writer, 87, 116–17
Mangu, Mongol khan, 31, 49–52, 55, 57–58
maps: Chinese, 141 i, 158–59, 160 i, 191, 194 i; Ricci's, 158–59, 158 i, 191, 192–93 i
Marcy, American secretary of state, 336
mare's milk, see *cosmas*
Marignolli, *see* John of Marignolli
Mark and Sauma, Nestorian Christians in China, 91, 94; *see also* Mar Yahballaha
Martin, W. A. P., American president of Peking University: on siege of Pei T'ang, 393; in "siege costume," 397 i; on Boxer Rebellion, 397–98
Mar Yahballaha (Mark), Nestorian patriarch, 91, 105; letter from, 105 i; *see also* Mark and Sauma
Ma T'ang, eunuch, 177–78, 179 i, 180, 184
matches as Chinese invention, 88
Maugham, W. Somerset, on China, 409
Mayer, Karl August, on Chinese College in Naples, 263–64
McCarthy, Michael F. G., on Boxer Rebellion, 395
McGhee, R. J. L., on British sack of Peking, 353–54
medicine, Manchu, 270–71
Memoirs of an Aesthete (Acton), 409
Mendoza, Juan Gonzalez de, historian, on d'Andrade expedition, 134, 136, 138
Meridian Gate, *see* Wumen
Mezzabarba, cardinal and papal legate, 276, 281–82
Middle Kingdom, concept of China as, 141 i, 208
Ming Cheng-te, Ming emperor, 144
Ming court in exile, 222–24
Ming dynasty: establishment of, 123; modeled on T'ang, 123; foreign policy of, 123, 131–32, 146–47; overseas voyages during, 123–24; degeneration and decline of, 175, 176, 179, 195, 202; depredations of eunuchs under, 176; fall of, 221; exiled court of, 222–24; architecture of, 356

Ming voyages, 123–24
missionaries, missions, *see* Dominicans, Franciscans, Jesuits, Protestant missionaries; *see also* Christianity in China
Mohammed, 109, 289
Mohammedans, *see* Moslems in China
money, *see* paper money
Mongol archer, 60 i
Mongol attitudes toward foreigners and the West, 29–31, 58, 79, 82–83, 104, 123
Mongol (Yuan) dynasty, *see* Mongols and Mongol dynasty
Mongol empire in 1290, map of, 69 i
Mongol idols, 38 i
Mongol princess and suite, 34 i
Mongols and Mongol (Yuan) dynasty, 29–31, 77–78; conquests of, 28, 30–31, 69 i; embassies sent from Europe to, 29–30; European captives of, 30, 51, 53; customs of, 34–38, 40, 41, 46, 77–79; under Kublai Khan, 68, 77–78, 82–83; change in culture of, 77–78; condition of China under 82–83; under Timur, 95; political and religious policies of, 95; downfall of, 123; *see also* Mongol attitudes toward foreigners and the West
Mongol yurts, *see* yurts
Montecorvino, *see* John of Montecorvino
Morrison, John Robert, English trader, 316–17
Morrison, Robert, English missionary, 316
Morse, Hosea B., American historian: on opium trade, 323; on Boxer Rebellion, 392, 393–94, 396
Moslems in China, 83, 95, 185, 237, 238
Mountney, Nathaniell, English seaman and trader, 210, 215
Mundoff, Ondrushka, Russian envoy to Ming, 196, 197
Mundy, Peter, English diarist and artist, 210; Chinese sketches by, 212 i; quoted, 213
music, Chinese, Ricci on, 183–84

Nan-ch'ang, Ricci in, 172–73, 175
Nanking: as Ming capital, 124, 172; Ricci in 159, 175–76; Semedo and Vagnoni in, 197–99; Treaty of, *see* Treaty of Nanking
Naples, Chinese College in, *see* Chinese College in Naples
Naples and the Neapolitans (Mayer), 263–64
Nationalists, 418
Natural Foot Society, 366
naval battle of Opium War, 325 i
Nean, Dominus, interpreter with Macartney embassy, 310 i
Nerchinsk, Treaty of, *see* Treaty of Nerchinsk
Nestorian crosses, 26 i, 27
Nestorian monk from Hermenia, 49, 52, 57
Nestorians, Nestorianism; 19–20, 21, 22, 23–27, 91, 124; William of Rubruck on, 42, 47–48; degradation of, 47–48; in India, 92; and Mongol hierarchy, 95; and John of Montecorvino, 96, 97–98; Western Liao kingdom of, 97–98
Nestorian Stone (Tablet of the T'ang), 18 i, 18–19, 21, 23, 24, 208
Nicholas, member of William of Rubruck's mission, 33
Nicholas of Pistoia, Franciscan priest, 93
Nieuhof, John, Dutch traveler in China, 227; reports on Dutch embassy to Manchu court, 228–32, 235
Noretti, Pablo, agent of Portuguese and Chinese in Canton, 214–15
Not Bind Foot Society, 369
Nurachi, Manchu khan, 202

observatory: in Nanking, 159; Verbiest's, in Peking, 243 i
"ocean devils" as epithet for foreigners, 144, 166, 202
"Ode to the Supernatural Horse," 117
Odoric of Pordenone, Franciscan priest, 107–20, 110 i; early life of, 107–8; in India, 108–9, 111; describes martyrdom of Franciscans at Tana, 108–9; and becalmed ship, 111; in China, 111–17; in Canton, 111; in Zayton, 112; in Hangchow, 112–13; in Yangchow, 113–14; in Peking, 114–16; as travel writer, 114; describes khan's hunting parties, 115; and "valley of death," 116; death and posthumous "miracles" of, 118–19; beatified, 119; tomb of, 119 i
officials, Chinese, ranks of, 155 n
Ogotai, Mongol khan, 31
Ogul-Gaimish, Mongol female regent, 29–30
On a Chinese Screen (Maugham), 409
opera, *see* Chinese opera
Ophthalmic Hospital, Canton, 329–32
opium and opium trade, 297–99, 300 i, 316–17, 319, 320, 321 i, 323–25, 328, 328 i, 358, 359 i
opium bales, 328 i
opium den, 300 i
opium ships, 321 i, 359 i
Opium War, 322–25; naval battle of, 325 i
Opus Majus (Roger Bacon), 59
oracle bones, 51
ordu (Mongol encampment), 36–37

Palace of the Four Barbarians, *see* Hui-t'ung-kuan
Palmerston, British prime minister, and Opium War, 324

Pantoja, Jesuit priest in China, 181, 183
papal bull dividing East and West, 129, 151
papal bull on Chinese rites, 281–82
paper money, 53, 82 i, 83
Paquette, European slave of Mongols, 51
Parement de Narbonne, 125 i
Parennin, Jesuit, on K'ang-hsi, 290
Parker, Peter, American missionary and diplomat, 329–41, 331 i; early life of, 329; and Ophthalmic Hospital in Canton, 329–32; as diplomat, 333–38; and Treaty of Wang-hia, 333–34; his project for colonization, 336; his plan for American annexation of Formosa, 336, 337
Parkes, Harry, British diplomat, 317, 319, 349–50, 350 i
Paul V, pope, and Mass in Chinese, 207
"pear garden," 24
Pearl River, 111, 136–37
Pedrini, Jesuit in Peking, 266, 267, 268, 281
Pegolotti, Francis Balducci, Florentine trader and author, 126–27, 317; his *Book of Descriptions,* 126, 126 i
Pei T'ang (Peking cathedral), 382 i, 385 i (plan); siege of, 379, 381–92; Western writers on siege of, 392–93
Peking (earlier named Ta-tu and Yen): described by Marco Polo, 77–78, 80–81; as Mongol capital, 77–78, 80–81; described by Odoric of Pordenone, 114–15; as Ming capital, 124, 183; described by Ricci, 177; described by Petlin, 196; captured by Manchu, 221; described by Ripa, 269–70; described by Bell, 279–80; described by Alexander, 299; in mid-nineteenth century, 345–47; seized by British and French (1860), 352–57; Summer Palace area of, 354 i; described by Mrs. Little, 363–64; in Boxer Rebellion, 376–79, 393–96; skeleton map of (1900), 380 i; first aerial photograph of, 393 i; described by Blofeld, 410–11; described by Sitwell, 412–14
Peking cathedral, *see* Pei T'ang
Peking opera, *see* Chinese opera
Peregrine of Castello, bishop, 101, 102, 103
Pereira, Emmanuel, *see* Yu Wen-hui
Perestrello, Rafael, Portuguese mariner, 132, 138, 146
Perugia, Andrew of, *see* Andrew of Perugia
Peter of Lucolongo, Italian merchant, 92, 94, 95, 100
Peters, Marie, wife of Bell, 283, 284
Peter the Great, tsar, 196, 275, 282; Bell's service under, 275; Bell's gifts from, 283, 283 i
Petlin, Ivan, Russian envoy to Peking, 196–97
Pian de Carpini, *see* John of Pian de Carpini
piracy, 146, 155, 164–65; by Japanese, 131, 139, 146, 155; by Portuguese, 133, 146, 155, 164
Pirès, Tomaso, Portuguese envoy to China, 130, 131–48; describes Malacca, 133; describes Chinese, 133–34; his *Suma Oriental,* 133–34; in Canton, 136–38, 144–45; in Peking, 144; imprisoned, 144–45
Po Chu-i, T'ang poet, 25
Poivre, on China, 290
Polo, Maffeo, 65, 66–70, 75
Polo, Marco, 63–89, 64 i; estimate of his book, 63–64, 87–89; his boyhood and youth in Venice, 64–66; scenes from his travels, 67 i, 68 i; itinerary of his journey to Peking, 68–70; describes Kublai Khan's palace at Shangtu, 70–72; meets Kublai Khan in Shangtu, 73–74; describes Kublai Khan's hunting preserve, 73; describes Kublai Khan, 74–75; appointed official by Kublai Khan, 74; describes Kublai Khan's harem, 75–76; describes Kublai Khan's palace at Peking, 76–79; describes Peking, 77–78, 80–81; curious omissions from his book, 80, 87–88; sent on journeys by Kublai Khan, 84; describes Hangchow, 84–85; leaves China, 85–86; his later life, 86
Polo, Nicolo, 65, 66–70, 75
porcelain, Chinese, 279
portolan charts, 129
Portugal, Portuguese: explorations of, 127, 129; attitude of, toward Orient, 129; in China, 130, 131–39, 142–48, 210, 213–15, 216; in Canton, 130, 134–39, 142, 143–45, 210; seizure of Malacca by, 131, 138, 142, 143; Chinese relations with, 131–32, 134–36, 137–40, 143–44, 146–48, 213–14; piracy by, 133, 146, 155, 164; in Tamão, 134; seizure of Macao by, 145, 164; in India, 151–52; Oriental trade of, 152–53, 164–65; rivalry of, with Dutch, 201–2, 229, 232
Portuguese galleons, *see* galleons
postal system, posting stations, under Mongols, 83, 115–16
Prester John, 30, 70, 87, 97–98, 127 n
printing, 88
Problem of China, The (Russell), 401; quoted, 400–402, 408
prostitutes, Marco Polo on, 81
Protestant missionaries, 316–17, 319–20, 329–32, 334–35, 338–41, 364
Purchas His Pilgrims, 189

"redheads" as epithet for foreigners, *see* Hung Mao
Reed, William B., American commissioner in China, 338
religion, Chinese attitudes toward, 162; *see also* Christianity in China
religious rivalry among Westerners in China, 202, 229, 232–33, 326
Renaissance, 127, 129–30
Republic of China, 414, 417–18
Ricci, Matteo, Jesuit priest, 149–94, 174 i; early life and education of, 150–51; his voyage to the Orient, 151; in Goa, 151–52; in Macao, 153–54; in Chao-ch'ing, 155–67; his map of the world, 158–59, 158 i, 191, 192–93 i; as scholar of Chinese, 161, 163, 164, 169–70, 175; his missionary strategy, 161–62, 171; and Chinese rites, 163, 186–87; persecuted in Chao-ch'ing, 165–67; in Shao-chou, 167–72; his route to Peking, 168 i; his Chinese dictionary, 170 i; in Nan-ch'ang, 172–73, 175; acquires new social status, 173; and Chinese etiquette, 173; in Nanking, 175–76; in Peking, 175, 180–91, 194; imprisoned in Lin-ch'ing, 177–78; and eunuch Ma T'ang, 177–78, 184; imprisoned in Tientsin, 178; his views on ancestor worship, 186–87; his contribution to Western knowledge of China, 188–89; evaluation of his writings, 188–89; last illness and death of, 191–92; his geographical teachings and maps criticized, 208
"rice Christians," 376
Ripa, Matteo, Jesuit priest, 263, 264–75, 265 i, 284–87; his voyage to China, 264, 266; and Rites Controversy, 267; on Jesuit missionaries, 267–68; and K'ang-hsi, 268–69, 271–73; appointed court painter, 268; describes Peking, 269–70; on infanticide in China, 269–70; on foot-binding, 270; treated by Manchu physician, 270–71; describes K'ang-hsi at Jehol palace, 273; leaves Peking, 285; homeward journey of, 285; founds Chinese College in Naples, 285; death mask of, 287 i
rites and Rites Controversy, 157 n, 186–87, 197, 244–45, 267, 281–82
Rites Controversy, *see* rites and Rites Controversy
Robinson, Thomas, member of Courteen Association expedition, 210, 215
Roman Church, *see* Catholic Church
Royal Asiatic Society, London, 323
Rubruck, William of, *see* William of Rubruck
Ruggieri, Michele, Jesuit priest, 151, 153; in Canton, 154; in Chao-ch'ing, 155, 156, 162; returns to Macao, 163; suspected of treachery, 165; sent to Rome, 169
Russell, Bertrand, on China and the Chinese, 400–402, 408
Russia, Chinese relations with: embassy of Petlin and Mundoff, 195–97; Baikoff embassy, 225–27, 234; Spathary embassy, 250–62; Treaty of Nerchinsk, 262, 276, 294; Ismailoff embassy, 275–78; Treaty of Tientsin, 348; Boxer Rebellion, 377, 395; under Communists, 415, 419
Russian ambassador and suite, 251 i

Saint Francis of Assisi, *see* St. Francis of Assisi
Saint Thomas, *see* St. Thomas
Saracens, 46, 47, 49, 55, 65, 116, 183; Holy Land occupied by, 28
Sartach, Mongol chieftain, 29, 30, 33, 34; receives William of Rubruck, 42–43
Sauma, *see* Mark and Sauma
Savage-Landor, A. Henry, English journalist, on Boxer Rebellion, 392, 394–95
scapulimancy, 51
Scatay, Mongol chieftain, 35, 38
Schall, Adam (Adam Schall von Bell), Jesuit priest and astronomer, 197, 203, 233 i; his book in Chinese on the telescope, 203, 206; as leader of Jesuits in Peking, 221; casts cannon for Ming, 224; and calendar reform, 224; and Dutch embassy to Peking, 228–29, 232–33; his ancestors ennobled by Shun-chih, 237; imprisoned, 237, 240; death of, 238; stele of, 238 i
Schall von Bell, Adam, *see* Schall, Adam
Schreck, John, *see* Terrentius
science, Chinese, 159, 161, 172, 188, 244; *see also* science, Western, in China
science, Western, in China: promoted by Jesuits, 161–62, 171, 203, 206–7, 241–42, 243–44; as Jesuit strategy, 161–62, 187–88, 203, 206–7, 209, 224, 241–42; attacked by Chinese, 208, 237; Jesuit ambivalence concerning, 209; rejected under Ming, 216–17; *see also* science, Chinese
Secret History, Mongol chronicle, 46, 57
Semedo, Alvarez, Jesuit priest and historian, 19, 21, 153, 197, 207 i, 208; imprisoned in Nanking, 198–99
Shangtu (Xanadu), 70–73; Kublai Khan's palace at, 70–71; plan of, 71 i
Shao-chou, Ricci in, 167–69, 171–72, 179
Shaw, George Bernard, on China, 408
Shen Ts'ui, Confucian scholar and persecutor of Jesuits, 197–99
Shun-chih, first Manchu emperor, 224–25;

and Dutch embassy, 229–33, 234; death of, 237
Siberia, Sino-Russian clashes in, under Manchu, 225
siege of Pei T'ang, *see* Pei T'ang
Siege of the Legations (Peking), 372, 376–81, 379 i
Sie Wan-kwoh, patient of Peter Parker's, 331
silk and silk trade, 20, 88, 125–26
Simpson, B. L. (Putnam Weale), English writer, on Boxer Rebellion, 395
Sino-Japanese wars: of 1894–95, 372; of 1937–45, 400, 409, 418
Sisters of St. Vincent de Paul, in Boxer Rebellion, 384, 389–90
Sitwell, Osbert, on China, 411–15
slaves: European, in Mongol China, 30, 44, 51, 53; Mongol and Chinese, in Italy, 125; of Portuguese, in Goa, 152
snakes as food, 111
Society for the Diffusion of Useful Knowledge in China, 332
Soldaia, castle of, 33 i
Southern Sung dynasty, 53, 84
Spain, Spaniards, 129, 147–48, 154, 155, 229
Spathary, Nikolai Gavrilovich, Russian envoy to Peking, 250–62; background of, 250–51; his embassy to Peking, 251–61; and Board of Rites, 252–57; handwriting of, 253 i; and Verbiest, 255–58, 259–62; his audiences with K'ang-hsi, 258–59, 260; importance of his narrative, 261
sphericity-of-earth diagram, 194 i
spice trade, 127
Star Messenger (Galileo), 203
Staunton, George Leonard, secretary with Macartney embassy, 288, 295, 302, 304, 307 i
Staunton, George Thomas, page with Macartney embassy, 288, 295; as linguist and writer, 296, 300; on audience with Ch'ien-lung at Jehol, 303, 311; pages from diary of, 303 i; in Canton, 316; with Amherst embassy, 320, 322–23
steam engine, 243–44
St. Francis of Assisi, 29, 90
St. Thomas, martyred apostle, 92, 93
Suma Oriental, The (Pirès), 133
Summer Palace (Yuan-ming Yuan), Peking: Macartney embassy at, 300; Amherst at, 322; occupied by Elgin expedition, 350; burned by Elgin expedition, 352–54, 356; sketch map of area of, 354 i; ruins of, 355 i
Sung dynasty, 53, 83, 146
Sun Yat-sen, 417, 418
Tablet of the T'ang, *see* Nestorian Stone
t'ai chi ch'uan exercises, 346
Taiping Rebellion, 319, 323, 345, 372, 418; Gordon and, 352
T'ai Tsung, T'ang emperor, 21–22, 22 i; harem of, 24; and Buddhists, 26
Taku, Taku forts: attacked by Elgin expedition, 348; captured by Western powers in Boxer Rebellion, 377; razed after Boxer Rebellion, 397
Tamão, 134, 142, 143
Tana (Indian city), martyrdom of Franciscans at, 108–9
T'ang dynasty: civilization and culture of, 17, 20–21, 22–23, 24; Westerners in China during, 20–21; Chinese conquests under, 22; condition of China under, 22, 25; antiforeign and anticlerical phase of, 25–26; decline and collapse of, 25–27
T'ang figurines, 20–21, 20 i
Taoists, 24
Tao-kuang, Manchu emperor, 323
Ta-tu, *see* Peking
Taylor, James Hudson, founder of China Inland Mission, 317, 319
tea, 23, 88
telescope, in China, 203, 204 i, 206
Temple of Heaven, Peking, 182, 183
Tenduc, 70, 97, 116
Terrentius (John Schreck), Jesuit astronomer, 203, 206
terrestrial globes: made by Ricci, 159; made by Longobardi and Diaz, 204, 205 i, 206
Thomas of Tolentino, Franciscan priest and martyr, 101, 108–9
Tibet, Tibetans, 22, 48
T'ien An Men (Gate of Heavenly Peace), 181, 347
T'ien-ch'i, Ming emperor, 202
Tientsin: Ricci imprisoned in, 178; British and French troops in (1860), 348; in Boxer Rebellion, 377, 378, 381; Treaty of, *see* Treaty of Tientsin
Timur (Ch'eng Tsung), Mongol emperor, 94, 95
Toghon Timur Ukhagatu, last Mongol emperor, 104
Tonkin War, 372
Tractatus de Sphaera (Holyroods), 191
trade, *see* foreign trade of China, opium and opium trade
Treaty of Nanking, 324–25; results of, 326
Treaty of Nerchinsk, 262, 276, 294
Treaty of Tientsin, 348, 357, 358; concessions to Western powers under, 358
Treaty of Wanghia (Cushing Treaty), 326, 333, 334; Tyler Dennett on, 326; Peter Parker and, 334

tribute bearers at Peking, 139 i
tribute flag, 297 i
Trigault, Nicholas, compiler of Ricci's journals, 149 n, 203; quoted, 184, 191
Trumbull, Henry C., on Peter Parker, 330
Tsungli Yamen (foreign affairs ministry), 358, 359; and memorial on footbinding, 366; in Boxer Rebellion, 376, 377
Tun-huang, Buddhist caves at, 70
Tze-hsi, *see* Empress Dowager

Uighurs, 97; script of, 48; as source of Kublai Khan's harem, 76
United States, *see* America

Vagnoni, Jesuit priest in China, 197; imprisoned in Nanking, 198–99
Valignano, Alessandro, Jesuit visitor and vicar-general, 150, 152, 153; his ideas on converting the Chinese, 154, 161; his writings on China, 188
Valleat, John, French Jesuit in Peking, 232
Venice as medieval principality and home of Marco Polo, 65–66
Verbiest, Ferdinand, Jesuit priest, 237–49, 241 i, 255–60, 261, 262, 268; criticizes Yang Kuang-hsien's calendar, 238–39; early life of, 240; imprisoned in Peking, 240; his friendship with K'ang-hsi, 240–42, 245–49; his observatory, 243, 243 i; his scientific pursuits, 243–44; his writings, 243; on safari with K'ang-hsi, 245–47; death of, 247; achievements of, 249; and Spathary embassy, 255–60; his treachery against K'ang-shi, 256–60, 262
Victoria, queen of England: Lin Tse-hsu's letter to, regarding opium trade, 324, 332; and loot from Summer Palace, 353 n
Visconti, Tedaldo, *see* Gregory X
Volta, Milanese priest and physician in Peking, 272
Voltaire, on China, 289, 290

Waley, Arthur, on opium traders, 317, 319
Wang, doctor of letters, 347–48, 349–52, 356–57; and mother's illness and death, 347–48, 349–52; describes British and French seizure of Peking, 352, 356
Wanghia, Treaty of, *see* Treaty of Wanghia
Wang Yang-ming, philosopher and moralist, 127
Wan-li, Ming emperor, 175, 176, 180–83, 182 i, 185; corruption of government under, 176; and gifts from Ricci, 180–83; and clocks, 181–82, 183; gives party in his tomb, 182; donates burial place for Ricci, 194, last days of, 195; death of, 202
warlords, 418
watertight compartments, Chinese invention of, 129
Weale, Putnam, *see* Simpson, B. L.
Weddell, John, English sea captain in charge of Courteen Association venture, 210–11, 213–15; and Portuguese at Macao, 213–15; attacks Chinese ships at Macao, 214; expelled from China, 215
Western attitudes toward China, 59, 146–48, 315; in sixteenth century, 146–48; in seventeenth century, 235; in eighteenth century, 289–94, 314–15; in nineteenth century, 338–39, 359–60; at time of Boxer Rebellion, 371–72; in twentieth century, 400, 411
Western embassies to Peking, table of, 236
Western Liao, kingdom of, 97
Western writers on China: in nineteenth century, 364; in twentieth century, 400, 402, 408, 409, 410–11, 415
wild asses, see *culam*
William of Rubruck, Franciscan priest, 28–60, 43 i; his relations with Louis IX, 29–30; seeks information on Mongols, 29–30; his expedition to the Mongols, 30, 32–60; itinerary of, 32; describes Mongols and their customs, 33–38, 40–41; at Scatay's camp, 38, 40; at Sartach's camp, 42, 43; at Batu's camp, 43–44; his journey to Mangu's camp, 44–49; at Mangu's camp, 49–53; in Karakorum, 53–58; returns to Europe, 59; evaluation of his writings, 59–60
Williams, Samuel Wells, American missionary and scholar of Chinese, 332, 334, 339; on importance of medical missionaries in China, 332
winged baby, 27 i
world map, Ricci's, 158–59, 158 i, 191, 192–93 i
writing system, Chinese, *see* Chinese writing system
Wu, Ming general, 221
Wu, T'ang empress, 24
Wumen (Meridian Gate), 181
Wu Tsung, T'ang emperor, 25–26

Xanadu, *see* Shangtu
Xavier, Francis, 150; on China and the Chinese, 150, 182; missionary tactics of, 154
xenophobia, *see* antiforeignism as official Chinese policy, Chinese attitudes toward foreigners and the West

yaks, 39 i, 40
Yangchow: Marco Polo in, 85, 88; Odoric of Pordenone in, 113–14

Yang Kuang-hsien, Chinese Moslem astronomer, 238–40; his calendar criticized by Verbiest, 238–39
yang kuei tzu ("ocean devils") as epithet for foreigners, 144, 166, 202
Yangtze River, described by Marco Polo, 84
Yeh-hei-tieh-er, Chinese-Arab architect, 77, 78
Yeh-lu Ch'u-ts'ai, adviser to Genghis Khan, 54
Yellow River: described by Marco Polo, 84; described by Ricci, 177
Yen, *see* Peking
Yi, *see* Concubine Yi
Yuan dynasty, *see* Mongols and Mongol dynasty
Yuan-ming Yuan, *see* Summer Palace
Yueh Fei, twelfth-century patriot, 410
Yule, Henry, English historian: on value of William of Rubruck's narrative, 59–60; on Odoric of Pordenone, 118, 119
Yung-cheng, Manchu emperor, 419
Yung-lo, Ming emperor, 124, 172
Yung-lo Ta-t'ien, Ming encyclopedia in manuscript, 387
yurts (Mongol portable houses), 35–36, 35 i, 37, 37 i (plan)
Yu Wen-hui (Emmanuel Pereira), painter and Jesuit convert, 174

Zayton (Ch'uan-chou-fu), 102; Andrew of Perugia in, 102–4; map of, 103 i; described by Odoric of Pordenone, 112

Oxford in Asia Hardback Reprints

CENTRAL ASIA

Robert Shaw
Visits to High Tartary, Yarkand and Kashgar
Introduction by Peter Hopkirk

Francis Younghusband
The Heart of a Continent
Introduction by Peter Hopkirk

Francis Younghusband
India and Tibet
Introduction by Alastair Lamb

CHINA

L.C. Arlington & W. Lewisohn
In Search of Old Peking
Introduction by Geremie Barmé

Juliet Bredon & Igor Mitrophanow
The Moon Year
Introduction by H.J. Lethbridge

Nigel Cameron
Barbarians and Mandarins

E.H.M. Cox
Plant-hunting in China
Introduction by Peter Cox

James W. Davidson
The Island of Formosa, Past and Present

E.J. Eitel
Europe in China
Introduction by H.J. Lethbridge

Reginald F. Johnston
Lion and Dragon in Northern China
Introduction by Pamela Atwell

Owen Lattimore
Inner Asian Frontiers of China
Introduction by Alastair Lamb

C.P. Skrine
Chinese Central Asia
Introduction by Alastair Lamb

INDONESIA

Carl Bock
The Head-Hunters of Borneo
Introduction by R.H.W. Reece

William Marsden
A Dictionary and Grammar of the Malayan Language (2 vols)
Introduction by Russell Jones

William Marsden
The History of Sumatra
Introduction by John Bastin

J.W.B. Money
Java, or How to Manage a Colony
Introduction by Ian Brown

Johan Nieuhof
Voyages and Travels to the East Indies 1653–1670
Introduction by Anthony Reid

Urs Ramseyer
The Art and Culture of Bali

Thomas Stamford Raffles
The History of Java (2 vols)
Introduction by John Bastin

Thomas Stamford Raffles
Statement of the Services of Sir Thomas Stamford Raffles
Introduction by John Bastin

Alfred Russel Wallace
The Malay Archipelago
Introduction by John Bastin

Beryl de Zoete & Walter Spies
Dance and Drama in Bali

MALAYSIA

S. Baring Gould & C.A. Bampfylde
A History of Sarawak under its Two White Rajahs
Introduction by Nicholas Tarling

Odoardo Beccari
Wanderings in the Great Forests of Borneo
Introduction by the Earl of Cranbrook

Hugh Low
Sarawak
Introduction by R.H.W. Reece

Sherard Osborn
The Blockade of Kedah in 1838
Introduction by J.M. Gullick

Owen Rutter
The Pagans of North Borneo
Introduction by Ian Black

Mubin Sheppard
A Royal Pleasure Ground: Malay Decorative Arts and Pastimes

Walter W. Skeat
Malay Magic
Introduction by Hood Salleh

Spenser St. John
Life in the Forests of the Far East
Introduction by Tom Harrisson

SINGAPORE

Charles Burton Buckley
An Anecdotal History of Old Times in Singapore 1819–1867
Introduction by C.M. Turnbull

Song Ong Siang
One Hundred Years' History of the Chinese in Singapore
Introduction by Edwin Lee

THAILAND

John Crawfurd
Journal of an Embassy to the Courts of Siam and Cochin China
Introduction by David K. Wyatt

George Finlayson
The Mission to Siam and Hué 1821–1822
Introduction by David K. Wyatt

Simon de la Loubère
The Kingdom of Siam
Introduction by David K. Wyatt

Carol Stratton & Miriam McNair Scott
The Art of Sukhothai: Thailand's Golden Age

Frank Vincent
The Land of the White Elephant
Introduction by William L. Bradley